Cultural Techniques

Cultural Techniques

Assembling Spaces, Texts & Collectives

Edited by
Jörg Dünne, Kathrin Fehringer, Kristina Kuhn,
and Wolfgang Struck

DE GRUYTER

We acknowledge support by the German Research Foundation (DFG) and the Open Access Publication Fund of Humboldt-Universität zu Berlin.

ISBN 978-3-11-064456-2
e-ISBN (PDF) 978-3-11-064704-4
e-ISBN (EPUB) 978-3-11-064534-7
DOI https://doi.org/10.1515/9783110647044

This work is licensed under a Creative Commons Attribution 4.0 International License. For details go to: https://creativecommons.org/licenses/by/4.0/.

Library of Congress Control Number: 2020939337

Bibliographic information published by the Deutsche Nationalbibliothek
The Deutsche Nationalbibliothek lists this publication in the Deutsche Nationalbibliografie; detailed bibliographic data are available on the Internet at http://dnb.dnb.de.

© 2020 Jörg Dünne, Kathrin Fehringer, Kristina Kuhn, and Wolfgang Struck, published by Walter de Gruyter GmbH, Berlin/Boston

Cover image: porpeller/iStock/Getty Images Plus
Typesetting: Integra Software Services Pvt. Ltd.
Printing and binding: CPI books GmbH, Leck

www.degruyter.com

Contents

Jörg Dünne, Kathrin Fehringer, Kristina Kuhn, and Wolfgang Struck
Introduction —— 1

Spaces

Tom Ullrich
Working on Barricades and Boulevards: Cultural Techniques of Revolution in Nineteenth-Century Paris —— 23

Jörg Dünne
Cultural Techniques and Founding Fictions —— 47

Wolfgang Struck
A Message in a Bottle —— 61

Gabriele Schabacher
Waiting: Cultural Techniques, Media, and Infrastructures —— 73

Christoph Eggersglüß
Orthopedics by the Roadside: Spikes and Studs as Devices of Social Normalization —— 87

Hannah Zindel
Ballooning: Aeronautical Techniques from Montgolfier to Google —— 107

Texts/Bodies

Bernhard Siegert
Attached: The Object and the Collective —— 131

Michael Cuntz
***Monturen/montures:* On Riding, Dressing, and Wearing. Nomadic Cultural Techniques and (the Marginalization) of Asian Clothing in Europe —— 141**

Jörg Paulus
Self-Imprints of Nature —— 165

Jürgen Martschukat
Identifying, Categorizing, and Stigmatizing Fat Bodies —— 177

Kathrin Fehringer
Techniques of the Body and Storytelling: From Marcel Mauss to César Aira —— 187

Collectives

Bettine Menke
Writing Out – Gathered Up at a Venture from All Four Corners of the Earth: Jean Paul's Techniques and Operations (on Excerpts) —— 219

Nicolas Pethes
Collecting Texts: Miscellaneity in Journals, Anthologies, and Novels (Jean Paul) —— 243

Kristina Kuhn
Reading by Grouping: Collecting Discipline(s) in Brockhaus's *Bilder-Atlas* —— 263

Stephan Gregory
Patience and Precipitation: Two Figures of Historical Change —— 277

Katrin Trüstedt
The Fruit Fly, the Vermin, and the Prokurist: Operations of Appearing in Kafka's *Metamorphosis* —— 295

Christiane Lewe
Collective Likeness: Mimetic Aspects of Liking —— 317

List of Figures —— 331

List of Contributors —— 335

Name Index —— 337

Subject Index —— 341

Jörg Dünne, Kathrin Fehringer, Kristina Kuhn, and Wolfgang Struck

Introduction

1 A Scene of Beginning

When Robinson Crusoe finds himself shipwrecked on a deserted, unfamiliar beach, he has but barely managed to save his life. Spending a night freezing, hungry, and in fear of unknown dangers, he realizes that his survival means nothing more than a merciless reprieve – the prospect of a slow death from wasting away instead of a quick death in the sea. Nature, to which Robinson finds himself exposed, appears to offer no means of securing his existence; indeed, it confronts him as a dangerous force. Only the following day, when he discovers his ship stranded not far from the coast, is he able to escape this resignation. This is followed by days of hard work during which he recovers preserved food, weapons, and tools from the wreck, and finally various nautical instruments, writing materials, and books: objects of use that seem capable of ensuring his survival in the short, medium, and long term. These are objects that Robinson could not have produced himself but are products of complex, specialized practices of labor, or, in other words, that are the product not only of individual actions but of operational chains reaching far back into the (cultural) history of humankind. In order to be able to start his life anew on the island, Robinson must be able to connect to a cultural tradition that he does not carry within himself but that is condensed into material things. The shipwreck functions as a container of everything that makes human beings human. And the lonely island is the place where Robinson is able to (re)create himself as a human being by connecting to all these things.

Read this way, Daniel Defoe's literary island-experiment is not so much the fiction of a new beginning far removed from civilization. Rather, it draws attention to what precedes (or underlies) this beginning: objects and practices that, by entangling material and symbolic dimensions, create not only "culture" but also "the human being." It is only at first glance that there appears to be a sequence of steps from the objects that make survival possible during Robinson's first few days (food), to those that allow him to establish himself on the island in the medium term (weapons, tools), to those that can only be useful in the long term (various "mathematical instruments," paper and ink, books – specifically, several Bibles). Precisely these things, whose immediate benefits are difficult to discern, soon

---

Translated by Michael Thomas Taylor

Open Access. © 2020 Jörg Dünne, Kathrin Fehringer, Kristina Kuhn, and Wolfgang Struck, published by De Gruyter. This work is licensed under a Creative Commons Attribution 4.0 International License.
https://doi.org/10.1515/9783110647044-001

prove to be indispensable. For example, it is paper and ink that allow Crusoe to capture and record the thoughts spinning in his head amid his despair and to sort them, like an accountant, into columns of "debtor" and "creditor." It is only when he makes a list comparing "good" and "evil" that Robinson gains clarity about his chances of survival, which turn out not to be so bad.[1] An earlier scene in the novel already emphasized the necessity of a self-distancing induced by media. When Robinson and the captain disagree during a storm over which course to take, they retreat into the cabin and, "looking over the charts," come to a carefully-weighed mutual decision. When the next storm arrives, they are not able to take a step back from the situation. The crew panics and abandons the ship:

> In this Distress the Mate of our Vessel lays hold of the Boat, and with the help of the rest of the Men, they got her slung over the Ship's-side, and getting all into her, let go, and committed our selves being Eleven in Number, to God's Mercy, and the wild Sea.[2]

The hasty decision, as it turns out later, is a mistake. The small boat is unable to withstand the storm, while the abandoned ship survives with only slight damage.

An education in *sang-froid*, or a suppression of spontaneous reactions that the anthropologist Marcel Mauss described 300 years after Defoe as the decisive effect of culturally specific "techniques of the body," proves itself here – or fails.[3] Robinson's example also shows that these techniques of the body work in conjunction with specific spaces (the protected cabin as opposed to the storm-battered deck) and media (the map). Their use, moreover, presupposes further objects:

> And now I began to apply my self to make such necessary things as I found I most wanted, as particularly a Chair and a Table, for without these I was not able to enjoy the few Comforts I had in the World, I could not write or eat, or do several things with so much Pleasure without a Table.[4]

A chair and table are not absolutely necessary to eat, but they make eating a cultural act. Like the fence with which Robinson surrounds his home, they create

---

[1] Daniel Defoe, *The Life and Strange Surprizing Adventueres of Robinson Crusoe, of York, Mariner*, ed. Michael Shinagel (New York and London: Norton, 1994 [1719]), 49–50.
[2] Defoe, *Crusoe*, 32–33.
[3] "La principale utilité que je vois à mon alpinisme d'autrefois fut cette éducation de mon sang-froid qui me permit de dormir debout sur le moindre replat au bord de l'abîme. / The main utility I see in my erstwhile mountaineering was this education of my composure [sang-froid], which enabled me to sleep upright on the narrowest ledge overlooking the abyss." Marcel Mauss, "Les Techniques du Corps," *Sociologie et Anthropologie*, introduction par Claude Lévi-Strauss, 365–388 (Paris: PUF, 1985 [1950]), here 385; translated as "Techniques of the Body" by Ben Brewster, *Economy and Society* 2, 1 (1973): 70–88, here 86.
[4] Defoe, *Crusoe*, 50.

a sphere of culture that sets itself apart from the surrounding nature, thus also distinguishing Robinson as a human from the wild animals on the island. They create a body kept upright not only by muscles and bones but also by rules that are learned.

The plasticity of this depiction positions *Robinson Crusoe* within a series of founding narratives about acts that engender culture. The furrow that Romulus ploughs to separate what will become the city of Rome from the surrounding wilderness is such an act. With the plough, he establishes a connection to the agricultural cultivation of the soil and thus to a spatial and symbolic demarcation that is fundamental, at least for the West.[5] Even before Romulus "misused" the plough to mark the city boundary, farming had always produced not only food but also a space that was no longer that of nature. Yet it is not only the relationship of human beings to "their" environment and "their" bodies that is determined in such an entanglement of material and symbolic practices; rather, it is such practices that define human beings in the first place. It is this entanglement that Marcel Mauss first observed in his "techniques of the body" and that has since become both a premise and an object of research in anthropology, the history of science, and media studies.[6] Viewed from a *cultural-technical* perspective, Robinson's attempt to assert his humanity in solitude proves to be a paradoxical enterprise. It is only possible in operations that always include what the human being should not be: those technical-medial objects recovered from the wreck and the nonhuman animals whose domestication, after many failed attempts, will eventually provide Robinson with sustenance and company.

The "expulsion of the spirit from the humanities" with which Friedrich Kittler initiated a media-technical turn in German cultural studies forty years ago has, in recent years, affected broad areas of the international humanities (→ Bernhard Siegert).[7] Critical animal studies, new materialism, or research on artificial intelligence (along with many other suggestions for reshaping the field of cultural studies) are united by an interest in questioning the anthropological difference,

---

[5] See Bernhard Siegert, "Kulturtechnik," in *Einführung in die Kulturwissenschaft*, ed. Harun Maye and Leander Scholz (Munich: Fink, 2011), 95–118.
[6] For an exemplary treatment, see the works of Timothy Ingold (e.g., *The Perception of the Environment: Essays in Livelihood, Dwelling and Skill* [London and New York: Routledge, 2000]), Bruno Latour (e.g., "Visualisation and Cognition: Drawing Things Together," *Knowledge and Society: Studies in the Sociology of Culture and Present* 6 (1986), 1–40) and Bernhard Siegert (e.g., "Media After Media," in *Einführung in die Kulturwissenschaft, Media After Kittler*, ed. Eleni Ikoniadou and Scott Wilson [New York and London: Rowman & Littlefield International, 2015], 79–91).
[7] See Friedrich Kittler (ed.), *Austreibung des Geistes aus den Geisteswissenschaften: Programme des Poststrukturalismus* (Paderborn: Schöningh, 1980).

defined as the exaltation of the human being as a creator of culture.[8] What a cultural-technical perspective has to contribute to this endeavor is, first and foremost, the possibility of operationalizing the border crossing this research strives for – precisely by making it possible to recognize the borders to be transgressed as the product of specific operations linking human and nonhuman actors. In the process, such an approach can build on the field of media studies as shaped by Kittler, while at the same time expanding on what this approach describes solely in terms of a capacity of technical media to include the cultural practices that, as this cultural-technical perspective fundamentally assumes, precede the concepts and objects constituted in and through precisely these practices.

This volume aims to take up the impulse that has emanated thus far primarily from media studies and to test this approach in a broader disciplinary field. This is, however, by no means intended as a transfer in only one direction. History and literary studies, as well as the history of science (all of which stand here together with media studies), each have their own specific theoretical and historical expertise, which is what makes research into cultural techniques a comprehensive field in the first place. What repeatedly emerges in this research are situations in which people enter into constellations with (media)technical and natural objects that do not simply yield to their actions or interpretations. These are scenes in which human beings are transformed from anthropological constants into variables in operations that elude human mastery by chaining together diverse actors and making them interdependent. This would be *Robinson Crusoe*'s message: survival is not ensured by the – vain – attempt to assert oneself as *human being* over and against an overpowering nature. Attention must be directed toward the many practices that mediate between human beings and nature, and in which both humans and nature are transformed. To put it another way: Robinson learns that he is as much an environment of nature as nature is his environment. But in order

---

[8] On critical animal studies, see Philippe Hamman, Aurélie Choné, and Isabelle Hajek (eds.), *Rethinking Nature: Challenging Disciplinary Boundaries* (London and New York: Routledge, 2017), in particular the overview article in that volume by Roland Borgards, "Animal Studies," 221–321; on new materialism see Karen Barad, *Meeting the Universe Halfway: Quantum Physics and the Entanglement of Matter and Meaning* (Durham, NC: Duke University Press, 2007); the classic introduction to artificial intelligence was authored by Alan Turing, "Computing Machinery and Intelligence," *Mind: A Quarterly Review of Psychology and Philosophy*, vol. LIX (October, 1950): 433–460; see also Selmer Bringsjord and Navee Sundar Govindarajulu, "Artificial Intelligence," *The Stanford Encyclopedia of Philosophy* (Fall 2018), ed. Edward N. Zalta, https://plato.stanford.edu/archives/fall2018/entries/artificial-intelligence/ (visited on August 27, 2019). On the adaptation of human-machine-hybrids in the context of a posthuman theory of culture, see Donna Haraway, "A Cyborg Manifesto," in *The Cybercultures Reader*, ed. David Bell and Barbara M. Kennedy (London and New York: Routledge, 2000 [1985]), 291–324.

to discern this message, a specific kind of attention is required, which we propose here to understand as a cultural-technical attitude, and which we hope to make use of in order to cast more or less familiar key scenes in a new light.

If we assume that an operational concatenation of specific practices produces "culture" in the first place then the case study is an indispensable instrument for researching culture. This might apply to "grand narratives," such as the culture of writing or a revolution, but an approach attuned to cultural techniques will always dissolve these *grands récits* into the many *petits récits* ("small narratives") of concrete objects and operations, such as the manufacturing of ink and paper (which Robinson fails to achieve, forcing him to eventually abandon writing in his diary) or the construction of barricades (→ Tom Ullrich). In order to illustrate the specific theoretical application involved, however, it is necessary to begin with several more fundamental deliberations.

2 What Are Cultural Techniques?

Priority of Practices

As these introductory reflections on Daniel Defoe's novel have shown, Robinson does not start from scratch on his deserted island. In other words, he does not begin in some kind of *état de nature* ("state of nature") – even if Jean-Jacques Rousseau, in his educational treatise *Émile*, gives his pupil the book *Robinson Crusoe* to help him imagine such a state, which might have been one reason for the continuing fascination of the Robinson paradigm in the history of literature and culture.[9] Rather, Robinson is busy deploying the objects and supplies he has recovered from the shipwreck (or storing them for later possible use) in such a way that they become not only fragments of a lost civilization, but building blocks for new practices. He assembles these practices using his own body as well

---

9 The educator in *Émile* excepts *Robinson Crusoe* from the ban applied to all other novels that might give the pupil a detrimental image of civilization. See Jean-Jacques Rousseau, *Émile ou De l'Éducation*, in J.-J.R., *Œuvres complètes*, vol. IV, ed. Bernard Gagnebin and Marcel Raymond (Paris: Gallimard, 1959), 454ff; *Emile, or On Education*, trans., introduction, and notes Allan Bloom (New York: Basic Books, 1979), 184ff. On this fundamental supplementarity, which is thus inscribed precisely in the imagination of an original state, see Jacques Derrida, *De la grammatologie* (Paris: Minuit, 1967); *Of Grammatology*, trans. Gayatri Chakravorty Spivak (Baltimore and London: Johns Hopkins University Press, 1997).

as his organizational abilities, supported by media, to create what recent cultural studies research calls "operational chains."[10]

Engaging with Robinson Crusoe's arrival on his island in this way makes clear that, from a perspective concerned with cultural techniques, concrete practices and operations always precede the orders and concepts they constitute. This priority of practices is a perspective that allows research into cultural techniques to reformulate, in terms of practice and action, fundamental assumptions about the conditions under which cultural orders emerge. There is no *human being* independent of cultural techniques for becoming human; there is no *time* independent of cultural techniques for measuring time; and, above all, there is no *space* independent of cultural techniques of spatialization.

It is cultural techniques that historically produce, reproduce, shift, and hybridize the fundamental distinctions of a civilization with its ideas, representations, concepts, and actions – not least of which is the opposition between nature and culture. Formulated grammatically, thinking in terms of cultural techniques thus means thinking primarily in verbs. Hence, when cultural techniques take the place of the predicate, the roles of things (the object) and people (the subject) trade places: things become observable as acting subjects and humans become observable as objects.[11]

Which Cultural Techniques?

Until now, research into cultural techniques has focused on what has been called elementary cultural techniques, such as reading, writing, arithmetic, and image making,[12] as well as on modern information and communication technologies.[13] Even our cursory glance at two "founding scenes" (Romulus's Rome, Robinson's

---

10 See specifically, with reference to André Leroi-Gourhan, Marcel Mauss, and André Haudricourt: Erhard Schüttpelz, "Die medienanthropologische Kehre der Kulturtechniken," in *Archiv für Mediengeschichte* 6 (2006): *Kulturgeschichte als Mediengeschichte (oder vice versa?)*: 87–110.
11 See Cornelia Vismann, "Kulturtechniken und Souveränität," *Zeitschrift für Medien- und Kulturforschung* 1, 1 (2010): *Kulturtechnik*: 171–182, here 171: "If media theory were or had a grammar, this power to act would be expressed in the fact that objects assume the grammatical position of the subject and cultural techniques represent verbs. Persons (meaning: people) then move to the place in a sentence that is reserved for the grammatical object. This reversal of positions is perhaps the most obvious characteristic of a theory of cultural techniques, considered from the perspective of media."
12 See Sybille Krämer and Horst Bredekamp (eds.), *Bild – Schrift – Zahl* (Munich: Fink, 2003).
13 See Christian Holtorf and Claus Pias (eds.), *Escape! Computerspiele als Kulturtechnik* (Cologne, Weimar, and Vienna: Böhlau, 2007); Valérie November, Eduardo Camacho-Hübner, and Bruno

island) has shown that it is worth investigating the entanglement of these reflexive cultural techniques with the agricultural origins of the concept of cultural techniques. This agricultural origin of the concept of culture, which also remains present in the concept of cultural techniques, among other things, can be rejected as an unmodern legacy of the past. But in it, we can also recognize a particularity that allows us to ask new questions about cultural techniques and the founding fictions associated with them. For example, one can ask what happens when the soil to be cultivated in an (aquatic) landscape is unable to sustain any distinction between (stable) ground and figure (→ Jörg Dünne).

After (post)structuralist cultural theory, which attempted to understand culture primarily in analogy to sign systems,[14] research into cultural techniques since the turn of the millennium[15] has begun reflecting on the operationality of cultural processes, as these were fundamentally described by Actor-Network Theory.[16]

Considering the history of research into cultural techniques that can be used for this purpose, we can distinguish three different approaches[17]: The term "Kulturtechnik" (1) has existed in Germany in the context of agriculture since the end of the nineteenth century to denote procedures developed in science and engineering, as relevant dictionaries indicate. The term goes back to an understanding of culture (lat. *colere, cultura*) that has referred since antiquity to techniques for cultivating the land and for settling the earth with dwelling places and cities. From the 1970s on (2), the primary school subjects of reading, writing, and arithmetic are referred to in Germany as "elementary cultural techniques." In the 1980s, the term was also applied in Germany to the use of television and, since

---

Latour, "Entering a Risky Territory: Space in the Age of Digital Navigation," *Environment and Planning: Society and Space* 28 (2010): 581–599.

14 For an exemplary treatment of cultural semiotics, see Roland Barthes, *Mythologies* (Paris: Seuil, 1970 [1957]); for an examination of the question of the "legibility" of culture against the background of research into cultural techniques, see Gerhard Neumann and Sigrid Weigel (eds.), *Lesbarkeit der Kultur: Literaturwissenschaften zwischen Kulturtechnik und Ethnographie* (Munich: Fink, 2000).

15 See especially the contributions in *Zeitschrift für Medien- und Kulturforschung* 1, 1 (2010): *Kulturtechnik* (edited by Lorenz Engell and Bernhard Siegert), as well as in *Theory, Culture & Society* 30, 6 (2013): *Cultural Techniques* (edited by Geoffrey Winthrop-Young, Ilinca Iurascu, and Jussi Parikka).

16 See the anthology by Andréa Belliger and David J. Krieger (eds.), *ANThology: Ein einführendes Handbuch zur Akteur-Netzwerk-Theorie* (Bielefeld: transcript, 2006); Mike Michael, *Actor-Network Theory: Trials, Trails and Translations* (Thousand Oaks, CA: Sage, 2016).

17 See Siegert, "Kulturtechnik," 79–101, and referring to Siegert in his overview of the history of German research into „Kulturtechniken," see Geoffrey Winthrop-Young, "Cultural Techniques: Preliminary Remarks," *Theory Culture & Society* 30, 6 (2013): 3–19.

the 1990s, to the use of information and communication technology. However (3), the use of the term "cultural technique(s)" that has persisted until today, with its increasingly international usage, comes from the media studies and cultural studies of the late 1990s. Here, the term continues to refer to the "elementary cultural techniques" in the sense just noted, while also pointing, with recourse to the older, agricultural meaning of the word, to those areas of *graphé*, writing, that go beyond the alphanumeric code.[18]

Spatialization

This recourse to the meanings of cultural techniques from agricultural engineering is crucial in terms of spatialization: it makes it more plausible to assume that every form of cultural-technical operationality already implies an elementary form of spatialization possessing a special quality compared to categories such as subjectivity and temporality. On a spatial level, it is not only the processual character of cultural-technical operations (which is, in itself, barely observable) but also associated processes such as subjectivization and temporalization that become visible and describable in a particular way.

Cultural-technical operational chains do not simply inscribe themselves into an existing physical space. First and foremost, as in the case of Robinson Crusoe's island, they constitute the accessible and thus adaptable spatiality of the island setting in which Defoe's novel unfolds, through the exploration of the terrain, the fencing in of pasture land, or agricultural development. In a cultural-technical sense, then, spatiality cannot simply be described as the prerequisite for an inert, natural "container space" in which cultural-technical operations take place,[19] even if novels such as *Robinson Crusoe*, resort to already stabilized ways of charting space and to the symbolic occupation of topographical spaces.[20]

---

[18] For an exemplary treatment, see Wolfgang Schäffner, "Topographie der Zeichen: Alexander von Humboldts Datenverarbeitung," in *Das Laokoon-Paradigma: Zeichenregime im 18. Jahrhundert*, ed. Inge Baxmann et al. (Berlin: Akademie-Verlag, 2000), 359–382; and Bernhard Siegert, *Passage des Digitalen: Zeichenpraktiken der neuzeitlichen Wissenschaften 1500–1900* (Berlin: Brinkmann & Bose, 2003).
[19] See Stephan Günzel for an overview of this belief as found in in recent research into space: *Raum: Eine kulturwissenschaftliche Einführung* (Bielefeld: transcript, 2017), especially 60–68.
[20] On cartography in this context, see especially the exemplary work of Robert Stockhammer, *Kartierung der Erde: Macht und Lust in Karten und Literatur* (Munich: Fink, 2007); and Jörg Dünne, *Die kartographische Imagination: Erinnern, Erzählen und Fingieren in der Iberischen Welt der Frühen Neuzeit* (Munich: Fink, 2011).

Spatialization through cultural techniques thus means, first of all, the establishment of spatial relations between actors that can be described topologically. Examples are operations such as differentiating between inside and outside by drawing boundaries. Such operations can also be used to process techniques of temporalization or subjectivization in spatial form. For example, the measuring of time by sundials or calendars,[21] the establishment and stabilization of anthropological difference,[22] the distinction between figure and ground,[23] or between word and number[24] can all be described through different types of spatial articulation or spatialization. In addition to the spatialization operations mentioned above, which are based on distinctions and demarcations, the operationality of less settled cultural practices, based on the habitualization and regulation of practices of movement, must also be taken into account.[25] Since critical revisions and elaborations of Actor-Network Theory have begun to focus on the concept of assemblage,[26] operations of assembling and condensing have attracted increasing interest, generating spaces and places that are not, however, based on demarcations. Hence the investigations of this volume begin with a cultural-technical view of processes for assembling spaces, texts, bodies, and collectives.[27]

Techniques of the Body

In his attempt to define techniques of the body, Marcel Mauss primarily examines habitualizations and regulations of movement. His lecture "Les Techniques du Corps" (1934) examines techniques of walking, swimming, washing, or giving birth, the theorization of which is based on the observation and narration of such concrete techniques and practices (→ Kathrin Fehringer).[28] Yet Mauss does

---

[21] See also Thomas Macho, "Zeit und Zahl: Kalender- und Zeitrechnung," in *Bild – Schrift – Zahl*, ed. Sybille Krämer and Horst Bredekamp (Munich: Fink, 2003), 179–192.
[22] See Siegert, "Kulturtechnik," 115f.
[23] See Manfred Sommer, *Von der Bildfläche: Eine Archäologie der Lineatur* (Berlin: Suhrkamp, 2016), especially 66ff; as well as the problematization of the basic distinction between ground and figure by Timothy Ingold in *Lines: A Brief History* (London: Routledge, 2007).
[24] See Denise Schmandt-Besserat, *Before Writing* (Austin: University of Texas Press, 1992).
[25] For an exemplary introduction to mobility studies, see Mimi Sheller and John Urry, "The new mobilities paradigm," *Environment and Planning A* 2, 8 (2006): 207–226.
[26] See especially Manuel De Landa, *A New Philosophy of Society: Assemblage Theory and Social Complexity* (New York: Continuum, 2006), and Bruno Latour, *Reassembling the Social: An Introduction to Actor-Network Theory* (Oxford: Oxford University Press, 2007).
[27] See below, 3.
[28] Mauss, "Les Techniques du Corps," 365–388; "Techniques of the Body," 70–88.

not stop at a general observation of techniques of spatialization. Rather, he sees techniques of the body as a motor for the formation and connection of (concrete) places and spaces through cultural techniques – a question that is also explored in this volume.

It is no coincidence that, since Mauss's fundamental considerations, techniques of the body have been at the forefront of more recent research into cultural techniques,[29] given that the body is one of the first sites where the operationality of techniques is articulated at all, in the form of heterogeneous spatial links. However, the body remains a resistant, ambivalent actor in the field of cultural techniques.[30] This can be seen, for example, in Eadweard Muybridge's photographs of movement from the end of the nineteenth century, which assembled techniques of the body such as walking, standing up, or jumping into series of images of movement sequences, thus making them observable in the first place. However, such studies of the body only become significant – in the case of Muybridge, for instance, as a deviation from the norm for bodies regarded as healthy – in the context of an interest in controllable techniques of the body that produces, in the course of the twentieth century, additional (classifying) cultural techniques (→ Jürgen Martschukat).

Techniques of the body thus play a central role in the constitution of cultural techniques (and vice versa), since, as techniques of spatialization, they can render concrete spaces describable in a particular way. This is the case even if, and especially as, research into cultural techniques increasingly advocates for positions that point beyond the anthropocentric reference associated with the question of spatial constitution: while the view of techniques of the body emphasizes the specificity of certain collectives (or "cultures"), it remains fundamentally focused on human practices. The engagement with cultural techniques therefore raises the question of the extent to which the spatialization associated with these techniques is already inherent to relations among different species of

---

29 Mauss's techniques of the body also form an important basis for work of his student, André Leroi-Gourhan, in paleoanthropologicy; see *Le geste et la parole* (Paris: Albin Michel, 1964); *Gesture and Speech*, trans. Anna Bostock Berger (Cambridge, MA, and London: MIT Press, 1993). See, however, Gilbert Simondon's detachment of technical objects from the human body and the formation of independent technical-geographical milieus: *Du mode d'existence des objets techniques* (Paris: Aubier, 1989 [1958]); *On the Mode of Existence of Technical Objects*, trans. Cecile Malaspina (Minneapolis: Univocal, 2017 [1958]).
30 For example, despite all supposed increases in performance, the body ultimately eludes the dynamics of accumulation ("everything always gets better") that the consideration of cultural techniques might occasionally imply, but which techniques of the body "presumably elude permanently"; see Erhard Schüttpelz, "Körpertechniken," *Zeitschrift für Medien- und Kulturforschung* 1, 1 (2010): *Kulturtechnik*: 101–120, here 113.

things (between solid, liquid, and gaseous bodies above and below the earth's surface, such as between humans and animals, plants, and bacteria), as well as the question of which relations between animate and inanimate objects trigger and shape spatializing processes.[31] Such considerations lead to questions about processes of assembling.

3 Assembling Spaces, Bodies, Collectives

Assembling

A "thinking in verbs," as Cornelia Vismann asserts in examining what cultural techniques are and what they perform, focuses on those processes, operations, and practices that make it possible to describe not only material infrastructures but also the entanglement of actors and objects, bodies and media, and not least of all symbolic orders (→ Katrin Trüstedt).[32] This volume aims to present and develop three central aspects of this insight: processes of spatialization, the assembling of bodies, and techniques of the collective. The following elucidations of the concept of assemblage make it clear, however, that the three terms in the title of this volume cannot be separated from each other: bodies are collectives just as they form spaces, collectives are composed of bodies that extend in space. Although the groupings of contributions in the subsections of the book thus focus on the constellations defining these respective points of emphasis, they should ultimately always be thought within this triad. What connects these three terms is the fact that they allow collective techniques of spatialization (and temporalization) to come to the fore that are determined by idiosyncratic bodies, both human and nonhuman.

In *A New Philosophy of Society: Assemblage Theory* (2006), Manuel De Landa has proposed the concept of assemblage to describe these kinds of processes. Building on sociological concepts, he understands these processes to generate "a wide range of social entities, from persons to nation-states ... constructed through very specific historical processes" – what he defines as "inorganic, organic and

---

[31] See Michel Serres, *Hominiscence* (Paris: Flammarion, 2003); Donna Haraway, *When Species Meet* (Minneapolis: University of Minnesota Press, 2008); Anna Lowenhaupt Tsing, *The Mushroom at the End of the World: On the Possibility of Life in Capitalist Ruins* (Princeton and Oxford: Princeton University Press, 2015).
[32] Vismann, "Kulturtechniken und Souveränität," 171.

social assemblages."³³ The parts of the assemblage "do not form a seamless whole"; assemblages form wholes in a completely different way: as "wholes whose properties emerge from the interactions between parts."³⁴ As a concept, assemblage thus allows us to see and describe, in a particular way, processual relations between diverse bodies and their integration into differentiated networks (→ Bernhard Siegert).³⁵ In a critical reading of Félix Guattari's and Gilles Deleuze's *Mille Plateaux* (*A Thousand Plateaus*, 1987 [1980]) De Landa proposes that we understand assemblages as a fluid connection of bodies and their individual components by considering effective practices that produce historically significant processes and make them analyzable. Guattari and Deleuze used the concept of *agencement* to investigate processes of coding and stratifying bodies, with the aim of describing processes of territorialization and deterritorialization and thus the complexity that De Landa attributes to assemblages.³⁶ De Landa then develops a theory of the collective by expanding Deleuze and Guattari's critique of a social ontology and arguing that "nested" assemblages such as language, science, nature, and culture have been erroneously considered to be composite, homogeneous wholes and must be reinterpreted from a theoretical perspective that he calls "neo-assemblage theory."³⁷

De Landa extends Guattari's and Deleuze's reflections, developing them into a theoretical approach that emphasizes, through a confrontation with Bruno Latour's concept of the network, the heterogeneity of what comes together in assemblages: according to De Landa, parts of the assemblage remain autonomous, i.e., independent of their network, and are not defined by the whole but by a function of the immanent material interrelations of their parts. Since each assemblage is itself historically singular and individual, De Landa sees the primary function of describing and theorizing it in analyzing the individual processes that found the assemblage in its complexity.

What is decisive here from a cultural-technical point of view is that assemblages consist of parts whose autonomy can never be completely dissolved, with the consequence that they should never be thought of as solid structure, for

---

33 De Landa, *A New Philosophy of Society: Assemblage Theory*, 3.
34 De Landa, *A New Philosophy of Society: Assemblage Theory*, 4–5.
35 See also Latour, *Reassembling the Social*.
36 The French word *agencement* literally means an arrangement or *Anordnung*, as a collection of things (that are gathered together, assembled); see Gilles Deleuze and Félix Guattari, *Mille plateaux: Capitalisme et schizophrénie*, vol. 2 (Paris: Minuit, 1980), especially 112; *A Thousand Plateaus: Capitalism and Schizophrenia*, transl. Brian Massumi (London and New York: University of Minnesota Press, 1987), 86.
37 De Landa, *A New Philosophy of Society: Assemblage Theory*, 4.

example, as dispositives.[38] Rather, they are dependent on constant stabilizations, on those "processes of assembling," in other words, that "thinking in verbs" attempts to comprehend.

In understanding techniques of the body as a "series of assembled actions," Marcel Mauss fundamentally ascribes to all cultural techniques, such a collecting, serial character that condenses in operational chains.[39] As habitualizations, these actions are thus also understood as collective, technical knowledge and are subject to effective processes of (de)coding, (de)terriorialization, and (de)stabilization. Seen in this way, the teachable body and all cultural techniques are, on the one hand, accumulations and condensations of such historically significant, technical knowledge, which makes them media.[40] And on the other hand, these techniques can themselves be described in terms of significant processes of assembling that underlie every cultural technique and highlight precisely those idiosyncratic objects that shape and are shaped by places and spaces, as well as human practices.

Spaces, Bodies, Collectives

One such idiosyncratic assemblage is the *montage* of horse and rider: not only does the body of the rider change to match that of the horse, but also vice versa. This assemblage, equally central and successful, demonstrates not only an interspecific history of techniques but also the emergence of interspecific network actors requiring a long period of preparation in which human beings and animals came to resemble each other (→ Michael Cuntz). The term "assembling" makes it possible to describe, from the perspective of cultural techniques, the combination of pieces of clothing and equipment (such as bridles with tack, and saddles with stirrups) with the gestures and techniques of the body required for riding to develop. A comparable operation is also performed by the writing and drawing pencil, and in this regard those (still unanswered) questions that attempt to

---

38 In contrast to the concept of dispositive as it has generally come to be used in media studies, however, Michel Foucault (from whom this concept originates) understood "dispositif" (usually translated as "assemblage") to mean an "ensemble résolument hétérogène" ("thoroughly heterogeneous ensemble") composed of discursive and nondiscursive practices. See the interview "Le jeu de Michel Foucault," in *Michel Foucault: Dits et écrits*, vol. III: 1976–1979, ed. Daniel Defert and François Ewald (Paris: Gallimard, 1994), 298–329, here 299; "The Confession of the Flesh," in *Power/Knowledge: Selected Interviews and Other Writings, 1972–1977*, ed. and trans. Colin Gordon et al. (New York: Pantheon Books, 1980), 194.
39 "série d'actes montées," Mauss, "Les techniques du corps," 372; "Techniques of the Body," 76.
40 See Schüttpelz, "Körpertechniken."

distinguish techniques of the body from cultural techniques by means of the instrument used in practice must be asked in a different way. Body and pencil cannot be separated from each other inasmuch as they represent an assemblage that is indeed similar to that of the horse and rider. Such effective, collective practices of assembling – and hence also practices such as arranging, collecting, selecting, directing, commuting – intrinsically connect diverse materials and/or bodies (for example textiles, leather, or paper) and operators (for example vectors that negotiate between one-, two- and three-dimensionality; → Wolfgang Struck) and thereby constitute concrete topographical, linguistic, cultural, and textual spaces.

The way in which culture and collective are oriented against each other is directly related to techniques and procedures of assembling – the accumulation, gathering, joining together, and building of text-, image-, and object-worlds that go hand in hand with the question of collection, collectivization, and the collective "use" of holdings and inventories, such as bundled textual bodies,[41] furniture for assembling,[42] or (museum) architectures.[43] The practices of collectivization endow topographic spaces with dimensions (for example, by piling or stacking), just as they fit into spatial structures (for example, in bindings or cupboards). As a body, the literary text in this respect derives less from textile fabric than from the convolute (lat. *convolutum*, "rolled up"). By means of cultural techniques, the flat parchment or the paper sheet undergo a change of dimension to become the massive three-dimensional body: by rolling, folding, flapping, layering, bundling, binding, and accumulating.[44] Paper develops the ability to bind objects and/or bodies in a way that goes beyond Latour's *immutable mobiles*, i.e., to do more than merely record them – for example, to merge with objects/bodies in the self-imprinting process of nature from the eighteenth and nineteenth centuries, to form operational structures that hold their contents together

---

[41] On the materiality of book and paper and the associated spatially constitutive operations, see Heike Gfrereis and Claus Pias (eds.), *Das bewegte Buch: ein Katalog der gelesenen Bücher* (Marbach am Neckar: Deutsche Schillergesellschaft, 2015).

[42] See Anke te Heesen and Anette Michels (eds.), *Auf/Zu: Der Schrank in den Wissenschaften* (Berlin: Akademie-Verlag, 2007).

[43] See Anke te Heesen and Margarete Vöhringer (eds.), *Wissenschaft im Museum: Ausstellung im Labor* (Berlin: Kadmos, 2014).

[44] Helga Lutz, "Folding Bodies into Books," in *Presence and Agency: Rhetoric, Aesthetics and Experience* of Art, ed. Caroline van Eck and Antje Wessels (Leiden: Leiden University Press, forthcoming); H.L., "Räume aus Falten, Falten aus Mustern, Muster aus Fäden: Interferenzen bildlicher und textiler Ordnungen an Beispielen burgundisch-niederländischer Kunst des 15. Jahrhunderts," in *Texturen von Bildlichkeit*, ed. Mateusz Kapustka, Martin Kirves, and Martin Sundberg (Emsdetten and Berlin: Imorde, 2018), 99–118.

(or stabilize them), endowing them with new dimensions (→ Jörg Paulus). The assembling of texts into collectives cannot therefore be described without considering correlating spatial constellations and spatial interventions, such as the dimensionality mentioned above and the media associated with it (text, image, map, book, body, etc.).

Several contributions to this volume explore the connection between literary and media practices and procedures, on the one hand, and cultural techniques, on the other. They ask, in other words, how collectives are assembled as/by textual worlds or how convolutes of texts are assembled as collectives. What implications does this initially philological ordering and rearranging have for a concept of cultural techniques and of the collective that builds on literary studies and media studies? Collective cultural techniques do not form classical symbolic systems, but they do participate in and modify the symbolic social communication. Inasmuch as symbols constitute a fundamental part of culture, literature opens up a linguistically coded cultural space in which symbol production can be described and demonstrated in an exemplary way – and within which this production again and again recursively takes up, or self-reflectively positions itself toward, its relationship to symbolic meaning.

Literature can be described as a collective that both deploys collective procedures and emerges from such procedures. It is worth recalling here the etymological origin of *legere* (lat. "reading," "collecting") of letters (*littera*) that is the foundation of literature. This etymology makes it clear that reading and writing are by no means always directed movements but can also be based on the automatic accumulation of letters, words, phrases, or entire parts of text, as in Baroque aleatorics – a nonintentional knowledge that is being suppressed, and more consistently so, by modernity's theories of the subject and their aesthetics of autonomy.[45] This deliberate forgetting entails a forgetting of the cultural techniques of literature and philology (→ Bettine Menke), which, in this fundamental sense, carry out and represent movements of collecting and dispersing.

Assembling – in the sense of accumulating and collecting – thus forms a starting point for a spatial-processual thinking of culture, whose processuality (leafing through, cutting out, excerpting, stapling to, tearing out, pushing or clapping open or closed, sifting, walking through, etc.) are rendered tangible and comprehensible by the procedures of the book, the reference library, the archive, the booklet, the supplement, and the sheet, i.e., by material practices and operations.

---

45 See Stefan Rieger, *Speichern/Merken: Die künstlichen Intelligenzen des Barock* (Munich: Fink, 1997).

Literature stands as a self-reflexive model not only for its own (eminently cultural-technical) procedures such as *reading* and *writing*, but also for other cultural techniques, which it seeks to experimentally imitate, subvert, and reproduce in its medium, at the level of its textual operations (not to mention the fact that many areas of knowledge, such as law, are, in any case, based on philological, i.e., exegetical and rhetorical procedures; → Katrin Trüstedt).

A consideration of cultural-technical procedures questions the phantasm of originality in several areas; the art of invention turns out, instead, to be a matter of digging something out again, of cropping or clipping and piecing together, of splicing and patching. Ultimately, the writing and collecting of texts is hardly distinguishable in operational terms, inasmuch as existing material is "only" rearranged and thus transformed (→ Nicolas Pethes): Here, too, concept and idea remain secondary phenomena that can subsequently appear. The potency of collecting and what is collected does not lie in its systematizing power but in its monstrous, grotesque exuberance, which yields (yet to be reconstituted) generic forms and allows them to proliferate (→ Nicolas Pethes, Bettine Menke).

Instead, innovations and new knowledge emerge – stealthily, in secret, as a shift or experiment from what is written out in advance, rather than being brought into the world by regulated provisions (prescriptions) or creative acts. This applies even to epochal phenomena such as the Enlightenment that spreads collectively and manifests itself less in a corpus of knowledge, in an attitude of consciousness or spirit (ideology), than in an infectious form of communication that "disseminates" itself through tricks, processes, and dynamics of its implementation, which are set in motion in the first place by textual techniques and their material, by letters, paper, the copyist and his excerpts, i.e., by cultural techniques (→ Stephan Gregory, Bettine Menke).

The cultural techniques of the text or the interference of cultural techniques and the intellectual transparency of a (mental) message decoupled from them become a theme in the literature of the eighteenth century itself; they become textwork. In the nineteenth century, in particular, practices of writing and the archive become institutionalized and institutionalizing practices of acquiring and generating knowledge (both in the philologies and the natural sciences),[46] with a significance that goes beyond any one epoch, as can be seen in the nineteenth century encyclopedia, whose accumulation and distribution of knowledge fluctuates between supplementation and systematization (→ Kristina Kuhn). Collectives,

---

46 See Kristina Kuhn and Wolfgang Struck, *Aus der Welt gefallen: Die Geographie der Verschollenen* (Paderborn: Fink, 2019), and Anke te Heesen and Emma C. Spary (eds.), *Sammeln als Wissen: Das Sammeln und seine wissenschaftsgeschichtliche Bedeutung* (Göttingen: Wallstein, 2001).

however, are also formed quite concretely through operations of collaging, which engender dimensions in the space of travel and the space of knowledge created by messages sent in bottles, and which not only accumulate the results of a scientific community but also themselves produce this community (→ Wolfgang Struck).

Cultural techniques, thus, are shared collectively and form collectives. They participate in the formation of communities and societies, for example, in the form of language, rituals, or religious practices. The collective can be understood as an equal and productive assembly of people and things or human and nonhuman actors.[47] It thus functions not only as a true-to-scale but also as a *relational* alternative to the concept of society.[48]

Collective cultural techniques/cultural techniques of the collective enable the temporary association (i.e., the combination and connection) of actors to form a collective. If collectives in a certain sense always represent an "event of connection," then one must also ask with which media and cultural techniques (of spatialization, synchronization, cooperation, and assembly) these connections are established, stabilized, and then once again dissolved.[49]

The paradox that emerges in any movement of assembling (of objects, texts, people) consists in a unifying, analogizing tendency that transforms the collected into the whole, while the collected objects, in their very difference, become an individual "valuable" component (of an assemblage) – a paradox also shared by the "small forms" of a modern media landscape when, for example, they process a parallel crossing of similarity and difference with the like button on Facebook, thereby producing belonging and exclusion (→ Christiane Lewe).[50] The definition of the collective, its specific spaces and the media of its cohesion, play a decisive role in determining how cultural and social processes (of exchange) are

---

[47] Bruno Latour, *Politics of Nature: How to Bring the Sciences into Democracy*, trans. Catherine Porter (Cambridge: Harvard University Press, 2004 [1999]). See Georg Kneer, Markus Schroer, and Erhard Schüttpelz (eds.), *Bruno Latours Kollektive: Kontroversen zur Entgrenzung des Sozialen* (Frankfurt am Main: Suhrkamp, 2008); Michel Serres, *Le Contrat naturel* (Paris: Bourin, 1990); *The Natural Contract*, trans. Elizabeth MacArthur and William Paulson (Ann Arbor: University of Michigan Press, 1995); Philippe Descola, "From Wholes to Collectives: Steps to an Ontology of Social Forms," in *Experiments in Holism: Theory and Practice in Contemporary Anthropology*, ed. Ton Otto and Nils Bubandt (Malden, MA: Wiley-Blackwell, 2010), 209–226.
[48] Lorenz Engell and Bernhard Siegert, "Editorial Focus Collective," *Zeitschrift für Medien- und Kulturforschung* 3, 2 (2012): *Kollektiv*: 5–11, here 5, 9, 10.
[49] Urs Stäheli, "Infrastrukturen des Kollektiven: alte Medien – neue Kollektive?" *Zeitschrift für Medien- und Kulturforschung* 3, 2 (2012): *Kollektiv*: 99–116, here 111.
[50] See the research program of the eponymous research training group (*Graduiertenkolleg*), http://www.kleine-formen.de/forschungsprogramm/ (visited on September 12, 2019).

conceived. This applies not least of all to the question of cultural transformation and the possibility of social openness and cooperation.

Since the modern period, processes of collectivization have been tied to an unprecedented extent to technical instruments and interventions that link physical techniques to foundations of social identity, be it through the construction of barricades in nineteenth-century Paris or the modern barricades of exclusion anchored in antihomeless-devices (→ Christoph Eggersglüß) found on the streets of socially stratified cities. Cultural techniques that unfold space and structure it – physically, mathematically, geographically, geopolitically, infrastructurally – prove to be socially effective techniques of power, especially in the instrumentalization of technical media, such as the balloon, in the service of a (humanizing) scientification (→ Hannah Zindel). The space-constituting dimension of cultural techniques, however, always contains a temporal index, as can be seen from the double meaning of "waiting" that is intertwined, in infrastructure, with cultural-technical competences; here, waiting places both human and nonhuman actors in a kind of "organic" relationship to each other that takes their own temporalities into account (→ Gabriele Schabacher).

This volume does not clarify different types of spatial articulation or spatialization in opposition to each other. Rather, it thinks them in terms of process, in logics of assembling techniques, objects, and spaces. It presents and stages, sometimes spectacularly, processes of (de)stabilization, (de)coding, (de)terriorialization, and dimensioning that become spatially visible, observable, and theorizable through cultural techniques and techniques of the body that are inherent to processes of assembling. These are scrutinized here, as the contributions from various disciplines collected in this volume show, through analyses of diverse materials and heterogeneous case studies.

***

This volume builds on the work of the Laborgruppe Kulturtechniken ("Cultural Techniques Research Lab"), a cooperative project between the disciplines of history, literature, and media studies at the Universität Erfurt and the Bauhaus-Universität Weimar, which was funded by the State of Thuringia from 2015 to 2017, and which is now continuing its activities as the research group on spatialization and cultural techniques at the Universität Erfurt.[51]

The focus of the volume on the assembling of collectives as a core idea of cultural techniques and techniques of the body comes from the Laborgruppe's conference on "Cultural Techniques of the Collective," which was conceived and

---

51 See https://www.uni-erfurt.de/projekte/kulturtechniken/ (visited on September 12, 2019).

organized by Anne Ortner and Kristina Kuhn and took place in winter 2017 at the Universität Erfurt in cooperation with the Bauhaus-Universität Weimar.

The conference and the volume were funded and supported by The ProUni committee of the Universität Erfurt, Prof. Dr. Michael Cuntz (Bauhaus-Universität Weimar), Prof. Dr. Jörg Dünne (Humboldt Universität Berlin), Prof. Dr. Bernhard Siegert (IKKM Weimar) and Prof. Dr. Wolfgang Struck (Universität Erfurt).

We would like to thank Reed McConnell and Valentine A. Pakis for translating two contributions into English, Benjamin R. Trivers for editorial assistance with several of the translations, and in particular Michael Thomas Taylor for his great work on translating and editing the remaining contributions to this volume. Last but not least our warm thanks go out to Stella Diedrich and all the employees of De Gruyter who have patiently and competently accompanied the publication of this volume.

Spaces

Tom Ullrich
Working on Barricades and Boulevards: Cultural Techniques of Revolution in Nineteenth-Century Paris

> Paris ... is Paris only in ripping out its cobblestones.
> Louis Aragon, *Plus belles que les larmes* (1942)

> The rules of procedure reflect the state of the art. In making a statement about cultural techniques, there is thus no need to speculate about whether or not these instructions have been followed. The fact that they exist points to a certain practice.
> Cornelia Vismann, *Kulturtechniken und Souveränität* (2010)

1 The Work of Revolution

This essay takes up the question of how revolutions were made in Paris and what can be known about them by visiting the revolution at its construction sites.[1] The invested work might have to do with the construction of a barricade, a building, or a boulevard, or with a hoe and a cobblestone, or with pen and paper. And it requires us to take a closer look at the connection between mediality and processuality in this kind of revolutionary activity. My topic is thus the revolution *under construction*, radical change *in its making* – which is especially suitable for investigation by the approach offered by research into cultural techniques.

What is to be done in times of revolution? What or who is at work? Reporting on the revolutionary conditions, the republican conspirator and wine merchant Marc Caussidière writes in his memoirs about the construction of barricades in Paris that took place in February 1848:

> The *insurgent work* went ahead with an extraordinary activity, in silence, and without any military force intervening to oppose it. *Paris was a barricade construction site* from the Boulevard de Gand to the Bastille, from the Porte Saint-Denis to the Seine. The insurgent

---

[1] Construction sites are revealing sites and topoi of modernity, as was shown by the Paris exhibition "The Art of The Building Site: Construction and Demolition from the 16th to the 21st Century" (November 9, 2018, to March 11, 2019). See Valérie Nègre (ed.), *L'Art du chantier: Construire et démolir du 16e au 21e siècle* (Paris: Cité de l'Architecture & du Patrimoine, 2018).

Translated by Michael Thomas Taylor

Open Access. © 2020 Tom Ullrich, published by De Gruyter. This work is licensed under a Creative Commons Attribution 4.0 International License.
https://doi.org/10.1515/9783110647044-002

people went out on the streets with their workshop tools, before taking up arms the next day. And alas! they felled the beautiful trees of the boulevards; they tore down the gates of the monuments, the gas lanterns, the fountains, and huts, and everything that might serve to block the way of the troops. In the streets, people carried away the building materials from the construction sites: beams, stone blocks, planks, and carts. And all of this was integrated into huge cobblestone walls. ... Soon the barricades were occupied and guarded by sentries; and one could see groups of men crouched around cracking fires, melting bullets and quietly smoking their pipes at this bizarre bivouac in the middle of the big, ploughed city, so that they might plant freedom there.[2]

Caussidière, who became known to the public in 1848 because of his dedicated commitment as a "barricade prefect,"[3] testifies to the extensive construction work on the barricades, referencing the materials, tools, skills, and persons that were involved. This is, however, also a topos that is imagined and stereotypically depicted in contemporary prints (Fig. 1). If one poses the question of material and symbolic work *to* and *with*, *for* and *against*, barricades, Caussidière's testimony and its iconographic compression makes it possible to work out three central aspects for my analysis.

First, barricades initially stand out amid the modern metropolis as unusual construction sites: the revolutionaries make use of other buildings under construction and the material, tools, and workers available there. This is followed by a highly ambivalent refashioning of the urban environment, especially of the street surfaces, that literally rips apart the old conditions to establish a new, transitory, and revolutionary culture meant to allow a flourishing of political and social change. Caussidière's use of the grand concept of "liberty" to name the objective this refashioning in combination with agricultural metaphors (plough, plant) – following the practice of planting liberty trees[4] – is highly relevant for research into studies of cultural techniques, for this concept of "liberty" is clearly preceded by entire operational chains of sociotechnical processes that are expected to produce, in the first place, that very liberty in the media network of revolutionary work.

Second, this not all, because the revolution must be cultivated and maintained. After the February Revolution, for instance, Caussidière quickly took over the management of the prefecture of the police and hence the responsibility for organizing the restoration of public order. He began by seeing to the repair of the

---

2 Marc Caussidière, *Mémoires de Caussidière, ex-préfet de police et représentant du peuple* (Paris: Michel Lévy frères, 1848), 47f.; my emphasis. All translations from the French are mine, with Michael Thomas Taylor.
3 Alexandre Herzen, *Passé et méditation*, vol. 4., ed. Daria Olivier (Lausanne: Éditions l'Age d'homme, 1981), 339.
4 See Emmanuel Fureix, "Freiheitsbaum," in *Lexikon der Revolutions-Ikonographie*, vol. 2, ed. Rolf Reichardt (Münster: Rhema, 2017), 940–964.

Fig. 1: Janet-Lange, *Barricade de la Rue St. Martin*, Paris (February 23 and 24, 1848).

road pavement torn up by the construction of the barricades.[5] In doing so, the "barricade prefect" reveals an understanding of himself that marks an interesting point of tension with other postrevolutionary repair work, because this phase of restabilization and reordering following the revolutionary barricades proves to be an ideological battlefield. The debate about who is to administer the changes and consequences of a revolution (in carrying out repair work), of when and by what right, is thus quite revealing: political positions and actions are ambivalent and range from the realization of revolution to counterrevolutionary restoration, or what has been called the "Haussmannization" of Paris in the second half of the nineteenth century.[6]

Third, all of Paris is euphorically declared a "barricade construction site." The construction sites of the revolution, shaped by practices of improvisation, routine, or skill, quickly became the object of conspiratorial planning fantasies (as we find in the writings of Auguste Blanqui); and during the uprising of the Paris Commune in 1871, these practices took on the bureaucratic character of an

---

5 See Caussidière, *Mémoires de Caussidière*, 75f.
6 Clément Caraguel, "Bulletin," in *Le Charivari*, January 7, 1865.

official construction site, as if the intention had been to outdo the large construction sites of the boulevards and magnificent architectures of the Second Empire. A peculiar genealogy becomes visible here: while the *communards* trusted in an "aura of invincibility,"[7] they also relied upon the more recent model of an authoritarian imagination of order and disorder, as well as on the media-technical repertoire of their political opponents.

Viewing the all-too-often glorified *Paris révolutionnaire*[8] from the perspective of media and culture studies, through an approach based in cultural techniques, allows us to go beyond previous approaches, most of which have been rooted in one discipline only. Through a case study of selected constellations of construction sites in the context of historical authoritarian urban planning, the following essay will show how engaging with the Parisian barricades of the nineteenth century can contribute to research into cultural techniques.

2 Researching Barricades: Source Materials and Research Methods

As one of the most curious phenomena of the nineteenth century, barricades and their (mass) media distribution were situated at the interface of multiple political, social, and industrial revolutions. Given the high frequency of revolutionary events in Paris between 1827 and 1871,[9] one can assume that a significantly large part of the constantly increasing urban population came into contact with barricade construction in some way, if only through acquaintances and relatives. Beyond the concrete techniques of resistance and protest, the "barricade" also left its mark linguistically and visually, which makes it all the more difficult to uncover its sociotechnical foundation apart from the metanarratives of history. The historiographic, administrative, artistic, and popular cultural knowledge of barricades that is widely available in libraries, archives, and museums proves to be both helpful and challenging. It forces us to pose hard questions from several perspectives in confronting many highly varied sources: printed sources, such as literary narratives and journal reports, conspiratorial and autobiographical writings, or military handbooks; archival documents such as parliamentary

---

[7] Friedrich Engels, "Die revolutionäre Bewegung in Italien," *Neue Rheinische Zeitung* 156, November 30 (1848): 78.
[8] Godefroy Cavaignac et al., *Paris révolutionnaire*, 4 vols. (Paris: Guillaumin, 1833–1834).
[9] There were minor uprisings or major revolutions in Paris in 1827, 1830, 1832, 1834, 1839, 1848, 1849, 1851, and 1870/1871.

debates, police reports, court records, and files from the Paris municipal administration or national ministries; and visual representations such as paintings, prints, photographs, cartoons, and, above all, sketches, plans and maps.

The diverse references in these sources document and perform the vehemence of the controversies of the day: barricades were a favorite topic of newspaper reports; they were frequently invoked as magical objects, or fetishes (what Engels called "spells"), of revolutionary groups;[10] and they were also the object of obsessive preoccupation of governments concerned with strategies to prevent and combat them.

Both quantitatively and qualitatively, barricades in the "capital of revolution"[11] of the nineteenth century appeared as a ubiquitous phenomenon that we must analyze as a historically unique excess, treating, as equally important, both discursive and nondiscursive aspects of a comprehensive knowledge of the barricades.

Barricades "draw together";[12] they assemble and mediate persons, things, and signs in an extreme situation of open conflict. They contribute to the formation of a revolutionary identity and collective by offering categories for locating oneself in a political or social order. That is why they also massively demarcate and exclude, break up and keep apart, social strata, political opinions, and generations: they divide friends from foes, arranging them spatially and symbolically this side and that side of the barricade. Herein lies the ambivalent mediality of barricades as powerful media agents or milieus, which is not, however, identical with the permanent presence of the Parisian barricades that has been conveyed by (mass) media, as evidenced by the sources mentioned above. Historical and sociological research has almost completely ignored these aspects.[13]

Nor has there been any work on barricades in research on the history of media that investigates their discursivization and representation in the context of concrete techniques and habitualized practices.[14] Research into cultural techniques allows barricades to be understood as busy spaces of assembly: they

---

10 Engels, "Die revolutionäre Bewegung in Italien," 79.
11 Heinrich Heine, *Über Ludwig Börne* (Hamburg: Hoffmann & Campe, 1840), 143.
12 See Bruno Latour, "Visualisation and Cognition: Drawing Things Together," *Knowledge and Society: Studies in the Sociology of Culture and Present* 6 (1986): 1–40.
13 Alain Corbin and Jean-Marie Mayeur (eds.), *La Barricade* (Paris: Publications de la Sorbonne, 1997); Mark Traugott, *The Insurgent Barricade* (Berkeley: University of California Press, 2010). It is striking how the barricade is understood as a simple extension of human instrumental capacities and reduced here to a singular phenomenon determined by two antithetic poles: the military and the popular, the practical and the symbolic, the organized and the spontaneous.
14 See Olaf Briese, "Moment-Architektur: Die Kunst der Barrikade und die Kunst ihrer medialen Mythisierung," in *Berlin im 19. Jahrhundert: Ein Metropolen-Kompendium*, ed. Roland Berbig et al. (Berlin: Akademie-Verlag, 2011), 433–447.

create hybrid collectives and are held together only by the temporary relations of their elements. A perspective concerned with cultural techniques thus reveals those working contexts of human and nonhuman actors that, in their historically contingent network, produce *something* (e.g., an idea or an event) that can, in retrospect, be defamed as *trouble* or celebrated as *révolution*. In this sense, it is necessary to question the grand concepts themselves: there is no such thing as *the* revolution, *the* revolutionary, or *the* barricade as such independent of specific practices and techniques, of the agency of distributed labor and its intermeshing operational chains. All the same, these concepts do concern the big picture, in its entirety, as it becomes comprehensible through what is quite particular, namely, by looking at concepts and terms in light of their detailed enabling conditions and operative constitutions. The respective concrete work on the boulevards and barricades in Paris of the nineteenth century – the planned, discursive, and situatively executed acts, with their respective conflicting contexts of justification – becomes particularly visible where barricades are not fully stabilized, where they are controversial or plainly under construction.

3 Analyzing Barricades: Construction Sites of Revolution

Under Construction: Building Barricades between Conversion and Utopia (1827–1848)

Every barricade is inextricably linked to urban space determining the material conditions under which it is produced at great effort. This ecology of the barricade influences not only the origin of its builders, their location, and the symbolic power attributed to them, but also their choice of materials. The selection of a barricade's elements depends on three properties: "their ready availability, their selective mobility and their facility for being creatively combined into an unyielding mass."[15] For this reason, the barrel (*barrique*), from which the word "barricade" likely derives, is as iconic as it is pragmatic: it is present everywhere as medium for storage and transport, easy to roll, and filled at its destination, often with cobblestones, making it relatively stable. Seen thus, urban construction sites – with the building materials and tools they offer, which are

---

15 Mark Traugott, "Barricades as Material and Social Constructions," in *Disobedient Objects*, ed. Catherine Flood and Gavin Grindon (London: V&A Publishing, 2014), 26–33, here 28.

usually stored loosely and in a way that is easily accessible – represent a location factor that is practically irresistible.

When, after a partial success in parliament against King Charles X, an exuberant celebration of several republicans on November 19, 1827, led to two-day skirmishes with the police, isolated barricades were built again for the first time since the French Revolution of 1789. Multiple officials filed independent reports of massive, almost spectacular looting of nearby construction sites. Consider, for instance, the report written by a certain Canler, head of the local *Service de Sûreté*:

> They went down toward the Seine; when they came near the Grand-Cerf passage, they stopped in front of a building under construction, and a man with the cane cried out: To the barricades! At these words, his accomplices attacked the building, removed all the building materials and scaffolding, and a moment later, with the help of twenty clerks from the surrounding shops, had built a formidable barricade, which was immediately followed by a second.[16]

Cobblestones that had been set out to repair a road are stolen as well as the quarry stone that had already been polished for another residential building under construction. Some of these stones are also broken up specifically to be used as projectiles against the police.[17] The *statistique des barricades* attached to a barricade map to memorialize the July Revolution of 1830 reports in detail on the – sometimes gruesome – extent of such material transformations of public street space in the course of the events:

> One hundred and twenty-five thousand meters of cobblestone roads were used to build 4,055 barricades, not even considering those built with the trees on the boulevards, building materials from buildings, carts, furniture, and even human corpses.[18]

Also worth noting is an ego document by Charles Jeanne, the chief of one of the barricades built during the republican uprising of June 1832. In a letter sent from prison, Jeanne provides a detailed inside perspective on those who built the barricades, writing about their procurement management:

> A building under construction on Rue Aubry-le-Boucher, near Rue St. Martin, facilitated our execution of these defensive measures, the wooden beams and crushed stones, together with the cobblestones that were continually being broken loose, were piled up with

---

16 *Mémoires de Canler, ancien chef du service de Sûreté*, vol. 1 (Paris: F. Roy, 1882), 151ff.
17 See Annie Lauck, "Les troubles de la rue Saint-Denis ou le renouveau des barricades à Paris: Les 19 et 20 novembre 1827," in Corbin and Mayeur, *La Barricade*, 55–70.
18 Charles Motte, *Révolution de 1830: plan figuratif des barricades* (1830), Paris: Bibliothèque nationale de France, Département des cartes et plans, GE DD-5711.

extraordinary speed: a considerable quantity of plaster, carried in baskets that we found in the building, served to fill in the gaps and thus consolidate our work.[19]

The numerous building sites of the barricades attract people in an almost physical way; the often-chaotic activities associated with the barricades are also directly aimed at drawing the fence-sitters and undecideds in the middle to politically and spatially take a side. Those who choose to fight for the uprising most probably do so not (only) after a reasoned consideration of the viability of a revolutionary idea, but – assuming they are not forced to do so – out of the physical-sensual attraction to the collective that is "innervated" at the barricade.[20] This is because the ecology of the barricades is multisensual. It affects through cobblestones held in the hands, and through the proclamations, rhythmically repeated in the ear, of the people gathered together. The communal handling of the converted objects and tools creates an immersive barricade milieu in which, like the street, an old entrenched order is then broken up with relish and piled up to form a new revolutionary sign of life. It affects, too, through the sounds and the feeling of the nonhuman parts of this wild structure – "the din is the applause of objects," as Elias Canetti writes in *Crowds and Power*:

> There seems to be a special need for this kind of noise at the beginning of events, when the crowd is still small and little or nothing has happened. The noise is a promise of the reinforcements the crowd hopes for, and a happy omen for deeds to come.[21]

It seems obvious, however, that not only things migrate from the building site to the barricade, but that the personnel, too, changes from one building site to another. This is primarily comprised of simple workers, shopkeepers, and craftsmen who occasionally became spokespersons for the revolution.[22] Martin Nadaud was one such man. He came to Paris in 1830 as a fifteen-year-old seasonal worker

---

19 Charles Jeanne, *À cinq heures nous serons tous morts! Sur la barricade Saint-Merry, 5–6 juin 1832*, ed. Thomas Bouchet (Paris: Vendémiaire, 2001). In *Les Misérables* (1862), Victor Hugo famously produces, in a detailed description of the events of June 1832, a literary monument to such rebellious practices of radical conversion.
20 Walter Benjamin wrote in his *Passagenarbeit* and in his essay on the work of art of revolutions as "innervations of the collective" see W.B., *Gesammelte Schriften*, vol. 5, 2, 801, and *Gesammelte Schriften*, vol. 7, 2, 666.
21 Elias Canetti, *Crowds and Power*, trans. Carol Stewart (New York: Farrar, Straus, and Giroux, 1984 [1960]), 19.
22 It has often been pointed out how most of the great Paris uprisings and revolutions were made by the lower classes from a wide range of trades, even if, as in the case of July 1830 or February 1848, the bourgeoisie was able to write itself into history books and images as the actual revolutionary subject.

from the rural region of Creuse; like many of his fellow countrymen, he then hired himself out as a stonemason. At an early age, he became involved in secret republican societies, and he received loud applause in one such meeting after the bloody uprising of June 1832, when he assured the co-conspirators present that he knew exactly where to get hammers and boards for the next opportunity to build barricades.[23]

Urban construction sites are thus immensely significant for the actor-network building the barricades: apart from their material dimensions, this is the case additionally, and especially, with the semantization of the barricades as a site of imagination for resistance and protest extending, in some instances, to their sociotheoretical overinterpretation as sites of utopia. For instance, the collective work on barricade construction sites first served French social theorists such as Charles Fourier (1835/36) and Joseph Déjacque (1858) as examples illustrating their social visions of egalitarian cooperation, as models of free individuals and the meaningful division of labor throughout society.

Fourier elucidates his concept of "attractive labor" by comparing it with the construction of barricades during the July Revolution of 1830.[24] Déjacque even claims that barricades bring forth what is most noble in humans. He argues that barricades are the site of an ideal social situation evident, for instance, in the example of a relationship between the sexes that he sees as becoming more egalitarian and polite in times of revolution. For Déjacque, the barricade becomes a form of work that, after some "moments of passing anarchy," leads to complete harmony and a new human being.[25] At the same time, for him the nonhierarchical construction and defense of barricades represents the utopian model of just labor, inasmuch as he considers this work to be a natural and entirely just result of each individual contributing to the community according to their own abilities. He exalts the barricade as an anti-authoritarian tool:

> The various groups of workers are recruited voluntarily, as are the men in a barricade, and they are completely free to stay as long as they want, or to change to another group, or

---

23 Martin Nadaud, *Mémoires de Léonard, ancien garçon maçon* (Bourganeuf. A. Duboucix, 1895), 212f. When the revolution broke out in February 1848, the staunch socialist Nadaud had just become a foreman of a group of stonemasons repairing a construction site at the town hall of the twelfth arrondissement. From there, he and his colleagues quickly join the crowd that was simultaneously storming the Tuileries Garden. Afterwards, Nadaud continued his political career as a member of the French parliament.
24 Charles Fourier, *La fausse industrie: Œuvres complètes de Charles Fourier*, vol. 8–9 (Paris: Bossange, 1835–1836), 400f.
25 Joseph Déjacque, "Die Humanisphäre, anarchische Utopie" (1858), in *Utopie der Barrikaden*, ed. Theo Bruns (Berlin: Karin Kramer, 1980), 151.

to another barricade. There is neither a permanent nor a designated leader. The one who has the best knowledge or aptitude for this work naturally leads the others. Every man has a chance to take the initiative, depending on the abilities he claims. In turn, everyone expresses their opinions and listens to those of the others. Friendly agreement prevails; there is no authority.[26]

It is telling how much Déjacque – who took part in the Paris workers' uprising in June 1848, to then be driven into American exile in 1854 – was still willing to believe in a kind of natural and consensual organization of barricade work, while at the same time in Europe, Friedrich Engels and Auguste Blanqui were working to optimize the processes of barricade building and street fighting, confronted with the increasing number of barricade uprisings ending in defeat and an urban planning policy that was becoming increasingly repressive.

Postrevolutionary Street Work: Repairing, Renovating, Restoring (1848–1870)

In 1848, when Paris experienced two extremely violent clashes over a period of five months between workers, bourgeois, and military forces, a discourse on urban planning that had already emerged in the eighteenth century reached a point of culmination. The aim was to free the French capital from its medieval structures: "*easy circulation, hygiene and safety* in Paris's quarters" is what the architects and engineers Grillon, Callou, and Jacoubet demand;[27] and at the same time, the journalist Henri Lecouturier publishes a "Plan for a new Paris in which revolutions will be impossible."[28]

With the expanded power held from 1849 onwards by the new president, later Emperor Napoleon III, the political course was set for a fundamental transformation of the metropolis according to Saint-Simonian ideas. In response to the repeated unrest and serious sanitary deficits, a plan was drawn up by a Commission for the Beautification of Paris to manifest the new balance of governmental power in the cityscape. These priorities can be clearly read from the colored highlights of the draft: the new barracks and railway stations to be built are red, while

---

[26] Déjacque, "Die Humanisphäre, anarchische Utopie," 165.
[27] Edme Jean Louis Grillon, G. Callou, and Théodore Jacoubet, *Études d'un nouveau système d'alignements et de percements de voies publiques faites en 1840 et 1841, présentées au conseil des batimens civils, d'après l'invitation de M. le Ministre de l'Intérieur le 8 aout 1848* (1848), 28; emphasis in the original.
[28] Henri Lecouturier, *Paris incompatible avec la République: plan d'un nouveau Paris où les révolutions seront impossibles* (Paris: Desloges, 1848).

the wide boulevards connecting them with the city center and with each other are blue (Fig. 2). In the original and more explicit version of the accompanying report of Siméon, this was justified, among other things, by the need to ensure that the new "great streets of communication ... shorten distances and ... in case of insurrection, ensure immediate repression of attacks on public order."[29]

In 1853, Baron Haussmann was appointed prefect in order to quickly realize the plan. Appropriating the commission's plans, and with the full power of the state's technocratic apparatus, he set about to "provide the means to meet the needs of continually increasing circulation."[30] The politically decreed measures can be understood (following Michel Foucault) as governmental techniques intended to repair the notoriously "congested," "sick," and "unworthy" metropolis of Paris.[31]

In his lectures on governmentality, Foucault describes this discourse as that of "a good town plan"[32] that actively factors in the openness and insecurity of modernity instead of suppressing it – specifically, "no longer that of fixing and demarcating the territory, but of allowing circulations to take place, of controlling them, sifting the good and the bad."[33] The concept of "Haussmannization" can thus be translated in the following matrix as the sum of the practices for controlling flows by sorting them:

Symptom to be fought	Object of the desired circulation	Desired effect
Filth	Citizens and tourists	Beauty and uniformity
Narrowness	Commerce and traffic	Speed and consumption
Illnesses	Light and air	Health and morality
Uprisings	Troops and cannons	Security and stability

---

[29] Bibliothèque Administrative de la Ville de Paris, MS 1780, fol. 6. Cited from Casselle Pierre, "Les travaux de la Commission des embellissements de Paris en 1853," in *Bibliothèque de l'École des chartes* 155 (1997), 645–689, here 654.
[30] Georges Eugene Haussmann, *Mémoires du Baron Haussmann*, vol. 2 (Paris: V. Havard, 1893), 53.
[31] See Tom Ullrich, "Reparieren nach der Revolution: Kulturtechniken der Un/Ordnung auf den Pariser Straßen des 19. Jahrhunderts," in *Kulturen des Reparierens: Dinge – Wissen – Praktiken*, ed. Stefan Krebs, Gabriele Schabacher, and Heike Weber (Bielefeld: transcript, 2018), 373–399.
[32] Michel Foucault, *Security, Territory, Population: Lectures at the College De France, 1977–78*, trans. Graham Burchell (New York: Palgrave Macmillan, 2009), 35.
[33] Foucault, *Security, Territory, Population*, 93. On this process as an administration of the things of Paris and the work that Haussmannian media of bureaucracy carried out in planning the changes, see Antonia von Schöning, *Die Administration der Dinge: Technik und Imagination im Paris des 19. Jahrhunderts* (Zurich and Berlin: Diaphanes, 2018), 132–140.

Fig. 2: Map of the Siméon Commission for the "Beautification of Paris" (1853).

It is helpful here to consider the threat scenario, from a bourgeois perspective, of a city center inhabited by workers and craftsmen in unhygienic conditions and constant rebellion, as Haussmann again recounts much later in his memoirs (1893): his aim had been to "gut [*éventrement*] the old Paris, the quarter of the unrests and barricades, through a wide central road that gradually pierces this almost unusable labyrinth" and whose "entirely straight alignment is not suitable for the usual tactics of the local uprisings."[34]

Usually, however, the justifications for the government's goals and construction measures during the Second Empire are more implicit and less clearly separated. For example, the objectives of controlled circulation outlined above occur in alternating combinations, but always with the power to define what spreads how, and where, and to shape public urban space. In any case, this was no longer to include displeasure on the part of certain sections of the population expressed by barricading streets – an example of the power held by the idea of urban health, which, in the context of the hygienic discourse of the time, combined questions of physical well-being, morality, urban planning, and political strategy.[35] To argue with Foucault: the aim of this sorting process was

> ensuring that things are always in movement, constantly moving around, continually going from one point to another, but in such a way that the inherent dangers of this circulation are cancelled out.[36]

The security of the rulers lies in the security of the populace, whose unreliable and dissatisfied segments must be actively integrated, as Charles Merruau, Secretary General of Haussmann's Seine prefecture, announced retrospectively in 1875:

> It was no longer mobs of insurgents who crossed the city, but troops of bricklayers, carpenters, and other craftsmen who rushed to the various construction sites. If the pavement was torn up, it was not to build barricades out of cobblestones, but to enable the circulation of gas and water underneath the streets.[37]

---

34 Haussmann, *Mémoires*, 54f.
35 See Philipp Sarasin, "Die moderne Stadt als hygienisches Projekt: Zum Konzept der 'Assanierung' der Städte im Europa des 19. Jahrhunderts," in *Stadt & Text: Zur Ideengeschichte des Städtebaus im Spiegel theoretischer Schriften seit dem 18. Jahrhundert*, ed. Vittorio Magnago Lampugnani, Katia Frey, and Eliana Perotti (Berlin: Gebr. Mann, 2011), 99–112.
36 Foucault, *Security, Territory, Population*, 93.
37 Charles Merruau, *Souvenirs de l'hôtel de ville de Paris 1848–1852* (Paris: Plon, 1875), 496.

Fig. 3: Construction work to open up the Avenue de l'Opéra (1870–77) photograph by Charles Marville (excerpt).

The aim of redirecting this building activity and announcing it as a social program was not only to prevent epidemics (such as cholera in 1832 and 1849) but, in particular, to also prevent "revolution, such as the series of urban revolts."[38]

The construction boom during the Second Empire, often criticized by contemporaries for its all-encompassing dimensions, also generated a mass migration of seasonal workers within France – including, not least of all, Martin Nadaud's politicized fellow citizens – that provoked a great deal of mistrust.[39] Numerous caricatures (e.g., from Daumier) and photographs (e.g., from Marville) from the time depict the construction sites of the *démolition* as a scene full of shovels and pickaxes in action (Fig. 3).[40] A revolutionary prophecy, too, can nevertheless be formulated on the basis of these tools' ambivalent use. For example, the novelist Émile Zola writes in a newspaper article from June 1868, two years before the fall

---

[38] Michel Foucault, "Space, Knowledge, and Power," interview in *The Foucault Reader*, ed. Paul Rabinow (New York: Pantheon Books, 1984), 243.
[39] See Casey Harison, *The Stonemasons of Creuse at Nineteenth-Century Paris* (Newark, DE: University of Delaware Press, 2008).
[40] See Eric Fournier, *Paris en ruines: Du Paris haussmannien au Paris communard* (Paris: Imago, 2008).

of Napoleon III, about the massive interventions of Haussmann's project into the cityscape of Paris:

> If this prefect's army has pierced Paris in all directions, it may not want to lay down its weapons, its Homeric pickaxes that have razed half a city to the ground, and God knows what use it will make of them.[41]

As a matter of fact, in the spring of 1871, the government's construction sites once again turn into construction sites of revolution.

Organizing the Uprising: Barricades as Professional Construction Sites (1868/1871)

During the two-month uprising of the Paris Commune in 1871, the construction of barricades was officially institutionalized, subjecting some earlier considerations to a practical test. One exemplary early theorist of the barricade in its new guise of a professional construction site is Auguste Blanqui.

During the nineteenth century, the name Blanqui stood for the type of professional revolutionary whose concept of an overthrow through an energetic fight at the barricades, to be carried out by a small conspiratorial elite, was criticized by Marx and Engels as "project-spinning" (*Projektenmacherei*).[42] Blanqui, who was active in secret societies at an early age, learned his lessons from Philippe Buonarroti, the author of an important book on protosocialist movements: *Conspiracy for Equality* (1828). In a section on "the order of the insurgent movement," the book offers a detailed discussion of the necessary sequence of measures *before*, *during*, and *after* an uprising; street fighting on barricades is mentioned as an emergency option.[43]

Having gained much practical experience, in 1868 Blanqui now builds upon this written engagement with his clandestinely circulated *Instructions for an Armed Uprising*.[44] The text, extant today only as a manuscript, discusses the

---

41 Émile Zola, in *La Tribune* (June 21, 1868).
42 This is the pejorative description by Karl Marx and Friedrich Engels in a review in *Neue Rheinische Zeitung: Politisch-ökonomische Revue* 4 (April 1850): 273. See Markus Krajewski (ed.), *Projektemacher: Zur Produktion von Wissen in der Vorform des Scheiterns* (Berlin: Kadmos, 2008).
43 Philippe Buonarroti, *Conspiration pour l'Égalité dite de Babeuf* (Brussels: Librairie romantique, 1828), 192ff.
44 See the English translation of the *Instructions pour une prise d'armes* in the Blanqui Archive of Kingston University: https://blanqui.kingston.ac.uk/texts/instructions-for-an-armed-uprising-1868/ (visited on December 17, 2019), as well as my German translation "Auguste

challenges of revolutionary tactics under the new urban conditions of Paris.[45] The central thesis is that a successful uprising depends above all on solid planning, joint organization, and absolute discipline. There are extensive chapters devoted to constructing barricades and combat, with Blanqui sketching the ideal type of "regular barricade" and further indicating which materials, tools, and locations are suitable for barricades and how best to organize their assembly. The book calculates, for example, that for the "9,186 cobblestones" required for a barricade (three meters high, twelve meters wide), exactly forty-eight meters of a typical Parisian street would have to be stripped of its paving.[46] Since Blanqui factors in the enormous time pressure during an uprising, he focuses on optimizing the insurgent operational chains: the procedural succession of the structural, organizational, and communicative acts, together with their logistical networking – "with the care and in the same order as indicated above."[47]

Blanqui cannot participate in the uprising of the Paris Commune because he is arrested, although his supporters form one of the many interest groups within the movement. Surprisingly, the attacked government withdrew from Paris on the first day of the uprising, March 18, 1871, leaving the barricades that had been erected as usual to the insurgents. A little-known photograph shows one of these barricades as a typical heap of stones on a street stripped of its paving, which then serves the participants not for protection, but as a stage for photo self-presentation (Fig. 4).[48] The communards as well as a female *cantinière* pose for the camera, although on the right edge of the picture there is evidence for the work this barricade required, in two persons demonstratively holding a pickaxe and a cobblestone.

Since many communards only unwillingly comply with the request of the newly formed central committee to dismantle their barricades (the fear of a counterrevolutionary counterattack is too great, even though the revolution also depends on a suitable circulation of goods, news, and the National Guard), the construction of barricades is quickly institutionalized. As an official part of the strategy to defend the city, the construction of barricades is now assigned to a committee tasked with this purpose. In its constituent meeting of April 12, 1871,

---

Blanqui: Anleitung für einen bewaffneten Aufstand," *Zeitschrift für Medien- und Kulturforschung* 8, 1 (2017): 85–107.
45 See Tom Ullrich, "Kommentar zu Auguste Blanquis 'Anleitung für einen bewaffneten Aufstand,'" *Zeitschrift für Medien- und Kulturforschung* 8, 1 (2017): 108–120.
46 Blanqui, *Instructions for an Armed Uprising*.
47 Blanqui, *Instructions for an Armed Uprising*.
48 See Jeannene M. Przyblyski, "Revolution at a Standstill: Photography and the Paris Commune of 1871," in *Yale French Studies* 101 (2001): 54–78.

Fig. 4: Photographic staging at a barricade during the Paris Commune (March 18, 1871).

the "Barricades Commission" defines responsibilities, plans, and standard measures to guide the systematic erection of defensive structures for the resistance throughout Paris.

Gaillard Père, a shoemaker who had achieved local fame as a speaker in socialist clubs, is appointed on April 30 as head of this extraordinary Barricades Commission. He is subsequently given command of a "battalion of barricade builders," consisting of about 800 workers and 35 officers.[49] In this way, the construction of the new barricades also comes to resemble the labor conditions of large construction sites, where workers are payed and organized in shifts under the guidance of architects and engineers. Uniformed soldiers guard the finished barricades, which now reflect – with their trenches, passages, and openings for cannons – the ballistic knowledge of military fortifications that informed their construction. Especially the barricade dubbed the "Castle of Gaillard" (*Château Gaillard*) in the Rue de Rivoli attracts a great deal of attention. The progress of this construction site was intensively discussed and pictured in newspapers (Fig. 5).

---

49 Marcel Cerf, "La barricade de 1871," in Corbin and Mayeur, *La Barricade*, 323–335.

Fig. 5: Print depicting the construction work on the Château Gaillard by Vierge, in *Le Monde illustré* (April 29, 1871), with a focus on shovels and pickaxes as well as curious onlookers from the bourgeoisie.

Here, there is an insistence on the term "barricade" in its functions for offensive mobilization and collective psychology. At the same time, a maximum contrast emerges to all barricade construction sites that this essay has previously examined: first, to the disorderly cobblestone heaps from March 18, 1871; and second, to the utopian idea of collective labor that is found in the writings Déjacque and Fourier (which was not at all based on financial compensation). Moreover, the care with which the construction is carried out (because the leading protagonists of the Barricades Commission believe they had sufficient time, materials, and labor) and, last but not least, the narcissistic ambition of Gaillard Père, produce a kind of activity at the construction site that is completely different from what we find in Blanqui's instructions.

Gaillard Père's untimely overzealousness and boundless overestimation of himself – all of the structures he built were bypassed or captured without a fight by the troops in the bloody battles for Paris at the end of May 1871 – gave his contemporaries plenty of reasons to ridicule him. Yet for the question about the ambivalence of barricade construction sites that I have articulated here in terms of cultural techniques, this case is extremely telling. Gaillard Père gives himself the title of "general barricade director, commanding the battalion of barricade workers," which he usually employs to sign his letters, together with a specially made red stamp (Fig. 6).[50]

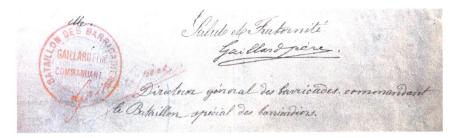

Fig. 6: Letter from Gaillard Père to the editor-in-chief of the magazine *La Commune* (excerpt, May, 1871): Père's signature as "directeur général des barricades, commandant le bataillon spécial des barricadiers" is visible in the letter.

Maxime du Camp, a close friend of the writer Gustave Flaubert and chronicler of Paris as a city enthralled by progress, reports on the peculiar impetus with which Gaillard Père sets up his headquarters at the Grand Hôtel du Louvre and, like an entrepreneur, establishes a bureaucratic apparatus for himself: letterhead paper with a letterhead in five variants corresponding to five different paper formats, as well a kind of blank power of attorney in the form of a preprinted letter allowing him, by virtue of his office, to confiscate all materials deemed necessary for the building of barricades.[51]

This constellation around the year of 1871, unique in many ways, marks the ultimate excess of the Paris barricades as a popular means of revolution that seemingly revolves only around itself. In a fatal misjudgment of its own situation, the Barricades Commission attempts to synchronize the mythical power of an outdated technology of local uprising ("action immediate, révolutionaire") with the modern idea of an industrialized order ("procédé regulier"), with the aim of defending a metropolis. The proceedings from the first meeting of the Commission April 12, 1871, reveals the entire paradox of the enterprise in a single sentence: "the barricades must be methodically studied and executed in a revolutionary way."[52]

---

[50] Raymond Huard, "Napoléon Gaillard chef barricadier de la Commune, 1815–1900," in Corbin and Mayeur, *La Barricade*, 311–322.
[51] Maxime du Camp, *Les Convulsions de Paris,* vol. 2 (Paris: Hachette, 1881), 231.
[52] Commission des barricades, "Séance du 12 avril 1871," in *Journal officiel de la République française* 103 (April 13, 1871): 237.

At the center of this historical experiment stands the unfortunate shoemaker and barricade director Gaillard Père, who, like many other communards, loves being photographed in front of his constructions.[53] His obsession with erecting particularly magnificent barricades, rather than strategically distributing as many barricades as possibly throughout the area of the uprising, comes close to that of Baron Haussmann: he also succumbs, for instance, in his very own way, to the fascination of an "administration of things" in which technology and imagination irresolvably reproduce each other.[54]

A few years later, the author and draftsman Albert Robida parodies these conditions in his futuristic novel *Le Vingtième Siècle (The Twentieth Century*, 1883), presenting the revolution, which has become a planned event taking place every ten years, as a large bourgeois folk festival in the year 1953 (Fig. 7).

In depicting a lively barricade construction site, Robida succeeds in ironically exaggerating the entire discourse on revolutionary cultural techniques of the nineteenth century as a caricature framing a page of text: in the foreground on the left, a *professeur de barricades* is bent over construction plans, offering "free lessons" in barricade construction; and on the right, zealous citizens are busy expanding the site with hoes, shovels, stones, and sandbags. A sign asks "passers-by" to contribute "a cobblestone" to the barricade ("un pavé en passant"). Above it, children are playing about on cannons that have already been put in place. Further up, three gentlemen study a map while others are drinking wine. The whole scene is being photographically documented by two large format cameras above and below the barricade. In the background, the facade of the house is solemnly decorated with posters, ultimately establishing the construction site of the revolution as a civic amusement park of a society of the spectacle-to-be (Guy Debord): "Vive le future gouvernement!" (Long live the future government!")

4 Conclusion

During what was known as the "night of the barricade" on May 10, 1968, several of the Parisian students took possession of a construction site in Rue Gay-Lussac,

---

[53] See Michaela Giebelhausen, "The Studio Collard and the Barricades of 1871: A Challenge Not Only to the Architecture of Paris," in *Nineteenth-Century Photographs and Architecture: Documenting History, Charting Progress, and Exploring the World*, ed. M.C. Nilsen (Farnham: Ashgate, 2013), 105–119.
[54] Antonia von Schöning, *Die Administration der Dinge*.

Fig. 7: "Construction des barricades," illustration, in Albert Robida, *Le Vingtième Siècle* (Paris: G. Decaux, 1883), 249.

erecting dozens of barricades in the street that had been carved out a century earlier by the prefect Haussmann. In December 2018, yellow jackets protesters – distinguishable, in their yellow warning vests, from the regular street workers only by the willingness some of them showed to commit violence – pile up construction site barriers and cobblestones loosened from the pavement on the Champs-Élysées in Paris – an action that spreads virally as #barricades.

Until now, what these recent protests represent from a perspective concerned with media and cultural techniques has been explored only to a limited extent.[55] Situating these events in Paris within a wider historical context not only makes it possible to better understand certain practices, for example, between protesters and the state. From the point of view of research into cultural techniques, barricade protests cannot be considered independently of the many tools, communications, and techniques of the body through which they are realized. Focusing on the materiality of the mediation work that barricades carry out allows us to break down both these historical and contemporary events into operations that go beyond individual actors and situated practices and would thus stand at odds to absolute categorizations of barricade building as a mere military tactic or revolutionary symbol.

By contrast, barricades are to be seen more as a specific constellation in which people, things, and signs are stabilized in a network of actors. In this sense, for example, the agency of a cobblestone is an extremely relevant factor, both in a planned uprising following Blanqui's instructions and for the urban planner Haussmann, who implemented new methods for paving roads. This has become clear in my essay's analysis of selected barricade construction sites and boulevards, and in the participation of human and nonhuman actors circulating between them in the period between the July Revolution of 1830 and the Paris Commune of 1871.

The results of this examination and of other work that would build on it allow us to write a different cultural history of "revolutionary Paris" – one that engages with the perspectives of mediality, processuality, and relativity, instead of repeating a "history of the barricade" that promises knowledge about immutable entities.[56] Such a cultural history as media history is characterized by an exploration of the remainders and margins of all different kinds of knowledge that seeks to find stories and histories in what is impure and mixed, in the in-between and simultaneity of things that are not simultaneous. For the confrontation of archival

---

[55] See Anna Feigenbaum, "Resistant Matters: Tents, Tear Gas and the 'Other Media' of Occupy," *Communication and Critical/Cultural Studies* 11, 1 (2014): 15–24.
[56] Eric Hazan's *History of the Barricade* (2015) is an account of the "myth of the barricade" that is rich in sources, but primarily anecdotal and theoretically lacking.

materials shows: both *in situ* and *in actu*, it is nearly impossible to distinguish social and technical, cultural and natural, processes in revolutionary situations. No consensus was ever reached on what barricades actually do or should do, or on what they definitely represent. It is only *a posteriori* that, for instance, a clear differentiating operation is often claimed; this distinction is by no means naturally given but rather an effect of its specific mediality. However, the Parisian barricades are in no way impervious; they have gaps and passages that connect and keep apart. They are always being rebuilt and repaired, abandoned and removed, controlled and exaggerated, observed and presented. Barricades were and are an important laboratory of modernity, and yet they play no role in existing cultural and technological histories of the construction site.[57]

If neither the human being as such nor time as such exist without the cultural techniques of hominization and of counting time that constitute them, then it is also true that "space as such does not exist independently of cultural techniques of spatial control ... The list is to be extended."[58] My proposal for expanding the list is as follows: neither the revolution, nor the revolutionaries, nor the barricades as such exists independently of the media and cultural techniques of the revolutionary through which they are constituted.

The task is thus to write their history as a genealogy of operations of stabilization and destabilization in nineteenth-century Paris. In this respect, barricades and boulevards do not emerge as closed formations in a static opposition (for example, as subversive disorder vs. monarchic/bourgeois order); rather, they become visible as agents themselves in those concrete techniques and operations that first create and determine these relations. Revolutionary cultural techniques of the barricade are – in Paris and elsewhere – nothing more than chains of interrelated negotiations between an "art of street blocking"[59] and an "art of the government."[60] In this way, the "insurgent work" of which Caussidière speaks becomes culturally observable and describable as an "art" of revolutionary change based on specific instructions for action.

---

[57] See Nègre, *L'Art du chantier*.
[58] Bernhard Siegert, "Kulturtechnik," in *Einführung in die Kulturwissenschaft*, ed. Harun Maye and Leander Scholz (Munich: Fink, 2011), 95–118, here 99.
[59] *L'illustration: Journal universelle* (May 13, 1871): 268.
[60] Michel Foucault, "Space, Knowledge, and Power," 240.

Jörg Dünne
Cultural Techniques and Founding Fictions

If one wants to take part in the current discussion about cultural techniques from the standpoint of literary criticism, an approach that deals with writing itself as a cultural technique would seem most suitable. This line of inquiry has certainly proven its relevance over the past few years, especially with regard to the operationality of writing,[1] which also unfolds in literary nexuses.[2] However, such an approach cleaves to the widespread exclusion of what have been called "primitive" cultural techniques from the field of inquiry of the humanities: while operations of writing would be counted among the reflexive "second-order" cultural techniques, "first-order" techniques – especially agricultural cultural techniques – would lack this reflexive potential.[3]

If the Cultural Techniques Working Group has set itself the goal of questioning this exclusion, and especially of assessing the agricultural sources of the concept of cultural techniques,[4] an undertaking of this kind poses special challenges to literary studies, as it even more decidedly transcends the borders of semiotic difference as the foundation of literary criticism than does the question of the operationality of writing. To the extent that literary analyses do not aim to limit themselves to understanding how cultural techniques can be described in literature on the level of content, the path from agricultural operations to literary techniques initially seems to be relatively long. This apparent incompatibility between literary studies and research into cultural techniques takes on a different appearance, however, as soon as one gives up the limit of applying findings about cultural techniques to literary texts and instead, conversely, begins to consider how research on cultural techniques invokes operations that, for their part, have literary character, or that can be regarded, alternatively, in a new light with an eye trained for reading literature in order to constitute objects of research.

---

[1] For example, see Sybille Krämer, "Operationsraum Schrift: Über einen Perspektivenwechsel in der Betrachtung der Schrift," in *Schrift: Kulturtechnik zwischen Auge, Hand und Muschine*, ed. Gernot Grube, Werner Kogge, and S.K. (Munich: Fink, 2005), 23–57.
[2] See Rüdiger Campe, *Spiel der Wahrscheinlichkeit: Literatur und Berechnung zwischen Pascal und Kleist* (Göttingen: Wallstein, 2002).
[3] Thomas Macho, "Second-Order Animals: Cultural Techniques of Identity and Identification," *Theory, Culture and Society* 30, 6 (2013): 30–47.
[4] See Bernhard Siegert, *Cultural Techniques: Grids, Filters, Doors, and Other Articulations of the Real*, trans. Geoffrey Winthrop-Young (New York: Fordham University Press, 2015), 9.

Translated by Reed McConnell, with Michael Thomas Taylor

More specifically, what follows will start from the assumption that in order to be able to describe "primitive" cultural techniques at all, one must imagine primal scenes that are necessary for setting into motion operational chains of cultural techniques, and that are fundamentally related to what are often called "founding fictions" in literary studies. By putting into play a connection of this type between research into cultural techniques and literary studies that questions the constitution of space through cultural techniques and literary operations of fictionalization, I would like to initiate a movement of double transfer: on the one hand, this approach will deal with the way that literary founding fictions can be more precisely described by means of research into cultural techniques; and on the other hand, I would also like to pursue the question of which implications of "primitive" cultural techniques become clear when placed in the context of literary research that would otherwise remain hidden.

1 "Ground-Laying" Techniques and Geopolitical Naturalizations of the Ground

Among the spatialization processes[5] constitutive of cultural techniques, the paradigm of the agricultural is especially well-suited to describe fundamental operations like the drawing of lines.[6] The "ground-laying" aspect implied in these operations should absolutely be understood in a literal sense: cultural techniques – such is the conjecture that I would like to pursue here – are not preceded by a naturally predetermined "ground" *upon* which they can develop a specific cultural technique; instead, such a ground is only constituted concomitantly through these operations. Giving primacy to practices over and against the orders that are constituted through them[7] raises the question of how the distinction between "ground" and "figure" is constituted in the first place: the differentiation of the terms of this pair of opposites, which originally stem from gestalt psychology, can, from a genealogical perspective, be traced back to specific cultural techniques such as dealings with textiles,[8] but also to the tools for cultivating land.

---

[5] On this point, see the introduction of this volume, especially, 8f.
[6] See Cornelia Vismann, "Kulturtechniken und Souveränität," *Zeitschrift für Medien- und Kulturforschung* 1, 1 (2010): *Kulturtechnik*: 171–182, here 171.
[7] See Erhard Schüttpelz, "Die medienanthropologische Kehre der Kulturtechniken," *Archiv für Mediengeschichte* 6 (2006): *Kulturgeschichte als Mediengeschichte (oder vice versa?)*: 87–110.
[8] See Lorenz Engell and Bernhard Siegert, "Editorial," *Zeitschrift für Medien- und Kulturforschung* 6, 1 (2015): *Textil*: 5–10.

Hence, "ground" and "figure" proceed from a primordial contact zone founded via specific operations, in which the two terms are not separated from the outset in the sense of a passive material, on the one hand, and an active form, on the other.

Such differentiations between "ground" and "figure" are of special pragmatic relevance wherever the "ground" is identified with the soil, or accordingly the earth in a geospatial sense. Although not nearly all operations of cultural techniques are based upon this construction, geospatially "earthed" ground-layings are of paramount importance for a history of the global consequences of "primitive" cultural techniques: from the perspective of the historical *longue durée*, one can surely say that linking cultural technique operationality to diverse manifestations of the "terrestrial" ensured the genesis of decisive thrusts in the powerful implementation of specific human forms of living and ruling, from the Neolithic Revolution with the origins of agriculture to colonial occupation in the time of terrestrial globalization.

Especially in the context of terrestrial globalization and colonization, these historically momentous founding figures are based upon the articulation of "paperwork" and the manipulation of the material world "out there," which Bruno Latour describes as the "making flat" of reality in his well-known remarks on the "immutable mobiles."[9] This occurs through forms of inscription, in which the two-dimensionality of a medial space of inscription is put in an operational relationship with the inscribed space. In Latour, the effect of "making flat" is especially conceived of with regards to the space of inscriptions and the scholarly practices associated with it, which allow the cartographer, for instance, to gain control on a two-dimensional surface over the places and operations the map represents; from a cultural history perspective there is, however, a reversal effect on the mapped space itself that must be considered, to the extent that this becomes the epitome of exactly the phenomena that are visible in its mapping: in this way, especially colonial maps[10] form potent apparatuses of perception and also generate political operations that foster the adjustment of geophysical space based upon the stipulations of space's cartographical controllability. One name for this adjustment is "geopolitics."

Carl Schmitt's remarks in *The Nomos of the Earth*,[11] have made clear – *inter alia* in the example of the famous Tordesillas Line – how intensely colonial international law regimes are predicated upon map-based colonization. There is

---

[9] Bruno Latour, "Visualisation and Cognition: Drawing Things together," *Knowledge and Society: Studies in the Sociology of Culture and Present* 6 (1986): 1–40, here 19.

[10] See William Boelhower, "Inventing America: A Model of Cartographic Semiosis," *Word & Image* 4, 2 (1988): 475–497.

[11] Carl Schmitt, *Der Nomos der Erde im Völkerrecht des Jus Publicum Europaeum* (Berlin: Duncker & Humblot, 1997 [1950]); *The Nomos of the Earth in the International Law of the Jus Publicum Europaeum*, trans. G. L. Ulmen (New York: Telos, 2006).

however an important difference between Schmitt's position and an approach concerned with cultural techniques. In regarding the earth as "mother of law,"[12] Schmitt hypostatizes the ground as always-already existing and legitimizes colonial settlement as a simple acceptance of this natural "motherly" offer. From the perspective of cultural techniques, on the contrary, it is only in the act of inscribing that the ground first begins to detach itself from what is inscribed.

As one can see in Schmitt, the geopolitical assumption of a pre-existing "ground" for operational inscription is also regularly tied to the adoption of a "primal state" from which the inscriptions of cultural techniques proceed. My thesis is that such thinking in "primal scenes," which act as "founding" regulative fictions,[13] is still a given in most approaches to cultural techniques that deal with the operationality of "primitive" cultures, even where these approaches eschew the adoption of a predefined territoriality as passive matter, which is the prerequisite for the active inscription of a form.[14] In the following, I would like to examine a historical example from Argentinian cultural and literary history – namely, the writings of Argentinian writer and politician Domingo Faustino Sarmiento – in order to show that such primal scenes are intensely literarized.

2 The Pampa and the River Delta: Sarmiento's Double Founding Fiction of the Argentinian Nation

The naturalization of the "ground" that develops through cultural techniques and results in geopolitically occupiable earth is not only the prerequisite for

---

12 Schmitt, *Nomos*, 13 (German), 42 (trans.).
13 On the significance of "regulative fictions" in political connections, see especially Albrecht Koschorke, "Macht und Fiktion," in *Des Kaisers neue Kleider: Über das Imaginäre politischer Herrschaft*, ed. Thomas Frank et al. (Frankfurt a.M.: Fischer, 2002), 73–84. My intention here is to more precisely articulate and further develop the considerations developed in these works, insofar as such fictions, at least in (geo)political connections, are related to a figure of territorial "founding." For this reason, I will hereinafter speak of founding fictions with a somewhat different, i.e. more strongly space-related accentuation of the term than Doris Sommer in her study *Foundational Fictions: The National Romances of Latin America* (Berkeley: University of California Press, 1991) on the Latin American novel of the nineteenth century, she investigates the allegorical relationship between the (nuclear) family and the nation in the fictional plots of novels.
14 For a critique of the hylemorphism implied in the requirement for a natural "earthy" ground, see Jane Bennett, *Vibrant Matter* (Durham: Duke University Press, 2010).

early modern colonial conquests but is also constitutive of the nation building in countries that arose from colonial regimes (for instance, in Latin America in the nineteenth century). In this context, the geopolitical formation of the Argentinian nation is a textbook case for the reversal effects that a medially constituted "flat" space of inscription has on material geopolitical operations with the territory constituted in this way: as counterpart to the flat space of medial inscription of the Argentinian nation, the pampa in the southwest of the capital city Buenos Aires was rendered an empty "desert" without positively describable characteristics.[15]

Such an overwriting of the geomorphology of the grassy landscape, which was admittedly actually generally flat, but was entirely manifold and above all fertile, and which is today mostly described as "pampa húmeda," via the imagining of a desert-like empty expanse, hinges not only upon cartographic, but also upon literary operations: this is evident in the work of Domingo Faustino Sarmiento, who was undisputedly the most important writer and politician of nineteenth-century Argentinian history and who, not least in his role as president from 1868 to 1874, had a decisive influence on the political fate of his country. As is characteristic for Latin American literature of this epoch, it is hardly possible to separate out literariness and political intervention in Sarmiento's texts, which also makes clear that an analysis of founding fictions does not require fictionality in a stricter sense.

In the following, I will begin with a short discussion of Sarmiento's well-known and oft-analyzed essay, in which the model of the desert unfolds in an exemplary fashion. I will then go into greater detail and aim to show that Sarmiento also designed founding fictions in his extensive work that are surely interesting from a cultural techniques perspective and that are different from those based upon the operation of "making flat" *qua* "desertization."

The Pampa and the Opposition between *civilización* and *barbárie*

The best-known model of a founding fiction for the Argentinian nation conceptualized by Sarmiento is familiar and will be presented here only in broad strokes: it has to do with the essay *Facundo*[16] written in 1845 in exile in Chile, a polemic against the dictatorship of Manuel Rosas in which Sarmiento establishes a momentous distinction, in the form of a sharp dichotomy, between urban

---
15 See Fermín Rodríguez, *Un desierto para la nación* (Buenos Aires: Eterna Cadencia, 2010).
16 Domingo Faustino Sarmiento, *Facundo* (Buenos Aires: Agebe, 2004 [1845]).

civilización after the European and North American models and indigenous or rural *barbárie*.[17] In doing so he locates not only the political opponent but the entire reach of the Argentinian nation south of the capital city of Buenos Aires on the degraded side of this oppositional relationship.

In *Facundo*, the pampa thus becomes, at least in the introductory reflections on the physical nature of the Argentinian republic,[18] an empty space whose only positive quality consists in its enormous expansiveness ("extensión"[19]). The figures of civilization that should be inscribed into this seemingly neutral ground include, for Sarmiento, not least a modern transportation infrastructure and especially a highly developed railroad network to connect the desolate spaces that it makes arable with the capital and metropole Buenos Aires.[20] The prerequisite for this is, however, a naturalization of the geospatial "surfaces of operation" into an empty, inert territory in which the organizing hand of the civilizer can act undisturbed. What is above all misappropriated in this situation is, as critical research on the history of the Argentinian nation has shown,[21] the fact that the emptying out of the desert results in the genocide of the indigenous population actually living in this "empty" space, which, toward the end of the 1870s, ultimately culminated in the military ventures of the so-called *conquista del desierto*.

It is not, however, this sharp geopolitical dichotomy of *Facundo*, which is nevertheless deconstructed in the text in many ways, that primarily interests me here. Rather, I am interested in an alternative model stemming from Sarmiento himself, with another geospatial setting that for its part has a great deal of similarity with research into cultural techniques and especially with the assumption of recursive chains of operations that proceed from the differentiation between figure and ground. As later on in Carl Schmitt, in Sarmiento aquatic spaces

---

[17] Of course, this literary stylization of Sarmiento's also has prototypes, among them Esteban Echeverría's epic poem *La Cautiva* (1837). On the literary archaeology of the pampa as "desert," see *inter alia* Jens Andermann, *Mapas de poder: Una arqueología literaria del espacio argentino* (Rosario and Berlin: Beatriz Viterbo and tranvía/Frey, 2000).

[18] See Sarmiento, *Facundo*, 22–35. A detailed reading, however, would show that over the course of the text, this empty space is consistently filled with different actors like the *gaucho*, who have a distinctly more complex relationship to their environment than that of simply moving through an "empty," purely passive space.

[19] Sarmiento, *Facundo*, 23.

[20] See Wolfram Nitsch, "La Argentina a finales de la época del caballo: Imaginaciones literarias de los medios modernos de transporte y de sus efectos culturales," in *Actas del VII Congreso Internacional Orbis Tertius de Teoría y Crítica Literaria*, ed. José Amícola (La Plata: UNLP/FAHCE, 2009), http://www.memoria.fahce.unlp.edu.ar/trab_eventos/ev.3579/ev.3579.pdf (visited on December 10, 2018).

[21] See especially David Viñas, *Indios, ejército y frontera* (México, D.F.: Siglo veintiuno, 1982).

already constitute the starting point of his alternative model. That said, Sarmiento's interest does not relate to the "model of the maritime," as with Schmitt and, building on Schmitt, with Gilles Deleuze and Félix Guattari.[22] This interest much rather relates to the transition zone between water and land in the delta of the Paraná river.

Carapachay and the Specter of Autopoietic Productivity

In comparison with his essay *Facundo*, Sarmiento's writings on the river delta collected under the title *Carapachay*,[23] which appeared starting in 1855 as scattered chronicles but above all in the national newspaper *El Nacional*, are relatively unknown. These texts are, however, especially worthy of attention insofar as they do not perpetrate the opposition between *civilización* and *barbárie* and the accompanying desertification of the *pampa*, but instead, from the perspective of cultural techniques, describe many more interesting forms of mediation between nature and culture, which, as will become clear, can be understood as an alternative founding fiction of the Argentinian nation.

Sarmiento's literary engagement with the Delta of the Paraná river, not far from Buenos Aires, is related not least to his house there, which he inhabited, with breaks, until shortly before his death. For him, however, the delta is above all a political space for the future of America: as early as 1850 he already imagines, in his essay with the title *Argirópolis*,[24] a confederation of the neighboring states of the Río de La Plata with a common capital city on the small river-island Martín García, with explicit similarities to the category of the island utopia that has circulated since Thomas More.

In his writings collected under the title of *Carapachay*, Sarmiento's report from his first trip into the river delta, which was described in a longer essay from 1857,[25] at first follows the most well-known model of the colonial occupation: together with General Mitre, who had just lost an important battle in the "desert" south of Buenos Aires against the Mapuche leader Calfucurá,[26] Sarmiento undertakes a "compensatory" replacement expedition into the Paraná Delta in 1855,

---

[22] Gilles Deleuze and Félix Guattari, *Mille plateaux: Capitalisme et schizophénie II* (Paris: Seuil, 1980), 597–602.
[23] Domingo Faustino Sarmiento, *Carapachay* (Buenos Aires: Eudeba e-book, 2012 [1855–1883]). In the following, all paragraph numbers refer to this edition.
[24] Domingo Faustino Sarmiento, *Argirópolis* (Buenos Aires: El Aleph, 2000 [1850]).
[25] Sarmiento, *Carapachay*, 557ff.
[26] See Liborio Justo's illuminating introduction to Sarmiento, *Carapachay*, 372.

which makes him in a certain sense into the conqueror, or more precisely, "inventor" of a new territory, in which case he compares himself explicitly with Amerigo Vespucci.[27] Like Vespucci, Sarmiento sees himself as inventor with regards to the river delta primarily because he gives the area the new name, "Carapachay," the origins of which I will discuss later.

However, the aforementioned repetition of the primal scene of the conquest of the Americas leads to several deferrals and reshufflings. Here, the clearly gendered model of a male/active "inscription" in the inert female/passive materiality of a territory is transformed into a distinctly more complex description of an extensive autopoietic cultural scenario of emergence. The river landscape, namely, possesses in Sarmiento's description a specific active materiality that has the entire world arise from the sediment of the river delta, without outward effect. This scenario is presented in the form of a report on creation.

The starting point, or, as Sarmiento puts it, the "first day"[28] of the creation story, which is faithful to the Old Testament model, is constituted by the moment when the sediment deposits appear on the water's surface and the vegetation begins with reeds ("junco"). This initial vegetation is complemented on the second day by bushes and trees that stabilize the process of silting up, before, on the morning of the third day, fruit trees appear that did not have to be planted by human hands according to Sarmiento. After briefly dealing with the fauna of these insular worlds on the fourth and fifth days, the sixth day of Sarmiento's history of creation introduces the first human being by describing him as follows:

> El sexto día de la creación de las islas, después de toda ánima viviente, apareció el carapachayo, bípedo parecido en todo a los que habitamos el continente, sólo que es anfibio, come pescado, naranjas y duraznos, y en lugar de andar a caballo como el gaucho, boga en chalanas en canales misteriosos, ignotos y apenas explorados, que dividen y subdividen el Carapachay en laberinto veneciano, nombre lógico que presta[n] al país los hombres que lo habitan, al revés de los otros países que dan su nombre al habitante, como de Francia francés, de España español. Aquí existía el carapachayo, sin que hubiera Carapachay, que nosotros hemos tenido que inventar, ya que nos ha cabido el honor de ser el primer Herodoto que describe estas afortunadas comarcas.

> On the sixth day of creation of the islands, once every soul was living, the *carapachayo* appeared, a biped looking in all ways like those of us inhabiting the continent, except

---

[27] Sarmiento, *Carapachay*, 1280f. The comparison with Vespucci is of consequence here insofar as Sarmiento does not claim to be the first conqueror to have set foot on this fluctuating territory, but instead to have recognized its true significance as a later visitor, and to have given the land a new name. In this way he ascribes himself a similar function to that which the historian Edmundo O'Gorman ascribes to Vespucci with regard to America in his *La invención de América* (México: FCE, 1958).
[28] Sarmiento, *Carapachay*, 591f.

that it is amphibious, eats fish, oranges, and peaches, and instead of riding a horse like a *gaucho* it rows atop barges in mysterious, unknown, and barely explored canals, which divide and subdivide Carapachay into a Venetian labyrinth, a name that is logically given to the country by those who inhabit it, the reverse of the other countries that give their name to the people inhabiting them, like the French from France, the Spanish from Spain. Here, there the Carapachayan used to exist, without there ever having been Carapachay, which we had to invent, which afforded us the honor of being the first Herodotus to describe these lucky regions.[29]

In this passage, the narrator takes on the role of Herodotus, the "father of history," who wants to create a bridge between mythic transmission and historical truth based on witness accounts by renaming a country with the name Carapachay. The primary focus of the passage, however, lies in the amphibious mixed being that bears the name "carapachayo."[30] For Sarmiento, the word "carapachayo" clearly embodies a primitive state of human history that is, however, distinctly different from the "barbaric" state of nature seen in the aforementioned essay *Facundo*. Unlike the *gaucho* as a representative of *barbárie* who rides on horseback, the *carapachayo* use barges as a means of transportation. In comparison to what Sarmiento elsewhere presents as the desolate environment of the pampa, the fluvial topography of *Carapachay* thus appears as a more heavily engineered and simultaneously more "primordial" form of civilization – an impression that Sarmiento strengthens even further by means of the connection of comparative references to numerous other aquatic locations of civilized life, from the Nile to Venice to the Mississippi and the Hudson River.

From a standpoint concerned with cultural techniques, however, the most notable feature of Sarmiento's description in this passage is the relationship between the name of the country and that of its inhabitants: for Sarmiento, the existence of the *carapachayo* precedes the country of "Carapachay"; in contrast to inhabitants of nation states with fixed boundaries, the inhabitants of this land are not, according to his depiction, named after the country. Conversely, Carapachay comes into being gradually where the *carapachayo* stops and creates new land through its activity of cultivation. In his account of creation, Sarmiento thus seems to anticipate a central feature of cultural techniques that was mentioned above, namely the precedence of practices over the orders that result from them: in this way, the space of Carapachay proceeds above all from the natural operations and the interwoven human operations of *land-making* in the river delta,

---

29 Sarmiento, *Carapachay*, 617–623 (all translations from Spanish to English by Reed McConnell).
30 Sarmiento remarks that the term comes from the *Guarani* and refers to a man who is fatigued in his actions from the lasting effects of hard work. Whether his *carapachayo* actually bears indigenous traits remains open.

instead of constituting a mere surface for the projection of an active human *nation-building*, like the pampa.

To bring my engagement with Sarmiento's *Carapachay* to a close, I would like to explore the history of the land a bit further: Sarmiento quickly connects the invented origin myth of the river delta with the modern world in his later essays;[31] and in this way, the creation story unexpectedly becomes a story of proto-capitalist production of economic added value through fruit growing. The river delta, as I have argued, is characterized by a remarkable fecundity that not only follows the tradition of specific biblical or mythological models but is instead connected very concretely to the present. The specific economic capital of the river delta consists in peaches ("duraznos"), which in Sarmiento's origin story take the place of the apple associated with the fall of humankind.[32]

Yet the product of the delta also must be delivered to the site of its consumption, meaning that the peach must initially be able to get to Buenos Aires and from there, to reach other port cities. Here, Sarmiento's interest in transport infrastructure is evident once more. Yet in his opinion, its establishment in the case of the river delta derives from nature itself, as one reads in his claim: "La naturaleza ha hecho del Carapachay el bello ideal de la viabilidad" ("Nature has made of Carapachay the beautiful ideal of viability").[33] In this context, one can once again view the amphibious nature of the *carapachayo* with his boat as a transitional figure between nature and culture: he constitutes the first link in a long chain of operations that open the river delta to economic circulation with Buenos Aires and beyond.

I cannot elaborate on the problems and aporias of this linkage of supposedly archaic forms of mobility on the river to modern engineered transportation infrastructure,[34] or on the logistics of peach transport – described in detail by Sarmiento – in baskets woven out of the reeds that grow at the river itself. All I can say here is that the baskets with which the peaches are transported become the starting point, in Sarmiento's mind, for an economically liberal fantasy of globalization that allows these and other fruits of the river delta to circulate through the entire world:

---

[31] See especially Sarmiento, *Carapachay*, 1481ff.
[32] Sarmiento intentionally withholds the fact that the peach is not an autochthonous plant of the Americas, but was instead imported from China via Europe with the first missionaries and settlers to America starting in the sixteenth century.
[33] Sarmiento, *Carapachay*, 694f.
[34] On this topic, see Wolfram Nitsch, "Insondables vías navegables: El Delta del Tigre en la literatura argentina" [forthcoming].

A este humilde instrumento de locomoción se debe hoy un comercio de millones de pesos, que no sólo provee a Buenos Aires de frutas exquisitas, sino que llega hoy a Río de Janeiro, donde, entre mangos, abacates, ananás, granadillas y extrañas frutas tropicales, se ostenta el durazno amarillo de las islas que derrota a todos los productos tórridos, salvo honorables excepciones, y se ha introducido en las costumbres fluminenses, no faltando el durazno, las peras y las manzanas de las islas en el postre de las familias menos acomodadas. Estas conquistas las ha hecho el canasto sacramental de las islas.

Due to this humble means of transport there exists today a trade worth millions of pesos that not only provides Buenos Aires with exquisite fruits, but that also reaches Río de Janeiro, where the yellow peach of the islands, which defeats all of the torrid products, save honorable exceptions, shows off among mangos, avocados, pineapple, passion fruit, and strange tropical fruits, and it has entered the customs of Río, where the peach, the pears, and the apples of the islands now do not lack in the dessert of families of lesser means. These victories have been won by the sacramental basket of the islands.[35]

In my short analysis of Sarmiento's *Carapachay* I hope to have shown that the river delta – in contrast to the empty "desert" of the pampa, which is in need of a firm intervention by a heroic initiator of civilization – is depicted in Sarmiento's work as quasi "autopoietic" space of fecundity and productivity, where the first act of creation consists not in the occupation of existing land, but in the development of a fruitful "ground" in the bottomland of the river. In *Carapachay*, with a quite remarkable sensitivity for geohistorical processes, Sarmiento replaces the geopolitical operation of *founding* on a desert-like mainland, which goes hand in hand with the dichotomous division of a territory into a subspace of *civilización* and one of *barbárie*, with a different form of ground-laying, which in a spatial sense also takes into account the vertical level of sedimentation and in a temporal sense takes into account the *longue durée* of geological land formation processes. However, by ultimately subordinating his primal scene of the co-emergence of ground and figure from the fertile river mud of Paraná entirely to his protocapitalist understanding of civilization, he ultimately once again sets the deep-historical opening of his version of the history of civilization that I have indicated into a much narrower spatiotemporal frame:[36] he functionalizes the primal scene, or the creation story that he tells, into a narrative of accretion through expanded market economy.[37]

---

35 Sarmiento, *Caparachay*, 1490–1494.
36 On "scale framing" as a way of dealing with differing forms of temporality in literary texts, see Timothy Clark, *Ecocriticism on the Edge* (London: Bloomsbury, 2015), 71–96; on scale framing in Argentinian literature see also my essay "Escribiendo el tiempo profundo: Ficciones fundacionales y el Antropoceno," *Orbis tertius* [forthcoming].
37 Other primal scenes of fluvial creation in Argentinian literature refer, for their part, to Sarmiento but ascribe, in these cases, a much more pessimistic scenario of founding violence

3 Founding Fictions and Research into Cultural Techniques

In closing I would like to return once again to the double transfer between research into cultural and literary studies, mentioned at the start, that is afforded by a reading of the essays of Domingo Faustino Sarmiento. In a certain sense, one can designate Sarmiento as a precursor of current research into cultural techniques: at least in his writings on the delta of Paraná under the title of *Carapachay*, he does not simply take the terrestrial "ground" of operations of cultural technique as a given, but instead observes its coming into being and the gradual differentiation of the relationship between ground and figure via "land-forming" chains of operations that cannot be traced back to an ur-act of the human will. Even when he seems to adopt a quasi "autopoietic" power of nature itself for the original dynamic of this differentiation, he emphasizes above all the constitutive importance of the agricultural paradigm for the further development of the river delta – and in any case he proceeds from the start from an excessive productivity that indicates that he *de facto* thinks in terms of the accumulative logic of commodities management in the age of industrialization and not in terms of an agricultural logic of sustainable land use. And related to this thought, Sarmiento develops a highly detailed scenario of transportation infrastructure connecting water and land in his further remarks on *Carapachay*, where he connects primitive forms of locomotion in the river delta with modern transportation techniques.

Hence, the model of a founding fiction in an insular delta-landscape that is based on fruit cultivation is set out much less dichotomously than the military model of the conquest of the pampa (which ultimately serves that of the creation of pastureland and consequently another form of agricultural use). Yet, it is in no way less laden with preconditions with regard to the founding fictions and the symbolic orders into which it inscribes itself: it propagates, in place of colonial settlement, an expansive model of economic growth.

Beyond this analogy between Sarmiento's *Carapachay* and current research into cultural techniques, a common principle can be found in the many different founding fictions of Domingo Faustino Sarmiento that provides information about semiotic operations of "founding" which are necessarily implied whenever research is dealing with "primitive" cultural techniques. What Sarmiento's imagining of the river delta has in common with scenarios of cultural techniques that mean to describe the differentiation between ground and figure is the fact that

---

to the specter of economic productivity characterizing the experience of the Argentinian military dictatorship. I think above all of Juan José Saer's *Nadie, nada, nunca* (1980).

in both cases, an original state must be imagined in which this differentiation is not yet a given. In this sense, research into cultural techniques that ventures to address fundamental questions is always, to a certain extent, also a history of creation, or at least it contributes to the invention of primal scenes of civilization.[38] However, the unavoidability of thinking in primal scenes or founding fictions does not in any way devalue the project of researching cultural techniques as such, insofar as this unavoidability inspires one to consider *alternative* founding scenarios and chains of operations, which contrast with the naturalization of the ground in the form of the colonial geopolitical way of thinking, while also contrasting with economically liberal specters of unceasing growth. In this sense, the remarkable versatility of Domingo Faustino Sarmiento with regard to the design of founding fictions can serve as a stimulus for research into cultural techniques. The task of creating alternative founding fictions based in a different set of cultural techniques could especially consist in clarifying whether – or to what extent – cultural techniques might be taken as a basis to conceptualize other founding discourses, not geopolitical in the conventional sense,[39] or whether non-accumulative chains of operations could be described as alternatives to economic logics of increase.[40] The extent to which contemporary literary founding fictions might do their part with regard to these questions deserve to be pursued further at the intersection of research into cultural techniques and literary studies.

---

38 On the link between foundations of culture and cultural techniques in island fictions, see also Gloria Meynen, "Die Insel als Kulturtechnik," *Zeitschrift für Medienwissenschaft* 2, 1 (2010): 79–91.
39 For an alternative conception of geopolitics, see concept of the "terrestrial" in Bruno Latour, *Où atterrir? Comment s'orienter en politique* (Paris: La Découverte, 2017).
40 See Erhard Schüttpelz, "Körpertechniken," *Zeitschrift für Medien- und Kulturforschung* 1, 1 (2010): *Kulturtechnik*: 101–120, here 114f.

Wolfgang Struck
A Message in a Bottle

A blank sheet of paper sets up the riddle around which an early short story by Bertolt Brecht revolves: "Die Flaschenpost" (The message in a bottle). Written in 1921 or 1922, but never published during Brecht's life, it tells the story of a young woman who had been left by her lover right after he proposed to her. On the "evening of decision" he had told her that he could not marry her immediately, but only after a journey to "the tropics," which might keep him away for several years. On his departure, he had given her an envelope and accepted her promise to open it only if, after three years, she had not received word from him.

> Ich öffnete nach drei Jahren, und ich fand ein leeres Blatt. Es ist weiß und dünn und völlig geruchlos, ohne einen Flecken.[1]
>
> After three years I opened the envelope, and I found a blank sheet. It is white and thin and odorless, completely immaculate.

The absence of writing directs one's attention to the paper itself. But this object, having no significant qualities, does not compensate for the lack. One elementary quality of the sheet remains, however, which becomes important in the woman's further attempts to solve the riddle: its flatness, its geometric two-dimensionality. Rejecting the possibility that her lover just played a cruel trick on her, she finds consolation in a kind of thought experiment, which not only introduces the message in a bottle from which the story's title is drawn but also performs an operation negotiating between one- and two-dimensionality:

> Eine Zeitlang beruhigte mich folgender Gedanke: Schiffer, die an der chilenischen Küste untergehen, übergeben in einer Flasche dem Meer Aufzeichnungen über ihre letzten Stunden und chilenische Fischer entkorken vielleicht nach 20 Jahren die Flasche, und obwohl sie in keiner Weise die fremden Schriftzeichen verstehen, erleben sie doch einen Untergang auf fremden Meeren nach. Wasser und Gischt haben die Schreiber vertan, aber die Schriftzeichen, frisch wie am ersten Tag, verraten nicht, wie lange es her ist. Wie lächerlich wäre die Botschaft, wäre sie lesbar; denn wie unmöglich ist es, in einem Leben ein Wort zu finden, das die Stille nicht stört, die nach Untergegangenem entsteht und irgend etwas sagt, das nicht böse ist![2]

---

[1] Bertolt Brecht, "Die Flaschenpost," in B.B., *Werke*, vol. 19, ed. Werner Hecht et al. (Berlin and Frankfurt am Main: Aufbau and Suhrkamp, 1997), 166–168, here 167.
[2] Brecht, "Flaschenpost," 167.

Open Access. © 2020 Wolfgang Struck, published by De Gruyter. This work is licensed under a Creative Commons Attribution 4.0 International License.
https://doi.org/10.1515/9783110647044-004

> For a while, I was comforted by the following thought: seafarers going down on the coast of Chile seal the record of their last hours in a bottle and commit it to the sea. Perhaps twenty years later, Chilean fishermen open the bottle, and – although they have no way of understanding the strange letters – they vividly experience a catastrophe on strange seas. Water and spray have tossed the writers aside. But the writing, fresh as on the first day, does not reveal how long ago this happened. How ridiculous the message would be if it were legible. For how impossible it is to find, in a life, a word that does not disturb the silence after a catastrophe, while still saying something that is not evil!

This "thought" does not really explain anything, but it offers comfort simply because it suspends the will to understand. It is not reading and understanding a text that establishes a community between fishermen and seafarers, but a shared maritime practice of sending messages in a bottle in the moment of catastrophe. One could read this resistance to (hermeneutic) understanding, as I have done elsewhere, as a metaphor for poetic practice itself – a metaphor that authors such as Celan and Adorno elaborated in the course of the twentieth century.[3] Here, however, I will follow a trace that leads to messages in bottles in a less metaphorical sense: as instruments of scientific research. Since the middle of the nineteenth century, (early) oceanographers made extensive use of bottles in order to track the ocean currents, and in the early 1920s bottle experiments were, though not without some objections, still considered to be useful tools of exploration.

In a first step to my reading, this practice may explain the – rather strange – analogy between the blank sheet found by the woman and the message in a bottle found by the Chilean fishermen, which is not blank at all but contains letters "fresh as on the first day." Their strangeness, however, not only subverts an understanding of the writing. It even applies to the order of the symbols, which – being found at the Pacific coast – need not be Latin but could also be, for example, Chinese or Arabic, requiring them to be read in different directions. The importance of the arrangement is further highlighted by a wordplay that the text performs within its own sphere of Latin letters. Fishermen and seafarers, the collective agents staged by the thought experiment, not only share a semantic assignment to the maritime world. In the German words *Fischer* and *Schiffer*, they also share alphabetic letters that transform into each other, as – nearly – an anagram.

If one draws a graphic representation of this transformation, it forms a plane between the two linear syntagmas of the words. Here, the addresses of the individual graphemes are voided and reorganized. And it is this (transitional) loss of

---

[3] Wolfgang Struck, "Ein Untergang auf fremden Meeren: Bertolt Brechts 'Flaschenpost,'" in *Seenöte, Schiffbrüche, Feindliche Wasserwelten: Maritime Schreibweisen der Gefährdung und des Untergangs*, ed. Hans Richard Brittnacher and Achim Küpper (Göttingen: Wallstein, 2018), 443–460.

address that leaves the letters floating free within the interspace and transforms them into "strange letters":

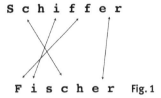

Fig. 1

The anagram creates a plane on which it is possible to perform certain operations – for example, by means of the vectors that point away from the direction of reading. In doing so, the anagram points to a process of change in the dimension between line and plane, which also marks a decisive difference between seafarers and fishermen: for seafarers, the sea is a space of transit between two harbors to be connected by the shortest possible line; for fishermen, it is a field they constantly plow, as it were, with peculiarities they know by heart from daily experience – not a "strange" sea, as for Brecht's seafarers, but part of a hybrid home; not divided by the shore but rather united. Fishermen appropriate this space through daily practice; they also segregate it, as their own fishing ground, from that of other fishermen. Seafarers, by contrast, cross borders; they claim the right of passage, and they can do so precisely because they do not intend to stay, because they do not expand the line of their passage into a plane. To put it slightly differently: both move on a plane, the surface of the sea. But sailors tend to contract this space into a line, while fishermen expand it into the space of the depths.

A similar transformation from line to plane and space takes place in marine sciences during the nineteenth century. And bottle messages are the most significant instruments of this transformation. An example where this becomes visible is the *Bottle Chart of the Atlantic Ocean,* drawn in 1843 with a second, updated edition from 1852, by the English oceanographer Alexander Becher, which he constructed on the basis of 121 found bottles and other drifting objects. It is no coincidence that this chart bears at least some small resemblance to my representation of Brecht's anagram (Fig. 2).[4]

---

[4] Alexander Bridport Becher, "Bottle Papers," *The Nautical Magazine and Naval Chronicle for 1843*: 181–184; Becher, "Bottle Chart of the Atlantic Ocean," *The Nautical Magazine and Naval Chronicle for 1852*: 568–572.

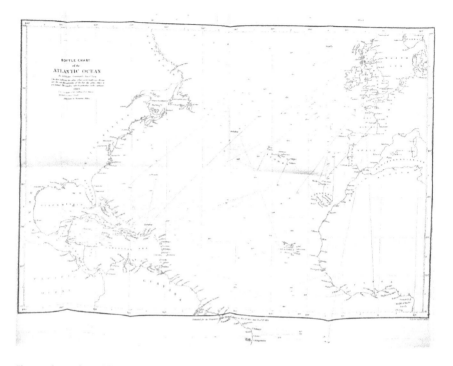

Fig. 2: Alexander Bridport Becher, *Bottle Chart of the Atlantic Ocean*, second edition (London, 1852).

The *Bottle Chart* was praised as a major breakthrough in marine sciences. But it was also rejected with equal vehemence, although – or probably precisely – because no one was in fact able to explain what it actually shows. The polar explorer Sir John Ross, for example, pointed out that the bottles cannot be observed during their voyage and, therefore, that the representation of their supposed routes of movement can be nothing but pure speculation, or, as he polemically called it: "bottle fallacy."[5] This is true but also unfair, since Becher himself noted that the lines in his chart are by no means meant to represent real trajectories. Yet he himself failed to give an answer to the question of what they show instead. One possible answer might refer to evidence based on the large number of bottle founds, i.e., in terms of probability. But this would not have been acceptable to nineteenth-century hydrography, which was eager to show that the oceans are structured not by amorphous flows but by clearly defined currents. To provide evidence in this sense, the many lines (as drawn in Becher's

---

5 John Ross, "The Bottle Chart," *The Nautical Magazine for 1843*: 321–323, here 321.

chart) would have to merge again into single lines – as is the case in charts drawn half a century later.[6]

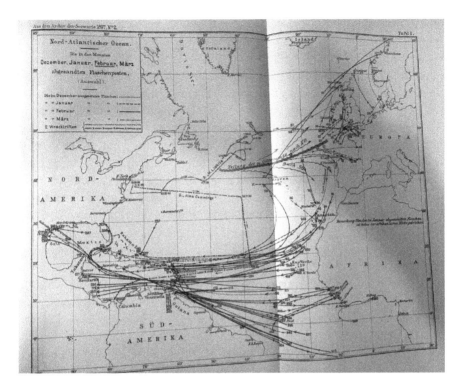

Fig. 3: Gerhard Schott, *Die Flaschenposten der Deutschen Seewarte* (Hamburg, 1897), plate 1: "Nordatlantischer Ozean: Die in den Monaten Dezember, Januar, Februar, März abgesandten Flaschenposten (Auswahl)" (North Atlantic Ocean: Messages in a Bottle Sent in the Months of December, January, February, March [Selection]).

In Becher's chart, however, something else and more fundamental happens. In their multitude, the lines make visible what is between the points of departure and arrival: not one single known, well-defined line but an unknown space, the space of not *one* actual passage but of *all* possible passages. Thus, the *Bottle Chart* carries out a paradoxical operation: transforming a line into a plane – or, if one considers the complex interactions of wind, surface and depth currents, into a space. The sea itself appears through this operation – not as a space of transit or

---

6 Gerhard Schott, *Die Flaschenposten der Deutschen Seewarte* (= *Aus dem Archiv der Deutschen Seewarte* 20, 2 [1897]).

as the medium of connectivity between the continents but as an epistemic object; or to again quote Becher, as a "fair field for discussion."[7]

In other words: what the chart makes visible are not lines of movement appearing as figures on the *ground* formed by a homogeneous surface of the sea but operators that let the sea itself appear as *figure*.

A striking embodiment of the reverse figure [*Kippfigur*] that inverts the relation of figure and ground can be found in a drift object that appears frequently in the debate about bottle messages: in 1892, the Canadian schooner *Fred B. Taylor* was rammed by the German express steamer *Trave* about 150 miles off the American coast and neatly cut into two pieces. Bow and stern both stayed on the surface for several weeks and drifted for nearly a thousand miles – but, to the puzzlement of the oceanographers, in opposite directions (as shown in the insert map on another one of Schott's charts, see Fig. 4). There have been several attempts to explain this fact (one half was more exposed to the wind, the other more to the currents ...), but apart from such explanations there remains something more fundamental. Like the bottles in Becher's chart, the two parts not only drifted into different directions, but into a different dimensionality. The collision with the express steamer, which continued undeterred on its course and arrived in New York twenty-four hours after the accident, transfers the schooner from the transit space of worldwide travel into a different space, a space defined no longer by the one, single line for traversing it but by the totality of all potential lines. The two opposite drifts are only two contingent actualizations of this potentiality.

Hence the drifting objects function as a medium enabling and compelling the sea to record and chart itself. In order to do so, however, bottle messages have to fulfill two preconditions. Their movements have to be relieved of human control (as happened with the two parts of the *Fred B. Taylor*). And they have to be relieved of the loquaciousness of humans who entrust not only the data desired by scientist to the bottles but also the individual stories they want to share with the world. To reduce such gossip, oceanographers invent forms that restrict the space for personal remarks to a minimum (see Fig. 5). Furthermore, they address their finders in different languages, thus distancing sender, relay, and receiver from their individual language and defining the sea as a space not so much of multilingualism (as was the case for so many ships in the nineteenth century), but beyond any individual language at all. As tools of scientific research, the standardized forms are operators of a reduction.

---

[7] Becher, "Bottle Papers," 181. In other words, the bottle experiments resemble what has been called an "open object" (see Lorenz Engell and Bernhard Siegert, "Editorial," *Zeitschrift für Medien- und Kulturforschung* 2, 1 [2011]: *Offene Objekte*, 5–9) or an "epistemic thing" (see Hansjörg Rheinberger, *Experimentalsysteme und epistemische Dinge* [Göttingen: Wallstein, 2001]).

A Message in a Bottle —— 67

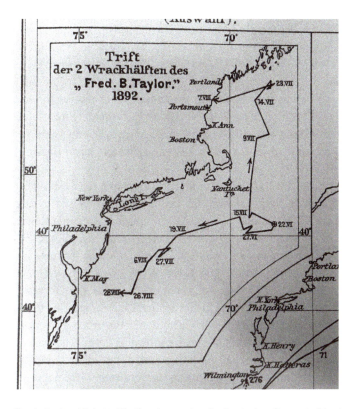

Fig. 4: Gerhard Schott, *Die Flaschenposten der Deutschen Seewarte* (Hamburg, 1897), plate 2 (extract): "Trift der zwei Wrackhälften des *Fred. B. Taylor*, 1892" (Drift of the Two Halves of the Wreck of the *Fred. B. Taylor*, 1892).

Fig. 5: Standardized form for messages in bottles used by the Hydrographic Office of the US Navy, 1893.

The illustration (Fig. 6) shows a form used by the German naval observatory (*Deutsche Seewarte)* in Hamburg, whose first director, Georg von Neumayer, was one of the most prominent researchers using bottle messages. The collection he initiated in 1867 was continued by his successors until the 1930s and contains more than one thousand messages, embedded in voluminous albums that reserve the left side of each double page for numerical data (which can be ultimately translated into the coordinates of a chart), while the original forms and all the other writings that accompanied them on their journey back to the observatory cover the right side, often in several layers.

In the example shown in Fig. 6a/b, Neumayer's assistant Gerhard Schott noted on the left-hand page that the form has been "delivered by the harbor master of Fort de France (Martinique)" ("eingeliefert von dem Hafenmeister in Fort de France"), to then extract only the data relevant for his research, namely direction and speed: "Drift in 264 days WSW/2W 2628 nautical miles, average of 10.0 nautical miles per day" ("Trift in 264 Tagen WSW/2 W 2628 Sm, durchschnittlich täglich 10,0 Sm"). The left-hand pages of the notebook thus increase the process of reduction already at work in the forms themselves. Opposite those pages, however, everything remains visible that is hidden in the process of reduction, and that will finally become invisible in the grid of a chart. The harbor master, for example, did much more than fill in the form according to the order and send it to Hamburg. He (mis)used the empty spaces for extensive remarks on the circumstances under which the bottle was found and how it made its way to him; he translated the English and German versions of its text into French (for whatever reason); he pointed out that the same year two other bottles had already been found in Martinique. And he himself drew up a very neatly laid-out map illustrating the "parcours" of the three bottles.

The abundance of the material kept in Neumayer's and other collections does not contradict the logic of reduction at work in the standardized forms. Even though they are manipulated and supplemented in multiple ways, they function as operators of a reduction transforming not only voyages into data but also voyagers – professional mariners as well as tourists – and coastal dwellers into disciplined collectors of such data. In practices of collecting and interpreting – and consequently in the final form of a map – amateur researchers such as the harbor master become participants in a collective scientific project, transforming the sea into an observable and measurable space. But this happens in the multiple operations performed by skippers, passengers, beachcombers, fishers, promenaders, harbor officials, and everyone else, who all have not only forwarded bottle messages but left their traces in the documents and thus created a maritime culture, as assembled in Neumayer's collection.

A Message in a Bottle — 69

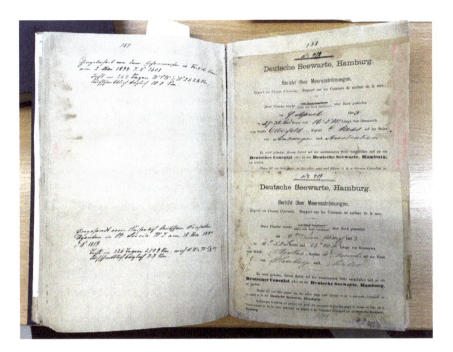

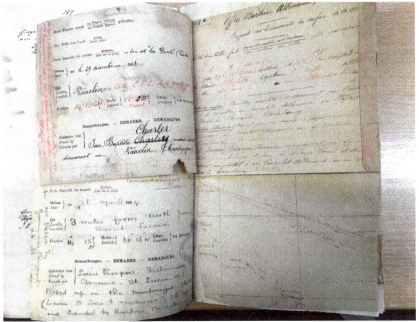

Figs. 6a/6b: Pages from Georg Neumayer's collection of messages in bottles.

Leaving an empty document before setting off to his expedition to the "tropics," Brecht's lost traveler may or may not take part in a similar scientific project. At the very least, he extrapolates a tendency vital for the scientific *persona* in the nineteenth century as it is illustrated by the research on and through bottle messages, and by its forms that continually reduce the possibility of individual expression, namely the self-erasure of the scientist as an individual subject. In order to produce objective knowledge, researchers must withdraw themselves from the field of study.[8]

Brecht's "Flaschenpost" translates this reduction into a space of *bio-graphy* and thus, in fewer than two book pages, revises the biographical model of the life path as it was cultivated by the *Bildungsroman*, probably the most successful genre in nineteenth-century German literature. In the voluminous novels of this genre, education is conceptualized as a journey through time and space, in which the protagonist's often intricate paths finally become legible as one single teleological progression. At the end of the century, when the Wilhelmine empire longed for colonial power, this model was adopted particularly by colonial novels and directed to the "tropics," now a prime subject for increasing imperial desire. Following this plotline, Brecht's male protagonist would set off on a journey supposedly forming a powerful subject. And he would enter a world where seafarers on strange coasts might easily turn out to be potential colonial aggressors. Here, in this political and literary context, is where the evil resides that threatens the silence after the catastrophe, and which only the illegible message might be able to exorcize.

But the blank paper does something more fundamental than avoiding a specific context. At least as it is read by the female protagonist, it counteracts the linearity of the paths of education and of life with a plane in which the consecutive logic of promise and fulfillment, engagement and marriage, decision and consequence, dissolves. Furthermore, this plane becomes visible as a constitutive precondition – albeit in contradiction to the linearity of narration – of every *graphy*.

Like the line, this plane is not simply there; it has to be constructed. *Tropes* and poetic figures such as the anagram are the operators that obstruct the singularity of one direction (for example, in the process of reading) and one meaning. Traveling to the "tropics," Brecht's protagonist enters a risky territory.[9] In the German imagination of the 1920s, the tropics are not only the aim of colonial desire. They are, like the sea, a space where an abundance of life erases any

---

8 See Lorraine Daston and Peter Galison, *Objectivity* (Cambridge, MA: MIT Press, 2007).

9 To borrow a phrase from Valérie November, Eduardo Camacho-Hübner, and Bruno Latour, "Entering a Risky Territory: Space in the Age of Digital Navigation," *Environment and Planning: Society and Space* 28 (2010): 581–599.

individual trace (any trace of an individual subject, that is). In literary studies, tropes are the operators of a fundamental ambivalence. Confusing and manipulating the various – material and symbolic – aspects of the process of signifying, they direct attention to this process itself as a chain of practices: collecting, arranging, selecting, assembling, directing ... All these practices are intrinsically related to specific materials (paper, for example) and operators (those vectors, for example, that negotiate between one- and two-dimensionality). This is exactly what biography does. Tracing a (single) line through the jungle of life, it is first of all a *graphy*, an act of writing that transforms the plane (of the paper, of a multitude of connections and relations) into a vector between cause and effect.[10] The idea that this transformation from one dimension to another is not irreversible is the final insight (or hope) of Brecht's female protagonist:

> Wie Sie wissen, gibt es sympathetische Tinten, die eine ganz bestimmte Zeit lang lesbar sind, dann verschwinden. Zweifellos sollte dasjenige, was aufzuschreiben sich lohnt, mit solcher Tinte geschrieben werden.[11]
>
> As you know, there are sympathetic inks, which remain legible for a certain time and then vanish. Without doubt, anything worth being written down should be written with such ink.

While the strange letters of the bottle message, "fresh as on the first day" but illegible, do not betray the writer's identity and thus, probably, liberate them from an evil history in which their own biographies are embedded, the blank paper fosters an awareness of the contingency of every text. Without value is writing that hides this contingency under an authoritative gesture suppressing all other possible writing. Only that writing has value, in other words, which implies the self-erasure of its author.

Such a reading does not solve the paradox – the riddle – of the empty sheet of paper, which – another paradox – only appears on a paper neatly covered with (Brecht's) writing. In order to appear in its flatness as a precondition of writing, the plane has to be cleansed of writing. That is the operation that Brecht's female protagonist performs in a line of thought experiments leading her to the "sympathetic ink," and thus to a conceptualization of writing that includes concepts of subjectivity, temporality, and of consecutive logic as the effects of specific scriptural operations. Stitching together writing and linearity, Brecht's "Flaschenpost" examines cultural techniques of connectivity. It exhibits not only the materiality of writing (that is, the arrangement of signs on a plane rather than on a

---

[10] For the fundamental cultural significance of linearity see Timothy Ingold, *Lines: A Short History* (London: Routledge, 2007).
[11] Brecht, "Flaschenpost," 168.

line) but the principal of linearity in a broader sense: in logic, in biography, in history ...

It may or may not be coincidence that Brecht designs his attack on the subjectivity of the *Bildungsroman* around a message in a bottle and thus an instrument scientifically used for – structurally – similar purposes. What is crucial is that he indeed has found an ally for combatting supposedly anthropologic universals with culturally and historically contingent cultural techniques – with quite specific practices that transform linearity and flatness into each other and process the difference between singularity and universality.

Following literature itself, the field of literary studies – at least insofar as it understands itself to be posthermeneutic – has developed a specific sensibility for the processuality, the proceedings [*Verfahren*] and the constitutive preconditions of writing. It is in this specific attentiveness to the material foundations of the symbolic that literary studies merges with research into cultural techniques. Both can profit from each other. Research into cultural techniques can profit inasmuch as it more fully considers those specific operations that take place in a vague space of potentiality – which in Brecht's "Flaschenpost" is the space of the thought experiment. And literary studies can learn to analyze poetic proceedings not only in relation to the self-referentiality of literary texts but also by developing a deeper interest in their epistemic effects. In this way, it may become possible to speak of a poetics of knowledge in a nonmetaphorical sense. Neumayer's atlas of bottle messages might thus become a *Bildungsroman* – in which, however, what is formed is not an individual person but a scientific collective.

Gabriele Schabacher
Waiting: Cultural Techniques, Media, and Infrastructures

In discussions on cultural techniques and in reflections on infrastructures, we come across the fact that both concepts are described by reference to materiality, interconnectedness, and processuality. However, research into cultural techniques and the field of infrastructure studies do not explicitly refer to each other. Taking this diagnosis as a starting point, the following paper explores the connection between cultural techniques and infrastructures. Such an approach is guided by the assumption that it can be productive to confront research on cultural techniques with insights from infrastructure studies, in order to elaborate on the socio-technical-discursive dimension of cultural techniques and infrastructures, respectively. Especially with regard to the notion of media, infrastructure studies could benefit from integrating a perspective on cultural techniques, just as research into cultural techniques may gain insights by taking up arguments from infrastructure studies. And even beyond that: media theory can be seen as the field where cultural techniques and infrastructures have a common ground.

The argument will unfold in four steps. In the first part, I will briefly sketch differences as well as proximities between research into cultural techniques and studies of infrastructures. In the second part, I will concentrate on the cultural technique of waiting to elaborate on its entanglement with infrastructural settings. In the third part, I will illustrate this idea by focusing on historical reflections on the upkeep of steam engines and railway maintenance in the nineteenth century. In the fourth part, I will briefly summarize important aspects and provide further perspectives.

1 Cultural Techniques and Infrastructures

Research into cultural techniques as well as infrastructure studies represent comparably young research agendas. The cultural-techniques approach originates from what has been labeled "German media theory."[1] With respect to media

---

[1] See Geoffrey Winthrop-Young, "Cultural Studies and German Media Theory," in *New Cultural Studies: Adventures in Theory*, ed. Gary Hall and Clare Birchall (Edinburgh: Edinburgh University Press, 2006), 88–104; Eva Horn, "Editor's Introduction: 'There Are No Media,'" *Grey Room* 29 (2008): 6–13.

Open Access. © 2020 Gabriele Schabacher, published by De Gruyter. This work is licensed under a Creative Commons Attribution 4.0 International License.
https://doi.org/10.1515/9783110647044-005

and communication studies in North America, German media theory was on a "collision course," as it abandoned "mass media and the history of communication in favor of those insignificant, unprepossessing technologies that underlie the constitution of meaning."[2] After its "antihermeneutic" phase from the early 1980s to the late 1990s, and in its "posthermeneutic" phase from the late 1990s to the present, media theory conceptualized media as cultural techniques.[3] In contrast to posthumanist thought in North America, which tends toward biological concepts and, therefore, toward focusing on the relation between humans and animals, the "nonhuman" stance of the German cultural-techniques approach always relates to techniques, technologies, and machines.[4] Thus, the concept of *Technik* is core: "Its semantic amplitude ranges from gadgets, artifacts, and infrastructures all the way to skills, routines, and procedures."[5] Understood in this way, the notion of *Technik* already provides a link between the dimension of practice and know-how on the one hand, and of stabilized material formations on the other. The field of infrastructure studies took up ideas from science and technology studies, from the large technical systems approach, from Actor-Network-Theory, and from practice theory in order to discuss, in particular, the challenges brought about by information infrastructures since the late 1980s.[6] The insights of infrastructure studies concerning information, energy, or transport infrastructures have been addressed with respect to their potential for media theory.[7]

---

[2] Bernhard Siegert, *Cultural Techniques: Grids, Filters, Doors, and Other Articulations of the Real* (New York: Fordham University Press, 2015), 3.
[3] Siegert, *Cultural Techniques*, 6.
[4] See Siegert, *Cultural Techniques*, 8. For an introduction to cultural techniques for an Anglo-American audience, see also Geoffrey Winthrop-Young, "Cultural Techniques: Preliminary Remarks," *Theory, Culture & Society* 30, 6 (2013): 3–19.
[5] Geoffrey Winthrop-Young, "Translator's Note," in Siegert, *Cultural Techniques*, xv.
[6] See Susan L. Star and Karen Ruhleder, "Steps Toward an Ecology of Infrastructure: Design and Access to Large Information Spaces," *Information Systems Research* 7, 1 (1996): 111–134; Paul N. Edwards et al., *Understanding Infrastructures: Dynamics, Tensions, Designs: Report of a Workshop on 'History & Theory of Infrastructure: Lessons for New Scientific Cyberinfrastructures' January 2007*, https://deepblue.lib.umich.edu/bitstream/handle/2027.42/49353/UnderstandingInfrastructure2007.pdf (visited on July 17, 2019); Geoffrey Bowker et al., "Toward Information Infrastructure Studies: Ways of Knowing in a Networked Environment," in *International Handbook of Internet Research*, ed. Jeremy Hunsinger et al. (Dordrecht and London: Springer, 2010), 97–117.
[7] See Gabriele Schabacher, "Mobilizing Transport: Media, Actor-Worlds, and Infrastructures," *Transfers* 3, 1 (2013): 75–95; Gabriele Schabacher, "Time and Technology: The Temporalities of Care," in *Hardwired Temporalities: Media, Infrastructures, and the Patterning of Time*, ed. Axel Volmar and Kyle Stine (Amsterdam: Amsterdam University Press, forthcoming). On the relevance of Actor-Network-Theory for media theory see Tristan Thielmann and Erhard Schüttpelz (eds.), *Akteur-Medien-Theorie* (Bielefeld: transcript, 2013); for respective reflections on the work

However, to mention cultural techniques and infrastructures in one breath is by no means a matter of course. At first glance, differences are notable. Infrastructures and cultural techniques seem to be located at opposite ends of the structure-agency relation. Thus, cultural techniques seem to refer to practices, infrastructures to materialities. Second, cultural techniques seem to articulate a premodern, even nonmodern, bias, while infrastructures, at least in the narrower sense, clearly seem to belong to the realm of modernity and only emerge in the horizon of industrialization during the nineteenth century. Third, cultural techniques appear to relate to (sequential) operations, while infrastructures seem to refer to (standardized) systems. Taking this into account, the question arises why it might make sense to relate the two concepts to each other. My argument is that both approaches are based on similar intuitions.

The first intuition concerns the fact that research into cultural techniques as well as infrastructure studies criticize the idea of hylemorphism, according to which subjects appear to be actors that intentionally shape the "material" of the cultural field.[8] This idea is opposed to the notion of "heterogeneous engineering"[9] by different actors that form socio-technical-discursive embroglios. Both perspectives are thus based on the idea of a fundamental interconnectedness of the actors involved (be they stones, people, or regulations). Research into cultural techniques, however, primarily has actor-networks and their trajectories in the sense of ANT in mind; it underlines the priority of the *chaîne opératoire* over the resulting artefacts and concepts.[10] In the heuristic sense, this priority means that "the task of an archeological reconstruction of technical operative chains ... is prior to *all* the involved, all reconstructed, found, documented, explored entities."[11]

---

of Susan L. Star, see Sebastian Gießmann and Nadine Taha (eds.), *Susan Leigh Star: Grenzarbeit und Medienforschung* (Bielefeld: transcript 2017).
8 For the critique of the hylemorphistic model and the idea of a different understanding of materiality, see Tim Ingold, "Toward an ecology of materials," *Annual Review of Anthropology* 41 (2012): 427–442.
9 John Law, "Technology and Heterogeneous Engineering: The Case of the Portuguese Expansion," in *The Social Construction of Technical Systems: New Directions in the Sociology and History of Technology*, ed. Wiebe E. Bijker, Thomas P Hughes, and Trevor Pinch (Cambridge, MA: MIT Press, 1987), 111–134.
10 Taking recourse to André Leroi-Gourhan, as well as to Marcel Mauss and Antoine Hennion, Erhard Schüttpelz reconstructs the relevance of these "operational chains" for the analysis of media; see Erhard Schüttpelz, "Die medienanthropologische Kehre der Kulturtechniken," *Archiv für Mediengeschichte* 6 (2006): 87–110, here 91ff.; E.S., "Die Erfindung der Twelve-Inch der Homo Sapiens und Till Heilmanns Kommentar zur Priorität der Operationskette," *Internationales Jahrbuch für Medienphilosophie* 3, 1 (2017): 165–182.
11 Schüttpelz, "Die Erfindung," 166; my translation.

Thomas Widlok's analysis of cracking Mangetti nuts in northern Namibia would be an example of such a *chaîne opératoire*.[12] Infrastructure research, on the other hand, understands the idea of interconnectedness more strongly as a type of (historical) layering, which is expressed in the concept of the path dependency of large technical systems.[13] Edwin Hutchins' analysis of distributed cognition in ship navigation can be understood as drawing together these two perspectives. Hutchins shows how the order of the steps in which something has to be done, the organization of the career cycle of navigation practitioners, and the spatial setting on board large ships are interrelated.[14] This example illustrates an aspect that is repeatedly emphasized in Susan Leigh Star's reflections on infrastructures: large-scale technical systems exist only in concrete constellations of use ("communities of practice"[15]) that determine how systems perform. Cultural techniques and infrastructures are thus both characterized by networked structuring, be it sequential chains of operations and "meshworks"[16] in the case of cultural techniques, or systems of different "communities of practice" based on distributed and standardized procedures in the case of infrastructures.

A second intuition that both research perspectives share concerns a strong interest in the invisible and implicit operations that work on a preconceptual, habitual, or indirect level to guarantee the functioning of things. Thus, cultural

---

12 Thomas Widlok, "Kulturtechniken: ethnographisch fremd und anthropologisch fremd: Eine Kritik an ökologisch-phänomenologischen und kognitiv-modularisierenden Ansätzen," in *Fremdheit – Perspektiven auf das Andere*, ed. Thomas A. Kienlin (Bonn: Rudolf Habelt, 2015), 41–59, here 45ff. Widlok also analyzes the "effective chain" to produce an arrow (Widlok, "Kulturtechniken," 53) within the process of hunting with bow and arrow as a whole as depicted by Marlize Lombard and Miriam Noël Haidle, "Thinking a Bow-and-Arrow Set: Cognitive Implications of Middle Stone Age Bow and Stone-Tipped Arrow Technology," *Cambridge Archaeological Journal* 22, 2 (2012): 237–264.
13 See Edwards et al., *Understanding Infrastructures*, 17ff.; Lawrence Busch, *Standards: Recipes for Reality* (Cambridge, MA, and London: MIT Press, 2011); for the concept of *momentum*, see Thomas P. Hughes, "The Evolution of Large Technological Systems," in *The Social Construction of Technological Systems*, ed. Wiebe E. Bijker, T.P.H., and Trevor Pinch (Cambridge, MA: MIT Press, 1989), 51–82, here 76ff.; on the "lock-in" effect of standards, see also Monika Dommann, "08/15, Querty, PAL-SECAM, Paletten und MP3: Standards als kulturelle Artefakte," in *Geltung und Faktizität von Standards*, ed. Thomas M. J. Möllers (Baden-Baden: Nomos, 2009), 253–260.
14 Edwin Hutchins, "The Technology of Team Navigation," in *Intellectual Teamwork: Social and Technological Foundations of Cooperative Work*, ed. Jolene Gallegher et al. (New York/London: Psychology Press, 1990), 191–220.
15 See Jean Lave and Étienne Wenger, *Situated Learning: Legitimate Peripheral Participation* (Cambridge: Cambridge University Press, 1991); Geoffrey C. Bowker and Susan L. Star, *Sorting Things Out: Classification and Its Consequences* (Cambridge, MA, and London: MIT Press, 1999), here 293ff.
16 Tim Ingold, *Lines: A Brief History* (London: Routledge, 2007), 80.

techniques research focuses on the preconceptual status of practices ("cultural techniques – such as writing, reading, painting, counting, making music – are always older than the concepts that are generated from them,"[17] such as "number" or "image"), on their embeddedness in architectural environments (one thinks of the operativity of architectural elements such as gate or door, wall, and corridor),[18] as well as on the appreciation of supposedly inferior cultural techniques (such as repairing).[19] This perspective on cultural techniques can be linked to the research on "invisible work,"[20] on "tacit knowledge,"[21] and on habitual routines[22] in the context of infrastructure analyses. For both approaches, the specific efficacy of practices is key, especially because they become visible only when networks fail. For this reason, moments of disruption (malfunction, accident, disaster) are highly relevant as knowledge-generating instances.[23]

---

[17] See Thomas Macho, "Zeit und Zahl: Kalender und Zeitrechnung als Kulturtechniken," in *Bild – Schrift – Zahl*, ed. Sybille Krämer and Horst Bredekamp (Munich: Fink 2003), 179–192, here 179.

[18] See Bernhard Siegert, "Door Logic, or, the Materiality of the Symbolic: From Cultural Techniques to Cybernetic Machines," in B.S., *Cultural Techniques*, 192–205; Stefan Trüby, *Geschichte des Korridors* (Paderborn: Fink, 2018); Markus Krajewski, *The Server: A Media History from the Present to the Baroque* (New Haven and London: Yale University Press, 2018), Chapter 1.

[19] See Gabriele Schabacher, "Im Zwischenraum der Lösungen: Reparaturarbeit und workarounds," *ilinx* 4 (2017): XIII–XXVIII. For studies on repair from different disciplinary backgrounds, see *Kulturen des Reparierens: Dinge – Wissen – Praktiken*, ed. Stefan Krebs, Gabriele Schabacher, and Heike Weber (Bielefeld: transcript, 2018); Stephen Graham and Nigel Thrift: "Out of Order: Understanding Repair and Maintenance," *Theory, Culture & Society* 24, 3 (2007): 1–25; Reinhold Reith and Georg Stöger, "Einleitung: Reparieren – oder die Lebensdauer der Gebrauchsgüter," *Technikgeschichte* 79, 3 (2012): 173–184; Steven J. Jackson, "Rethinking Repair," in: *Media Technologies: Essays on Communication, Materiality, and Society*, ed. Tarleton Gillespie, Pablo J. Boczkowski, and Kirsten A. Foot (Cambridge, MA, and London: MIT Press, 2014), 221–239.

[20] Susan Leigh Star and Anselm Strauss, "Layers of Silence, Arenas of Voice: The Ecology of Visible and Invisible Work," *Computer Supported Cooperative Work* 8 (1999): 9–30.

[21] Michael Polanyi, *The Tacit Dimension* (New York: Doubleday Company, 1966).

[22] Bowker and Star, for example, turn the attention to Sacks' "doing 'being ordinary'" (Harvey Sacks, *Lectures on Conservation*, vol. 2 [Oxford: Blackwell, 1992], 216) as a form of "naturalization" and "familiarity" in everyday work practice that desituates objects to the status of transparency (Bowker and Star, *Sorting Things out*, 299). With respect to work practice and information technologies, see also Paul Luff, Jon Hindmarsh, and Christian Heath (eds.), *Workplace Studies: Recovering Work Practice and Informing System Design* (Cambridge: Cambridge University Press, 2000).

[23] See Christian Kassung (ed.), *Die Unordnung der Dinge: Eine Wissens- und Mediengeschichte des Unfalls* (Bielefeld: transcript 2009); Gabriele Schabacher, "Staged Wrecks: The Railroad Crash between Infrastructural Lesson and Amusement," in *Infrastructuring Publics/Making Infrastructures Public*, ed. Matthias Korn et al. (Wiesbaden: Springer VS, 2019), 185–206.

2 Waiting

After having discussed differences as well as similar intuitions of the cultural-techniques approach and infrastructure studies, I would like to illustrate their relation to each other by focusing on the practice of waiting. Waiting is an almost unnoticed practice. This is the case for everyday life, as well as in terms of research. Only recently has waiting become a topic of interest. Primarily, the specific positioning of the subject, the affects waiting produces (such as boredom), and the time structure (delay) it goes with were regarded from philosophical, ethical, or anthropological perspectives.[24] However, waiting is also a low-threshold way of dealing with things. In this respect, waiting is neither a form of use nor a practice of invention or production. However, waiting *does* something to things. What this practice is about, and why we generally underestimate this type of handling of things, will be outlined in the following.[25]

The word *waiting* (in German *warten*) has two meanings: first, it refers to maintenance and upkeep, and second, it refers to the act of waiting and holding out. The connection between the two forms of waiting that are familiar to us today, temporal waiting and perseverance on the one hand, and procedures of maintenance on the other, is not immediately evident. If we follow the term's etymology in the most comprehensive dictionary of the German language, the *Deutsches Wörterbuch* by the Grimm brothers, the respective entry on the verb *warten* unfolds a panorama of relations between directed attention and the practices of guarding, watching, and caring – relations that are still visible in the German noun *Wärter*[26] (guard, warder). Only in the second half of the forty-two columns of Grimm's entry on *warten*, we do find the meaning of *warten* in sense

---

24 See Thomas Macho, "Waiting," in *Gregor Schneider: 7–8:30 pm 05.31.2007*, ed. Staatsoper Unter den Linden (Cologne: König, 2007), 31–39; Harold Schweizer, *On Waiting* (London and New York: Routledge, 2008); on the communicative relevance of delay, see Jason Farman, *Delayed Response: The Art of Waiting from the Ancient to the Instant World* (New Haven and London: Yale University Press, 2018).
25 For a more comprehensive analysis of the temporal dimensions of infrastructure with respect to the practices of repair, maintenance, repurposing, and abandonment, see Gabriele Schabacher, "Time and Technology."
26 "Wärter," in *Deutsches Wörterbuch*, vol. 27, ed. Jacob and Wilhelm Grimm (Leipzig: Hirzel, 1922), col. 2168–2170. Even earlier, in 1801, Johann Christoph Adelung claimed that the noun *Wärter* refers less often to the temporal dimension of waiting (for example, the gatekeeper at the gate) but primarily addresses the action of taking care for something or somebody; see "Wärter, der," Johann Christoph Adelung, *Grammatisch-kritisches Wörterbuch der Hochdeutschen Mundart*, vol. 4 (Leipzig: Breitkopf & Sohn, 1801), 1391.

of awaiting something to come.[27] In English, too, there is the dimension of guarding and care in the now obsolete meaning of the verb *to wait* in the sense of "to (keep) watch" as well as in the word *waiter*.[28] A waiter is thus primarily someone who observes carefully, and furthermore a "person who waits on or attends another" (be it as private servant or as employee in hotels and restaurants who waits upon the guests during the meals).[29] This correlation of keeping watch, care, and service is also central to the German verb *(be)wirten*, that is, to cater for somebody, which is said to stem from the basic meaning of caring ("pflegen") for somebody or something, making today's meaning of *(be)wirten* in the sense of providing hospitality only a special case (even though it is the oldest).[30]

Waiting is thus a specific form of care that consists of an attentiveness to things, people, or animals. It can be characterized as a practice that derives its temporality from the objects cared for, that is, from the things, people, and animals that it follows in order to preserve them.

Unlike the practice of repair, which can be said to react to a disruption or at least an irritation and, therefore, to work "retrospectively," the practice of waiting is directed prospectively towards the future of the objects being cared for. In doing so, waiting cultivates a specific type of concern – maintenance – that is directed towards preventing harm by caring for things, animals, and people on a regular basis. These acts of maintenance cover proper nutrition and hygiene measures for people, the dismantling, oiling, and cleaning of machines, or the replacing of wear parts. The lifetime of artefacts (and people) thus coincides with the process of their maintenance. Cost-intensive products such as buses or airplanes or infrastructures, as well as goods under conditions of scarcity such as clothes or furniture, tend to live "forever," if they are cared for properly. In the Global South, we can see such a prolongation of lifetimes of cars or bicycles, for example, that are sorted out in the first-use societies of the Western World.[31] However, the longevity that maintenance practices produce also indicates the fundamental "fragility" of

---

27 "Warten," in *Deutsches Wörterbuch*, vol. 27, col. 2125–2167, here col. 2149ff.
28 "wait, v.1," "waiter, n.," in *OED Oxford English Dictionary* (Oxford: Oxford University Press, 2019), https://www.oed.com (visited on June 22, 2019).
29 "waiter, n.," in *OED*.
30 "Wirt," in *Deutsches Wörterbuch*, vol. 30, col. 629–648, here col. 630. I thank Bernhard Siegert for the hint on the relation between *warten* and *wirten*.
31 See Hans P. Hahn, "Das 'zweite Leben' von Mobiltelefonen und Fahrrädern: Temporalität und Nutzungsweisen technischer Objekte in Westafrika," in *Kulturen des Reparierens: Dinge – Wissen – Praktiken*, ed. Stefan Krebs, Gabriele Schabacher, and Heike Weber (Bielefeld: transcript, 2018), 105–119.

matter.[32] Therefore, research on repair and maintenance is related to a general shift in perspective towards a "broken world thinking."[33] It allows to see that things do not "exist" in an easy way, but only because of unaccountable acts of work and care invested in them. This is the reason why the notion of care has been discussed recently also in science and technology studies.[34]

Nevertheless, in the context of infrastructures, maintenance work displays certain characteristics that adapt the general feature of care to what are generally considered modern conditions. As outlined above, the premodern care for things and people has a cyclic structure. Care work in this sense is repeated work that takes place on a more or less regular basis. The industrialization and urbanization of the nineteenth century bring about ideas of linear progress and permanent acceleration.[35] This also affects the cyclic structure of care, in that the maintenance of things is secularized and becomes itself scheduled according to industrial time regimes.

Two aspects should be noted here. First, the idea of so-called life cycle management, that is, the attempt to extensively exploit all the phases during the life of a product.[36] This can go along with planned obsolescence, as implanted limits for household and for consumer products (and the respective fashion cycles) allow for the exact calculation of the artefacts' life times. This illustrates that "maintainability"[37] is not the main priority when it comes to the capitalist logic of value creation. Second, maintenance processes themselves are organized according to specific regimes of timing that structure their execution, repetition, and control. Procedures such as customer services for household appliances, regular inspection for vehicles, as well as health check-ups and computer updates establish the repetition of certain maintenance activities according to specific intervals, and thus obey modern time regimes of logistics and management.

---

[32] Jérôme Denis and David Pontille, "Material Ordering and the Care of Things," *Science, Technology, & Human Values* 40, 3 (2015): 338–367, here 341.

[33] Jackson, "Rethinking Repair," 221.

[34] See Maria Puig de la Bellacasa, "Matters of Care in Technoscience: Assembling Neglected Things," *Social Studies of Science* 41, 1 (2011): 85–106; Annemarie Mol et al. (eds.), *Care in Practice: On Tinkering in Clinics, Homes and Farms* (Bielefeld: transcript 2010); Denis and Pontille, "Material Ordering."

[35] See Hartmut Rosa, *Beschleunigung: Die Veränderung der Zeitstrukturen in der Moderne* (Frankfurt am Main: Suhrkamp, 2005).

[36] See John Stark, "Product Lifecycle Management," in *Product Lifecycle Management (Volume 1): 21st Century Paradigm for Product Realisation*, ed. Stark (London et al.: Springer, 2015, third edition), 1–29.

[37] Denis and Pontille, "Material Ordering," 358.

3 The Maintenance of Steam Engines

For the adaptation of care practices to industrial settings, the history of the Technischer Überwachungsverein (TÜV) provides a good example.[38] Just as infrastructural improvements in the nineteenth century took place on the basis of a successful "learning" from accidents,[39] industrially supported maintenance activities also followed from disastrous failures of machines and infrastructures. In the nineteenth century, the most salient technical failure concerned the steam engine, or more precisely: accidents by exploding boilers. Therefore, the revision of boilers was a central task of supervision and inspection.[40] After another of these fatal accidents in a Mannheim brewery in 1865, twenty regional boiler operators founded the first inspection association for steam boilers in 1866, which later became part of the steam boiler inspection associations DÜV (Dampfkessel-Überwachungs- und Revisionsvereine). The accident prevention of these privately organized inspection associations was so successful that they were granted broader competences during the 1880s, which made them in turn politically influential, as all factories were affected by regulations of boiler inspection. Important questions that arose in connection with this reorganization of the entire steam boiler inspection business ("Dampfkesselrevisionswesen")[41] were inspection intervals and resulting costs, as the private associations prescribed boiler inspections more often than the state. Whereas the private associations demanded external inspections at least once a year, and internal inspections every two years, the state allowed longer intervals of two years for external inspections and even six years for internal inspections.[42]

In nineteenth-century publications on the topic of maintenance, the upkeep of steam engines is a prominent and recurring theme. Around 1850, explicit

---

[38] See Günther Wiesenack, *Wesen und Geschichte der Technischen Überwachungsvereine* (Cologne: Heymann, 1971).
[39] For this argument see Schabacher, "Staged Wrecks."
[40] See Wolfgang Ayaß (ed.): *Quellensammlung zur Geschichte der deutschen Sozialpolitik, I. Abteilung: Von der Reichsgründungszeit bis zur Kaiserlichen Sozialbotschaft (1867–1881)*, vol. 3: *Arbeiterschutz* (Stuttgart: Fischer, 1996); W.A. (ed.): *Quellensammlung zur Geschichte der deutschen Sozialpolitik, II. Abteilung: Von der kaiserlichen Sozialbotschaft bis zu den Februarerlassen Wilhelms II (1881–1890)*, vol. 3: *Arbeiterschutz* (Darmstadt: Wissenschaftliche Buchgesellschaft, 1998).
[41] "Report on the founding meeting of the central association of the Prussian associations of steam boiler revision," No. 41, June 21 (1884), *Zeitschrift des Vereins Deutscher Ingenieure* 25, in Ayaß, *Quellensammlung, II. Abteilung*, vol. 3, 142–145, here 143.
[42] "Memorandum of the chairman of the Steam Boiler Inspection Association in Pommern, Hugo Delbrück, for the Prussian Trade Minister, Otto Fürst von Bismarck," No. 37 (1884), in Ayaß, *Quellensammlung, II. Abteilung*, vol. 3, 124–128, here 126.

guidelines for the "Maschinenwärter" (the person operating and maintaining the machine)[43] provide preparatory remarks on the laws of combustion and the characteristics of steam, followed by chapters on how to heat boilers and keep them running as well as on their components and possible accidents.[44] The main aspects which should be kept in mind by the "Maschinenwärter" in order to maintain the functioning of the steam engine are listed A to Z. They underline the attention directed to the boiler and the necessity to select the right time for the respective actions: inspecting the water level several times a day, removing the ash if necessary, cleaning the boiler at the right times, observing and regulating the heat of the water, listening to the sound of the steam, lubricating the piston every hour, oiling the bearings at least every day etc.[45] The list is also given as a supplement in larger letters so that it can be pinned to the wall.[46]

In his guidelines *Der Führer des Maschinisten* ("The machinist's guide") from 1845, Ewald Friedrich Scholl considers the necessary attitude of the machine operator.[47] In the first part, Scholl addresses the relation of the machinist to his work and claims that besides possessing qualities such as health, strength, and endurance, as well as the ability to read and write and manual skills such as that of a blacksmith, locksmith, or carpenter, the machinist should above all pay special attention to the operation of the machine in order to notice the slightest deviation and error. He thus should display a "never resting care for the good operation of the machine."[48]

By the end of the century, the tone of the publications on maintenance work changes towards a more practical mastery of the already known malfunctions of steam engines (see Fig. 1) and their different components.[49] This development is

---

43 See "Maschinist," in *Meyers Großes Konversations-Lexikon*, vol. 13 (Leipzig and Vienna: Bibliographisches Institut, 1908), 390.
44 See Baumgartner, *Anleitung zum Heizen der Dampfkessel und zur Wartung der Dampfmaschine* (Vienna: Heubner, 1841).
45 See Baumgartner, *Anleitung*, 122–125.
46 See Baumgartner, *Anleitung*, supplement.
47 Ewald Friedrich Scholl, *Der Führer des Maschinisten: Anleitung zur Kenntnis, zur Wahl, zum Ankaufe, zur Aufstellung, Wartung, Instanderhaltung und Feuerung der Dampfmaschinen, der Dampfkessel und Getriebe: Ein Hand- und Hülfsbuch für Heizer, Dampfmaschinenwärter, angehende Mechaniker, Fabrikherren und technische Behörden* (Braunschweig: Vieweg und Sohn, 1845).
48 Scholl, *Der Führer des Maschinisten*, 7.
49 See, for example, Hermann Haeder, *Die kranke Dampfmaschine und erste Hülfe bei der Betriebsstörung: Praktisches Handbuch für Betrieb und Wartung der Dampfmaschine* (Duisburg: Selbstverlag von Hermann Haeder, 1899, second edition).

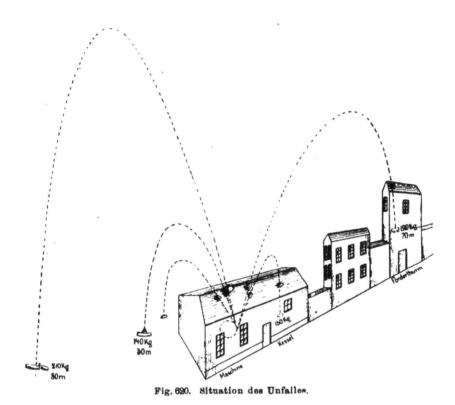

Fig. 620. Situation des Unfalles.

Fig. 1: Sketch of an accident indicating the impact of the explosion of a steam engine's fly-wheel (1899).

part of the establishment of training schools for machinists and other workers in this field.[50]

Questions of maintenance, however, are discussed at this time in relation not only to machines but also to large technical systems such as railway infrastructure. Here again, maintenance costs are an important subject. Interestingly enough, even at the end of the nineteenth century, fixed expenses for maintaining the railway as a whole were difficult to calculate, as there were no reliable data, only different opinions. "Yet I do not remember to have ever seen any attempt to estimate this cost, or to have seen it discussed," Marshall M. Kirkman argues in 1886 with respect to the "cost of maintenance and operating under *normal* conditions."[51] For this purpose, Kirkman discusses the deterioration of rails, ties,

---

50 See "Maschinenwärterschulen," in *Meyers Großes Konversations-Lexikon*, vol. 13, 390.
51 Marshall M. Kirkman, *Maintenance of Railways* (Chicago: Trivess, 1886), 21.

roadways, bridges, and so on due to climatic and other forces in correlation to the resulting costs for their maintenance (see Fig. 2). He argues that the problems in estimating these costs also depend on regionally different environmental settings.[52]

FIXED EXPENSES.

Percentage of the total cost of Operating that is due to Maintenance of Organization or that arises from Natural Decay of the property.

NAME OF ACCOUNT.	PERCENTAGE. (Fixed Charge.)
Renewal of Rails	2
Renewal of Ties	70
Repairs of Roadway and Track	57
Repairs of Bridges, Culverts and Cattle Guards	75
Repairs of Buildings	70
Repairs of Fences, Road Crossings and Signs	95
Repairs of Locomotives	8½
Repairs of Passenger Cars	9
Repairs of Freight Cars	10
Telegraph Expenses (Maintenance)	10
Agents	50
Clerks	25
Train Force	12½
Salaries General Officers and their Chief Clerks	50
Law Expenses	50
Oil, Waste and Tallow	1
Stationery and Printing	1
Contingencies (and Miscellaneous)	1
Insurance	10

In the case of a Railroad not in operation the expense would be { 5¾ / 6¾ / 9 }

In making these estimates, the wages of the force retained are reduced 50 per cent.

Fig. 2: Railway maintenance costs (1886).

In the German context, railway maintenance ("Bahnerhaltung") becomes prominent during the 1890s. Karl Hartmann, for example, discusses the disadvantages of the existing maintenance procedures that rely on gangs of workmen carrying out corrections wherever necessary, which makes them work in many places without an overall plan. He criticizes the limited reliability of such groups of workers, the lack of overview, the unfair distribution of responsibilities, and unnecessary additional costs.[53] In his opinion, one should adapt to the procedure

---

52 See Kirkman, *Maintenance*, 55.
53 Karl Hartmann, "Bahnerhaltung durch Haupt-Untersuchungen," *Organ für die Fortschritte des Eisenbahnwesens* (1892): 147–153, here 148f.

already in operation in France, where the gangs of workmen are situated near the respective section of the railway they are responsible for:

> Each section is paced off once a day. For this purpose, a workman is living at each end, whereas the others live near the middle section. The daily work of the two workers living at the end of the section begins with pacing off and examining the railway; if the worker meets the group, he reports to the foreman and joins the group.[54]

The most important suggestion concerns the introduction of regular general inspections. Hartmann distinguishes two types of causes that affect the state of the railway: regular effects and contingent effects. Whereas the contingent causes have to be removed immediately by "local repairs," the second type of change is brought about by regularly occurring weather conditions and normal operation and can only be detected by the new general inspections. They guarantee a close examination of the railway, in which respective damages are marked with chalk.[55] With his suggestion to introduce general inspections, Hartmann puts into practice an insight he already formulated in 1837: "the progression of damages can only be prevented by never interrupting the work of repair."[56]

As we can see from these examples, the discourse concerning the maintenance of steam engines in the mid-nineteenth century takes up notions of skill and care and underlines the necessity to be attentive to machines. At the end of the century, publications still refer to the skill of machinists and the type of labor they perform, but the focus now is not only on the guidance of the single worker but on the practical training of future machinists as a group. These texts therefore describe procedures of coping with malfunctions that are classified according to different categories, with the fixed expenses caused by maintenance and repair as well as the organization of regular inspection businesses (such as the DÜV in Germany). The question thus is no longer whether maintenance is important, but how to organize it. The focus shifts from the reliability of the single worker to the reliability of the inspection procedure. In this development, we find evidence for

---

54 Hartmann, "Bahnerhaltung," 150; my translation. See also "Bahnwärter," in *Enzyklopädie des Eisenbahnwesens*, vol. 1, ed. Victor von Röll (Berlin and Vienna: Urban & Scharzenberg 1912, second completely revised edition), 460–463.
55 Hartmann, "Bahnerhaltung," 151.
56 Karl Hartmann, *Praktisches Handbuch über die Anlage von Eisenbahnen, ihre Kosten, Unterhaltung und ihren Ertrag, über die Anfertigung und Prüfung guß- und stabeiserner Schienen, und die Einrichtung der Dampf- und anderen Eisenbahnwagen* (Augsburg: Jenisch und Stage, 1837), 352; my translation.

the influence of management ideas and bureaucratic organization brought about by the process of industrialization and especially the railroad.[57]

4 Conclusion

Practices of maintenance concerning devices, machines and large technical systems are in fact closely linked to basic processes within the realm of the organic: *cultura* in the sense of the Latin verb *colere* means to care for the soil in agriculture, which is fundamental for the growth and prosperity of life – plants, animals, and humans (think of the concept of education as letting a seed flourish). Maintenance in the sense of waiting is thus related to a specific temporality that demands that one waits (sometimes holds out), but always that one watches and lets things take their own time.

The notion of waiting thus points to the temporal dimension of cultural techniques, on the one hand, and infrastructures, on the other. As an agricultural, premodern practice, waiting implies a cyclic understanding of time, in which the act of waiting follows the time of the things it preserves. With industrialization and ideas of acceleration, progress, and optimization, the time of care in the sense of repair and maintenance gets industrialized, too. The natural rhythms and cycles of care become secularized as planned maintenance and repair intervals within the management of a product's life cycle.

In terms of the connection between cultural techniques and infrastructures, the activity of waiting (in the wider sense) is thus an example for a set of extremely underestimated practices of concern that are relevant to the continuity of culture, society, and technology. The relevance of these cultural techniques of care can only become visible, however, if they are not exclusively analyzed in relation to the present and in differentiated settings such as the hospital or the repair workshop. Rather, their historical development and premodern bias has to be taken into account, allowing one to locate the waiting of people and the waiting of things in one and the same field. In doing so, it is possible to show how cultural techniques of care (for things, people, etc.) represent indispensable practices within the framework of infrastructural systems, without which these could not exist and "survive."

---

[57] See James R. Beniger, *The Control Revolution: Technological and Economic Origins of the Information Society* (Cambridge, MA, and London: Harvard University Press, 1986); JoAnne Yates, *Control through Communication: The Rise of System in American Management* (Baltimore: Johns Hopkins University Press, 1989).

Christoph Eggersglüß
Orthopedics by the Roadside: Spikes and Studs as Devices of Social Normalization

Fig. 1: Planter with deterrents/armoring, New York, 2012.

1 The Subtle Circumstances of Technosocial Control

In June 2014, a few square meters of a building entrance on London's Southwark Bridge Road caused a sensation. The news about the niche (some called it an alcove) barely big enough for one person to fit in (which is why it was also problematic) quickly spread in the English-language media, triggered solely by a photo

---

Translated by Michael Thomas Taylor

∂ Open Access. © 2020 Christoph Eggersglüß, published by De Gruyter. This work is licensed under a Creative Commons Attribution 4.0 International License.
https://doi.org/10.1515/9783110647044-006

taken by a passer-by on their way to work.[1] Social policy, private householder's rights, and engineering know-how intersected in this apparently undesired, subsequently obstructed sleeping niche. Mayors, social workers, and dismayed citizens soon gathered around this barely visible spot. What was it about this unplanned, open shelter that caused people to make such a fuss? Why did people argue so vehemently about whether this smallest of places should offer refuge to individuals who were marginalized? Critics lamented the numerous, brightly cleaned, thumb-sized steel cones on the ground of the public sleeping space. They objected to what they saw as the clearly visible, inhuman architecture of a little area that would presumably deny the homeless their very last place of retreat.[2] What did their anger focus on, specifically?

Deterrents: spikes and studs, embedded and anchored in the concrete, arranged in rows in order to seal off a potential surface for lying down against any use for social exchange and make it permanently inhospitable. The intention was to make it difficult for the sensitive human body to settle down at this spot, to make it even less likely, because of this anti-homeless armoring, that rough sleeping (or sleeping on the street as a homeless person), something that is already exhausting enough, would happen. The uproar was directed, one might argue from the perspective of the more recent technosocial vocabulary of science and technology studies (STS) or Actor Network Theory (ANT), against an architectural, antisocial program realized by these anti-homeless devices.[3] The issue was the architecturally unfortunate continuations of a microstructural regulatory problem of public budgets in the registers of defensive microarchitecture, expressed here in the concrete, subsequently spiked sleeping niche: human beings and things – or more precisely: human beings and the ground – should no longer maintain lasting contact at this place. The relational structure was regulated by the disruptive third that became the focus of attention on the surface and thus decisively shaped the discourse of such antisocial architectures.[4]

---

[1] "Anti-homeless studs at London residential block prompt uproar," at *theguardian.com*, *The Guardian*, June 7, 2014, http://www.theguardian.com/society/2014/jun/07/anti-homeless-studs-london-block-uproar/ (visited on January 7, 2018).
[2] Josh Halliday, "Council urged to act over 'inhumane' use of spikes to deter homeless," *The Guardian*, June 8, 2014, http://www.theguardian.com/society/2014/jun/08/metal-spikes-london-flats-homeless/ (visited on January 7, 2018).
[3] See Robert Rosenberger, *Callous Objects: Designs against the Homeless* (Minneapolis: University of Minnesota Press, 2017), and Fredrik Edin, *Exkluderande Design* (Stockholm: Verbal, 2017).
[4] See Michel Serres, *The Parasite* (Minneapolis: University of Minnesota Press, 2007).

Orthopedics by the Roadside: Spikes and Studs as Devices of Social Normalization — 89

In July 2014, this "very important object," the *Kent Spike StudKSS/35/45* from Kent Stainless Ltd., 73 mm × 35 mm, with a highly polished, brand-new counterpart of *Spike Stud, 2014, Stainless Steel*, was acquired by the Victoria and Albert Museum as part of a new form of so-called rapid response collecting. Presumably because of the bad press, however, the product is no longer listed on the manufacturer's website (Fig. 2).[5]

Fig. 2: Kent Stainless Studs, 2014.

---

5 There, one reads: "The Kent Spike Stud is used to deter people from unwanted sitting areas such as window sills and wall tops. Its Spike like [sic] design makes it a good deterrent but also not a sharp and dangerous object. The stainless steel material is ideal for high-traffic pedestrian areas and does not adversely affect the aesthetics of the elements they are protecting." Kent Stainless, "Kent Spike StudKSS/35/45," http://www.kentstainless.com/our-products/stainless-steel-street-furniture/studs/kent-spike-stud, (accessed via Wayback Machine; the item is no longer listed on the Kent Stainless site). See also Lauren Collins, "Very Important Objects," *The New Yorker*, July 28 (2014), http://www.newyorker.com/magazine/2014/07/28/important-objects/ (visited on July 29, 2018).

Used on the road, these few, purposefully placed bolts thus change the properties of the ground and create a fait accompli.[6] Yet the defense practiced by the spike does not follow a merely reactive scheme in the sense of resistance. Rather, the ability to be one factor determining a course of movement can be seen as a potential to act, hence as an "initiative to act," and as an invitation to take (or make) other paths.[7] While the relationship between the spike and the human body was short-lived, the pointed disputes found enduring resonance. Just a few days later, a citizen's initiative stirred at the windowsill of a Tesco store: "Customers told us they were intimidated by antisocial behaviour outside our Regent Street store and we put studs in place to try to stop it," reported a spokesman for the supermarket chain. "These studs have caused concern for some, who have interpreted them as an anti-homeless measure, so we have decided to remove them to address this concern. We will find a different solution and hope this clears up any confusion."[8]

Yet this is not only an outline of how things played out in the mass media. Rather, we also have here several examples of how a misguided solution to a problem got people going by means of things. Ultimately, people transformed steps and ledges from a mere matter of fact to a matter of dispute, causing architecture to teeter as an (a)social problem.[9] Attention was focused

---

[6] One could also speak here of a reversal of the affordance of the formerly idle and unused surface, which now mediated an anti-affordance of not-lying; see James J. Gibson, *The Ecological Approach to Visual Perception* (Boston, MA: Houghton Mifflin, 1979). On the problems of thinking the furnishing of the world, see Tim Ingold, "Bindings against Boundaries: Entanglements of Life in an Open World," *Environment and Planning A* 40, 8 (2008): 1796–1810. And on the anthropological significance of the ground beneath one's feet, see Tim Ingold, "Culture on the Ground: The World Perceived Through the Feet," *Journal of Material Culture* 9, 3 (2004): 315–340.

[7] Erhard Schüttpelz, "Elemente einer Akteur-Medien-Theorie," in *Akteur-Medien-Theorie*, ed. Tristan Thielmann and E.S. (Bielefeld: transcript, 2013), 9–78, here 10.

[8] Josh Halliday, "Tesco to remove anti-homeless spikes from Regent Street store after protests," *The Guardian*, June 12 (2014), http://www.theguardian.com/society/2014/jun/12/tesco-spikes-remove-regent-street-homeless-protests/ (visited on July 1, 2018). The unwanted bad press motivated Tesco to immediately promise nothing less than an architectural remedy, whereby it failed to address the homeless situation. For this solution to allegedly antisocial behavior also operated within the registers of combating symptoms. According to the mayor's tweets, such architectural measures should not be so visible, anyway: "Spikes outside Southwark housing development to deter rough sleeping are ugly, self defeating & stupid. Developer should remove them ASAP," Josh Halliday and Haroon Siddique, "Boris Johnson calls for removal of anti-homeless spikes," *The Guardian*, June 9 (2014), http://www.theguardian.com/society/2014/jun/09/boris-johnson-calls-removal-anti-homeless-spikes/ (visited on July 1, 2018).

[9] See Theo Deutinger: *Handbook of Tyranny* (Zürich: Lars Müller Publishers, 2018), 85ff. Bruno Latour, "Why Has Critique Run out of Steam? From Matters of Fact to Matters of Concern," *Critical Inquiry* 30 (Winter 2004): 225–248.

not only on these small things. They defined the formerly marginal, smallest surfaces as places for "undesirables". In belatedly installing immobile street furniture and armoring, they also made it possible to even more narrowly outline certain groups of "undesirables" in the entryway to the building. It was not only that the small things on the surfaces open to all were successfully intertwined with disciplinary measures and regulatory mechanisms of social defense[10]: by influencing the network of relationships between people and things, the balance of power was newly arranged and the patterns of movement were adapted locally. The surfaces were moved from the infrastructural background of the city into the foreground and connected to a strategic way of thinking. The newspaper reports, pictures, and observations highlighted the urban background. They focused attention on what would have remained invisible in everyday life as long as it did not constantly break down.[11] Niches and street furniture, which had previously been accepted as such, no longer existed all throughout these reinforced surfaces, beneath the threshold of perception of an informed user. Rather, they clearly stood out.[12] The spikes and bolts completed a *gestalt switch* and brought the ground to the fore, together with its discourses and contexts:

> This inversion is a struggle against the tendency of infrastructure to disappear (except when breaking down). It means learning to look closely at technologies and arrangements that, by design and by habit, tend to fade into the woodwork (sometimes literally!).
>
> Infrastructural inversion means recognizing the depths of interdependence of technical networks and standards, on the one hand, and the real work of politics and knowledge production on the other. It foregrounds these normally invisible Lilliputian threads and furthermore gives them casual prominence in many areas usually attributed to heroic actors, social movements, or cultural mores.[13]

In the short film *Le Repos du Fakir* by Stéphane Argillet and Gilles Paté, this principle of inversion is not only illustrated; it even becomes a method. Here, it serves to demonstrate a politics of urban furnishing.[14] Paté sets out onto uncomfortable

---

10 See Michel Foucault, *Society Must Be Defended. Lectures at the Collège de France (1975–1976)* (New York: Picador, 2003).
11 See Stephen Graham and Nigel Thrift, "Out of Order: Understanding Repair and Maintenance," *Theory, Culture & Society* 24, 3 (2007): 1–25.
12 See Geoffrey Bowker and Susan Leigh Star, *Sorting Things Out: Classification and its Consequences* (Cambridge: MIT Press, 2000), 35, and Susan Leigh Star, "The Ethnography of Infrastructure," *American Behavioral Scientist* 43, 3 (1999): 377–391.
13 Bowker and Star, *Sorting Things Out: Classification and its Consequences*, 34.
14 Stéphane Argillet and Gilles Paté, *Le repos du Fakir* (Paris: Production Canal Marches, 2003), 06:20 min.

"furniture" of the Paris metropolis, traveling into the blocked off areas, the niches that have been obstructed and made unavailable, in order to make visible the design intentions and possibilities and their mechanisms of exclusion – even to those who would otherwise pass by (Figs. 3–5). With his body, he imitates a resistance that everyday inhabitants of these spaces, or even city users – as they are called these days in the positive relabeling that is, given the homeless, at the same time rather insidious – are unable to muster.

2 Orthopedic Surfaces and Their Media Theory

In the seemingly impenetrable, vast hustle and bustle of these kinds of forces of attraction and repulsion in public spaces, numerous steps of differentiation and delimitation take place between bodies. What perspective do these processes offer for the media-theoretical description of the entanglement of technology and politics that is operative in these places, which could be described, loosely following Michel Foucault, as a "social orthopedics?"[15] A closer look reveals two modes of normalization, of influencing, moving, and arranging bodies: one operates at specifics points of application, physically, i.e., orthopedically; and the other operates visually, meaning it regulates, reduces and strengthens relationships of visibility. What both regulating procedures or theoretical views have in common is that bodies are constantly dispatched, guided, diverted, adjusted, and placed in certain ways, that transitions and changes of actions take place in which individual stages and mediators of certain norms and policies can be identified, whether they be small, artificial proxies or other bodies guided by additional bodies:

> 'Follow the actors!' thus means, more precisely, 'Follow the mediators!' – with the focus here on the capacity of all entities and factors involved to allow other entities to take action, to initiate or delegate actions.[16]

As representatives, mediators, and delegates, the spikes regulate both the visibilities and the local compositions of the social structure: they not only explicate disciplinary measures; they also process ideal programs and ideas about norms, whether from engineers, politicians, neighborhood managers, or indeed owners.

---

[15] See Michel Foucault, *Discipline and Punish* (New York: Random House, 1977), and "Truth and Juridical Forms," in M.F., *Power – Essential Works of Foucault 1954–1984*, vol. 3, ed. James D. Faubion (London: Penguin, 2002), 1–89, here 17.
[16] Schüttpelz, "Elemente einer Akteur-Medien-Theorie," 19.

Orthopedics by the Roadside: Spikes and Studs as Devices of Social Normalization —— 93

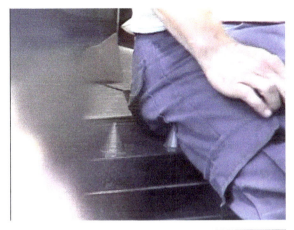

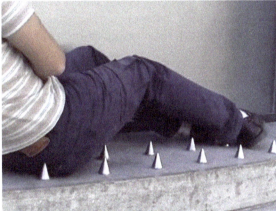

Figs. 3–5: *Le Repos du Fakir* (2003), stills.

They make distinctions, in the sense of a delegation, standing in for others and creating an asymmetry[17] between groups of people, based on selected needs and desired user behavior. And one can follow them on their decision paths.

Instead of grounding the constitution and explanation of such a differentiating (a)social milieu in its entirety on this milieu itself, as the result of superactors and "force fields of attraction,"[18] it is therefore possible to discern within them individual, small-scale and detailed impulses, in order to describe decision-making parameters and to tie them back to their originator. Certain developments in semipublic spaces can thus be linked to individual elements of the pattern of interaction. The aim is to shed light on the larger picture of a policy of inclusion and exclusion in its sociotechnical chains of operation, even as other entities and factors, such as consumer policy or social change, may still be in the background.[19] This might merely delineate the literal circumstances of blocked movements and impeded body techniques, or touch upon the mechanisms of exclusion this fosters; but it thus also opens these things up to further analysis in the first place:

---

[17] As Latour says: "An object stands in for an actor and creates an asymmetry between absent makers and occasional users," Bruno Latour, *Pandora's Hope: Essays on the Reality of Science Studies* (Cambridge, MA: Harvard University Press, 2001), 189.

[18] See Laura Frahm, "The Rules of Attraction: Urban Design, City Films, and Movement Studies," *Zeitschrift für Medien- und Kulturforschung* 5, 1 (2014): *Producing Places*: 85–99, here 95.

[19] At the same time, their individual historical strands of development cannot, for reasons of textual economy (among others), be deeply traced in these discourses in the history of knowledge. The aim here, however, is still to open them up for further investigation. See Michael Lynch, "Ontography: Investigating the Production of Things, Deflating Ontology," *Social Studies of Science* 43, 3 (2013): 444–462. The essay follows an ontographic procedure, which makes it possible to describe a scene and its relations, and to remain at this scene and be attentive to the connected facts of the inventory of a photograph and exploded view – without, however, insisting too much on explicating the history of each object; see Ian Bogost, *Alien Phenomenology, or, What It's Like to Be a Thing* (Minneapolis: University of Minneapolis Press, 2012), 35ff. Certainly, this is only snippet-like, situational, and hence only one way of tracing these things. However, it would be advantageous here to take stock of the situation in order to be able to pursue, concretely in local individual cases, the microphysics of configurations such as the sociotechnical constellation of modes of action, instead of getting lost in the world of larger periods of time. See Steve Woolgar and Javier Lezaun, "The Wrong Bin Bag: A Turn to Ontology in Science and Technology Studies?" *Social Studies of Science* 43, 3 (2013): 321–340. Moreover, one would have to separate the interlocking forces in terms of their relations and reattach them to actants, tracing their historical development and their coming together step by step, in order to avoid producing a linear total narrative favoring one factor either socio-deterministically or even techno-deterministically. See Schüttpelz, "Elemente einer Akteur-Medien-Theorie," 28f.

Orthopedics by the Roadside: Spikes and Studs as Devices of Social Normalization — 95

In other words, the presumption is that, whereas we cannot straightforwardly read-off the technical characteristics of the technology, we are somehow capable of reading off the (actual) social and/or political characteristics that apply.[20]

In the end, the implicit uncertainty of the techniques used to describe the complex processes of public spaces conditions the media-theoretical toolbox and vice versa. On the one hand, a supposedly clear conflict can be emphasized in the description between two adversaries vis-à-vis the representation of far-reaching connections between regulations and traditions, in order to keep a clear overview of the actors in a narrative.[21] On the other hand, relying on a network that is widely ramified and only partially describable holds the danger of not having to clearly select possible agents and intentions, i.e., establishing points of contact to explain movement and change in the arrangement of bodies. Social differences, as well as the hardening of urban surfaces, thus remained mere symptoms of the always-contingent coming together of a series of actors; instead, the task would be to investigate in more detail precisely where something might be awry within a chain of operations and a political line of decision-making.[22] Because of the demonstrative defensive stance of the spikes, special characteristics of usual surfaces can be examined independently of sheer outrage in order to place them in the forefront of public discussion and, ultimately, understand their sociotechnical peculiarities, i.e., the microarchitectural politics of their individual situations:

---

[20] Steve Woolgar and Daniel Neyland, *Mundane Governance: Ontology and Accountability* (Oxford: Oxford University Press, 2013), 39.
[21] Criticism from recent work in science and technology studies argues, for example: "In sum, the programme-antiprogramme schema neglects situations where there is no overt contestation, and yet where the objects and technologies may none the less be considered to be part of (social) ordering, regulation, and control." Woolgar and Neyland, *Mundane Governance: Ontology and Accountability*, 45–46, note 19.
[22] Latour's often all-too-conclusive narratives of actor networks and assemblages made it possible to not directly associate any characteristic of a thing, of an actor or actant, with an observed effect, and to rest the effects on a large whole, as Latour's argument about the specific "Berlin key" shows: "The straight forward [sic] essentialist account – the weight of the keys forces customers to leave them at the desk – has been replaced by a form of distributed essentialism." Woolgar and Neyland, *Mundane Governance: Ontology and Accountability*, 39, 45. See Bruno Latour, "The Berlin Key or How to Do Words with Things," in *Matter, Materiality and Modern Culture*, ed. P.M. Graves-Brown (London: Routledge, 2001), 10–21. The stability of the network is invoked without, however, having to search in detail for further conditions for the course of action.

Or, to put it in slightly less trenchant terms, readings of the governance implications of technology are occasioned. This means that a primary focus for analysis is the ways in which objects and technologies are made to 'do' political work.[23]

Ultimately, the niche's means of deterrence introduce and implement a pattern of technical mediation in the sense of "crossing the border between signs and things." Situation-specific guidelines for action can be traced; the functional deployment of things can be emphasized.[24] With Latour, it was the speed bump, the "sleeping policeman" made of concrete, that regulated the speed on the street – quietly, but materially assertive.[25] Here, it is bright and shiny spikes and knobs that are now encountered[26] as an inhuman disciplinary instrument multiplying the numbers of the "missing masses."[27] The flaw that must be admitted and addressed here clearly makes visible the asocial intention – which joins together with deterrents at this place of displacement – of owners, investors, and social engineers. These, too, nevertheless represent a choice of the (supposedly) lesser aesthetic evil,[28] since

---

23 Woolgar and Neyland, *Mundane Governance: Ontology and Accountability*, 39.
24 See Latour, *Pandora's Hope*, 185. Crossing the "boundary between signs and things" is the "fourth meaning of technical mediation." Latour describes this crossing in terms of the speed bump that is meant to prevent a car, and thus a driver, from exceeding the speed limit. This speed bump casts a regulation in concrete and delegates the function of ensuring that vehicles slow down. The delegation of a program of action occurs by means of engineers, and the speed bump no longer requires interpretation but compliance; see Latour, *Pandora's Hope*, 186–187. Delegation is a shift in the meanings of agents. In the case of the speed bump, delegation means a series of shifts of the actors, since it is neither a policeman standing guard nor a traffic sign "indicating" but a concrete threshold that intervenes into the events; the streets receive a new actant that is constantly present and fulfills its task. The engineers are absent: "An object *stands in* for an actor and creates an asymmetry between absent makers and occasional users," Latour, *Pandora's Hope*, 189.
25 Latour, *Pandora's Hope*, 186. Latour writes: "Techniques act as shape-changers, making a copy out of a barrel of wet concrete, lending a policeman the permanence and obstinacy of stone," Latour, *Pandora's Hope*, 189. "The speed bump is ultimately *not* made of matter; it is full of engineers and chancellors and lawmakers, commingling their wills and their story lines with those of gravel, concrete, paint, and standard calculations. The mediation, the technical translation, that I am trying to understand resides in the blind spot in which society and matter exchange properties" (Latour, *Pandora's Hope*, 190).
26 See Gordan Savičić and Selena Savić (ed.), *Unpleasant Design* (Belgrade: G.L.O.R.I.A., 2013), 123ff, and Cara Chellew, "Design Paranoia," *Ontario Planning Journal* 31, 5 (2016): 18–20.
27 See Bruno Latour, "Where Are the Missing Masses? The Sociology of a Few Mundane Artifacts," in *Shaping Technology/Building Society*, ed. Wiebe E. Bijker and John Law (Cambridge, MA: MIT University Press, 1992), 225–258.
28 To prevent undesirable behavior, they have the possibility of completely closing off a location, of making it appear unaesthetic in order to make it unlikely that it will be used, or – quite the opposite – of carrying out a kind of refinement to carry out a procedural social hygiene of

the later retrofitting and reinforcement of the surface was a makeshift solution to ultimately limit the space of action [*Handlungsspielraum*] that had not been originally envisioned in the construction plans. What became manifest in the spikes was both the implementation of and the insistence on a rigid guideline for private property, as one homeless person previously affected by the device commented in the *Guardian*: "Hard property jutting out against soft homeless bodies, saying: how dare you be poor in plain sight?"[29] The small things are means of constant displacement and intentional seizure of space: the aim is to force "undesirables" to gather at selected places (shelters, kitchens for the homeless, drinking halls), to keep them moving in the streets and, in this way or in general, make them controllable.[30] Conversely, the private entryway to the building needs to be kept "clean" and orderly, if only for fear of otherwise promoting the supposedly unstoppable rise in menacing criminality. At least according to the distorted causality of aesthetic standards, any visible sign of deficiency, such as a broken window, will always be followed by another.[31] Talk of "hardening" gained currency for speaking about this preventative "infrastructuralization", which is decidedly aimed against supposedly antisocial behavior and impedes spaces of refuge.

---

architecture by increasing the number of desired users; see Jan Wehrheim, *Die überwachte Stadt: Sicherheit, Segregation und Ausgrenzung* (Leverkusen: Opladen, 2006), 102ff.

29 Alex Andreou, "Spikes Keep the Homeless Away, Pushing Them Further Out of Sight," *The Guardian*, June 9 (2014), http://www.theguardian.com/commentisfree/2014/jun/09/spikes-homeless-london-metal-alcove-defensive-architecture-poverty/ (visited on July 1, 2018). With caution, one can certainly observe the precise aims and the decisiveness of supposedly historically posited forms of technodeterminism that have become established and persist as myths, for instance, in the debate around Langdon Winner, "Do Artifacts Have Politics?" *Daedalus* 109, 1 (1990): 121–136; Bernward Joerges, "Do Politics Have Artefacts?" *Social Studies of Science* 29, 3 (1999): 411–431; Steve Woolgar and Geoff Cooper, "Do Artefacts Have Ambivalence? Moses' Bridges, Winner's Bridges and Other Urban Legends in S&TS," *Social Studies of Science* 29, 3 (1999): 433–449.

30 The underlying principle follows a concept of order perfected in modern times of parceled-out and measured urban space, in which spaces can be assigned and individuals can thus be addressed. See Foucault, *Society Must Be Defended*.

31 See George L. Kelling and James Q. Wilson, "Broken Windows: The Police and Neighborhood Safety," *The Atlantic Monthly*, March (1982): 29–38, and Anna Minton, *Ground Control: Fear and Happiness in the Twenty-First-Century City* (London: Penguin Books, 2012).

3 Hard Times – Rough Sleeping

If one follows the semantics of a hardening of urban surfaces, one immediately encounters the contradictions between program and statement. Sometimes these surfaces are hard, sometimes they are soft, and sometimes they appear receptive and yet also repel. How does one measure the sociality of asphalt and concrete, or the hardness of granite and marble? What makes some of them hard and others soft, what allows them to enter into relationships and how do they make distinctions?[32] Everyday edges, steps, and ledges, marginal stone blocks and cornices in niches, become matters of concern of living between buildings.[33] "It is the edges of sitting surface that do the work, and it is the edges that should be made the most of."[34] How does one approach a political ontology of grounds and surfaces that now function to draw social boundaries?

As the example case shows, surfaces feature design tools for social relations and serve to regulate behavior and mixing, both operationally and semantically. Yet their intervention into the access point is neither unambiguous nor one-sided, as is the case with gates that open and close. One must adapt oneself rather than pass through them. Despite being largely devoid of objects, in comparison to windows and other openings, urban surfaces likewise function as "operative structure."[35] They nevertheless open and close less through a difference between inside and outside than by selecting certain patterns of behaviors, by enabling these behaviors or preventing them, or at least by promoting or hindering certain actions. With the hard surfaces there is, moreover, a shift in the thinking of inside and outside that is otherwise decided by lines, borders, and gates.[36] The possibility of entering into social relations is increasingly being opened and closed, topologically and in a less binary fashion, than one might want for an unambiguous description. The demarcation of social behavior takes place topologically, which means that boundaries are drawn in the social diagram or in the desired

---

[32] See Latour, "The Berlin Key or How to Do Words with Things."
[33] See Jan Gehl, *Leben zwischen Häusern: Konzepte für den öffentlichen Raum* (Berlin: Jovis, 2012 [1971]), and Bruno Latour, "Why Has Critique Run out of Steam? From Matters of Fact to Matters of Concern," *Critical Inquiry* 30 (Winter 2004): 225–248.
[34] William H. Whyte, *The Social Life of Small Urban Spaces* (Washington: Project for Public Spaces, 1980), 39.
[35] See Wolfgang Schäffner, "Elements of Architectural Media," *Zeitschrift für Medien- und Kulturforschung* 1, 1 (2010): *Kulturtechnik*: 137–149.
[36] See Georges Teyssot, "A Topology of Thresholds," *Home Cultures* 2, 1 (2005): 89–116; Laurent Stalder, "Turning Architecture Inside out: Revolving Doors and Other Threshold Devices," *Journal of Design History* 22, 1 (2009): 69–77, and Bernhard Siegert, "Türen: Zur Materialität des Symbolischen," *Zeitschrift für Medien- und Kulturforschung* 1, 1 (2010): *Kulturtechnik*: 151–170.

normal distribution rather than being directly and unambiguously implemented by visible mechanisms of exclusion (fences, barriers, walls, passports). The social hardness of a surface, which should not be confused with the actual hardness of the material, is determined via this potential of invisible/visible exclusion, i.e., gatekeeping. This can be seen upon a closer reading of the semantic hardening and charging of surfaces, on which their force of attraction and repulsion cannot be directly measured.[37]

Finally, a well-documented historical case for the instrumentalization of these mechanisms can be found in the middle of Manhattan, at Seagram Plaza, documented in the urban research project *The Street Life Project* by the self-trained sociologist William H. Whyte, who, in the 1970s, investigated the fluctuating (a)sociality of barren, hard surfaces of the city using both qualitative as well as quantitative, image-supported data.[38] "It takes real work to create a lousy place. Ledges have to be made high and bulky; railings put in; surfaces canted. Money can be saved by not doing such things, and the open space is more likely to be an amenable one."[39] Whyte's world perpetuated itself in such simply constructed causal chains; the point was to show that the built environment played a role, both as a symptom and a cause: "Fear proves itself."[40]

Whyte's observations provided a concrete knowledge of the sociality of urban spaces, a knowledge that is explicitly linked to the properties of surfaces. This knowledge thereby attempted to make suggestions about the selection and targeted occupation of seating in order to promote certain associations, hinder others, and thus regulate the degree of accessibility of certain spaces. Whyte argued vehemently against reinforcing seats, as happened in the example described above; instead, he proposed ostensibly open placeholders. The solution for avoiding the ubiquitous fear of the other, he argued, is simple and cost effective; if the seats were appealing in their appearance, this would already govern how people come together: "The best way to handle the problem of undesirables is to make a place attractive to everyone else. The record is overwhelmingly positive on this score."[41] Whyte emphasized that the "undesirables"

---

[37] For a more precise concept and an art-historical justification of the relationship between semantic charging and the operationality of surfaces, see Marion Hilliges, *Das Stadt- und Festungstor: Fortezza und sicurezza – semantische Aufrüstung im 16. Jahrhundert* (Berlin: Gebr. Mann, 2011).
[38] See Whyte, *The Social Life of Small Urban Spaces*.
[39] Whyte, *The Social Life of Small Urban Spaces*, 29.
[40] Whyte, *The Social Life of Small Urban Spaces*, 61. See Nan Ellin (ed.), *Architecture of Fear* (New York: Princeton Architectural Press, 1997).
[41] Whyte, *The Social Life of Small Urban Spaces*, 63.

were a problem that could be solved by technical means; that this problem could be recorded, scientifically counted and measured, calculated and managed; and that the "undesirables" could thus finally be displaced to other locations, as if they were changed in the process, by conceiving sufficiently harmonious arrangements of seats, people, and additional attractions, to then conclude: "Most of the undesirables have gone somewhere else."[42] The "humanitarian" William H. Whyte integrated similarly configured processes of control into his suggestions, as already existed in Oscar Newman's principle of natural surveillance or the neighborhood security technology of "eyes on the street" from the urban critic Jane Jacobs.[43]

Surfaces represent hinges of gathering together embedded in space. Instead of opposing assembly with barriers or other rigid and visible security mechanisms, they filter and regulate this activity more subliminally. The social hardness of the surface appears variable depending on the techniques of the body it enables or prevents. It renders steps, landings, and concrete blocks political.

Fortress L.A. by Mike Davis, which links the attractive New York of William H. Whyte from the 1970s with the postliberal L.A. of the 1990s in several main points of its theses, offers another key text and precursor to the theoretical illumination of an (a)social hardening of urban surfaces: "an unprecedented tendency to merge urban design, architecture and the police apparatus into a single, comprehensive security effort."[44] Davis argues that this development has ultimately even resulted in the destruction of public space and the spread of "sadistic street environments." On the one hand, this creates showcase plazas: "'soft' environments," equipped with "fountains, world-class public art, exotic shrubbery, and avant-garde street-furniture," some of which resulted from Whyte's studies.[45] On the other hand, the homeless and "undesirables" are thus forced out of the street niches into areas of the city designated for them. To do this, Davis argues, the city employs two decisive procedures of a sociospatial strategy: "increased police harassment and ingenious design deterrents."[46] And it is this strategy that produces the actual hardening, which Davis defines

---

[42] Whyte, *The Social Life of Small Urban Spaces*, 62.
[43] See Oscar Newman, *Defensible Space: Crime Prevention through Urban Design* (New York: Collier, 1972), and Jane Jacobs, *The Death and Life of Great American Cities* (New York: Vintage Books, 1961).
[44] Mike Davis, *City of Quartz: Excavating the Future in Los Angeles* (London: Verso, 2006), 224.
[45] Davis, *City of Quartz*, 226, 232.
[46] Davis, *City of Quartz*, 232.

even more precisely as "programmed hardening of the urban surface."⁴⁷ But what should be understood by the conscious "'hardening' of the city surface against the poor" that at least knows its counterpart or even its enemy? Davis cites so-called "bumproof benches" and "outdoor sprinklers" as concrete examples. He criticizes what he sees: that "the city is engaged in a merciless struggle to make public facilities and spaces as 'unliveable' as possible for the homeless and the poor."⁴⁸

Almost incidentally, he follows Whyte's argument that the quality of urban environments is correlated with that of their seating.⁴⁹ At the same time, even without expressly invoking Whyte, he succeeds in describing the effects of the regulating process of displacement that Whyte assumed in the case of his more or less closed semipublic spaces:

> They set up architectural and semiotic barriers to filter out 'undesirables.' They enclose the mass that remains, directing its circulation with behaviorist ferocity. It is lured by visual stimuli of all kinds, dulled by muzak, sometimes even scented by invisible aromatizers.⁵⁰

Davis accordingly sees a decisive mechanism in both architectural and semiotic barriers, which were at first "invisible signs warning off the underclass 'Other,'" but then could not be overlooked: "the semiotics of so-called 'defensible space' are just about as subtle as a swaggering white cop."⁵¹ Whereas ultimately, the homogenization of the desired mass is based on a selection process of targeted mechanisms for excluding "undesirables," for Davis the concrete effect of the architectural configurations seems to respond even more through their signs.⁵²

---

47 Davis, *City of Quartz*, 223.
48 Davis, *City of Quartz*, 232f. And finally, it is fear that he sees as the driving force: "The social perception of threat becomes a function of the security mobilization itself, not crime rates," Davis, *City of Quartz*, 224. Wherever the "security-driven logic of urban enclavization" (244) might not be applicable, have little success, or meet with too much opposition, more subtle strategies of the "fear of the crowds" (257) are found to accommodate the situation.
49 Davis, *City of Quartz*, 232.
50 Davis, *City of Quartz*, 257.
51 Davis, *City of Quartz*, 226.
52 A similar focus on semantics can be found in the "extremely functional aesthetics of a large-scale urban restructuring" that narrows the "sign system of exclusion [serving] the social 'purification' of space defined as public"; see Monika Wagner, "Die Privatisierung von Kunst und Natur im öffentlichen Raum: Die Plazas von Manhattan," in *New York: Strukturen einer Metropole*, ed. Hartmut Häussermann and Walter Siebel (Frankfurt am Main: Suhrkamp, 1993), 286–298, here 287. She argues that, in these "rooms of stone," refined with waterfalls, trees, and works of art, one finds above all "natural stone as cladding material"; Wagner prefers the visual and aesthetic characteristics and meanings of this environment: "Much more directly, by contrast,

Davis vacillates undecidedly between clear signs of displacement, defensiveness, and violence or, according to his definition, signs invisible to the general public that would only give warning of a particular group or class of *Others*.[53]

4 Programmed Environments – Instances of Gatekeeping

The fortified typologies of the building catalogues not only testify to the worries of building owners, urban developers, and investors. The method here is closing up and excluding: defining spaces for specific groups and uses without erecting high fences or establishing otherwise unattractive security mechanisms. Rather, the mechanisms used are intended to prevent specific human assemblies: of skaters, homeless people, bums, and loiterers.[54] These groups are to be denied certain spaces of activity by addressing and hindering supposedly characteristic techniques of the body that in turn define and identify these groups as marginal: body/cultural techniques, such as grinding and skating, sleeping and lying down, or resting and sitting are accordingly met with various devices, structures, and reinforcements of a hostile architecture. The means used for this purpose bear their intentions in their name: antiskate, antirest, antihomeless devices.[55] The intention is to prevent pointless hanging out, antisocial behavior, undesirable misuse of, and hence constantly feared damage to, street furniture and property value. Ultimately, things create access bound to a specific purpose;

---

the material unfolds its social function by articulating the level of aspiration of its surroundings, smoothing away the boundary between an object meant for daily use and art and thus contributing to both identification and social exclusion"; Wagner, "Privatisierung," 293.

53 With Whyte's motto, "fear proves itself," Davis concludes: "The social perception of threat becomes a function of the security mobilization itself, not crime rates"; Davis, *City of Quartz*, 224.

54 Steven Flusty, *Building Paranoia: The Proliferation of Interdictory Space and the Erosion of Spatial Justice* (West Hollywood: LA Forum for Arch & Urban Design, 1994); Ellin, *Architecture of Fear*; Michael Sorkin (ed.), *Indefensible Space: The Architecture of the National Insecurity State* (New York: Routledge, 2008); T. J. Demos, "The Cruel Dialectic: On the Work of Nils Norman," *Grey Room* 13 (2003): 32–53, here 41.

55 See Marcel Mauss, "Techniques of the Body," *Economy and Society* 2, 1 (1973 [1934]): 70–88; Benn Quinn, "Anti-homeless spikes are part of a wider phenomenon of 'hostile architecture,'" *The Guardian*, June 13 (2014), http://www.theguardian.com/artanddesign/2014/jun/13/anti-homeless-spikes-hostile-architecture/ (visited on July 1, 2018); Ocean Howell, "The Poetics of Security: Skateboarding, Urban Design, and the New Public Space," *Urban Action* 2001, http://bss.sfsu.edu/urbanaction/ua2001/ps.html (accessed via Wayback Machine; visited on July 1, 2018).

they produce a selection of desired groups and a certain order.[56] The demarcation of these groups, or more precisely of their patterns of behavior, is not made solely on the basis of a topographically defined level, through gates, doors, and barriers that everyone has to pass through. Rather, it takes place through the topological closure of a space by means of small, dispersed deterrents that process previously established normative boundaries at the behavioral level: the boundary is translated from the rigid line into a method and inscribes itself, as a model of terrestrial movement, into the procedures of production and processes of these spaces.[57]

Hence, these little things can only mark a social difference on and with the surface they make difficult to approach, thus implementing a sociotechnical distinction. Amid the rows of deterrents, spikes, and studs, an earthbound strategy of displacement thus rises up. These devices distribute power relations on the steps, window sills, and benches and direct the trajectories of the relationships that develop on these surfaces.[58] Bollards, railings, barriers, flower pots, benches, garbage baskets, and, above all, everything on the surfaces themselves, on the ground and on all other surfaces that are actually suitable for sitting, stands here for an instrument or at least a part of a sociotechnical relationship of control.[59] Even more than the regulations of the municipal administrations and the catalogues of street furniture manufacturers, projects developed in art contexts in recent decades have worked through such a network of relationships in urban space, thereby providing a reliable contemporary inventory of these semipublic conditions of endeavors of inclusion and exclusion. A well-known example is Nils Norman's *The Contemporary Picturesque*: images from an archive that has grown over the years form an inventory of microarchitectural configurations. Norman

---

[56] For more on this issue, see Shilpa Phadke, Sameera Khan, and Shilpa Ranade, *Why Loiter? Women and Risk on Mumbai Streets* (New Delhi: Penguin Books, 2011).

[57] See Sandro Mezzadra and Brett Neilson, *Border as Method, or, The Multiplication of Labor* (Durham: Duke University Press, 2013).

[58] These artificial measures of new urban design can be subsumed under the concept of defensible architecture: Recently the term "unpleasant design" has become established. The Survival Group from France, among others, documents sites where measures are targeted to exclude certain behaviors, which it defines with the term "anti-sites," and in recent years, the group has produced a catalogue of typologies of *mobilier urbain anti-SDF* (SDF: Sans Domicile Fixe, antihomeless urban furniture) in order to not only create an inventory of the fortified arsenal of street furniture but also to get a grasp of the partially abstruse formal language of these reinforcements; see Savičić and Savić, *Unpleasant Design*, 93ff.

[59] See Savičić and Savić, *Unpleasant Design*, 123ff; Cara Chellew, "Design Paranoia," *Ontario Planning Journal* 31, 5 (2016): 18–20, and Fredrik Edin, *Exkluderande Design* (Stockholm: Verbal, 2017).

sometimes gave them impressive titles reflecting their ability to bind themselves to humans in context-specific ways by preventing an action.[60] Yet many of the typologies produced by the building code and street regulations(s) that Norman so impressively captures are reminiscent of a completely different technological aesthetic. In their simple program and minimal aesthetics, the means of art and social engineering entirely coincide:

> Minimalism thus becomes a tool of urban discipline, its negativity reverses into instrumentalization, its anti-anthropomorphism becomes urban design's sadism. If minimalist structures once provided complex spaces of perceptual critique or offered ideal possibilities for phenomenological self-consciousness, they are now directed against the public body, transmogrifying into a logic of 'defensible space' and psychosomatic stimulation.[61]

A curious repertoire of social control mechanisms that attack at specific points and determine laminarly, connected to a specific surface technology for displacing undesirable behavior, comes together on the city's niches, steps, ledges and windowsills – though the pronounced and abstruse techniques of surface defense cannot be illuminated further at this point. These are mechanisms that, in recent years, have mostly been installed after the fact when social problems could not be solved in any other way, or when the limits of what is permitted were tightened or a desire for order became stronger.[62] Today, as Lorraine Gamman, professor of design and director of the Design Against Crime (DAC) Research Center (Central St. Martins College of Arts and Design) says, people have recognized that this "outdated fortress aesthetic" has once again become obsolete because it increasingly prohibits any kind of use. Instead, there is a profound desire for inclusive urban design constructed like the very communities that are to emerge from it.[63] The surface technologies serve to further the integrated, albeit not invisible, normalization of patterns of behavior and thus pursue a mix of disciplinary mechanisms and regulatory measures that lay like a net over the

---

[60] Norman lists, among others: "Studded Paving, King's Cross, London. Concrete pavement with mouldings to prevent people from lying or sitting down in front of shop windows"; "Surface Studs, City of London. Upmarket metal studs on marble prevent people from sitting on private property and adds [sic] a perverse yet strangely apt S&M flavour to the City's streets." Nils Norman, *The Contemporary Picturesque* (London: Bookwords, 2001, without page numbers). Today on Instagram and co., one can find additional collections of this kind, https://www.defensiveto.com/ among others (visited on July 1, 2018).
[61] Demos, "The Cruel Dialectic: On the Work of Nils Norman," 40–41.
[62] See George L. Kelling and Catherine M. Coles, *Fixing Broken Windows: Restoring Order and Reducing Crime in our Communities* (New York: Simon & Schuster, 1996).
[63] See Quinn, "Anti-homeless Spikes Are Part of a Wider Phenomenon of 'Hostile Architecture.'"

city, now refined by street furniture, in places where power relations were dense and social control was locally strengthened.[64]

5 Surface Technologies

What helps architecture and infrastructure to establish social differences, and what renders them political?[65] In Whyte's attraction models, the filter shifts into an (a)socially differentiating surface technology and thus into an environmental technology of social control. Social processes, triangulated with spectacles, now filter and displace, instead of obvious barriers, and decide upon those groups of individuals who want to take a seat.[66] Instead of rejecting everyone, they limit the area that desired groups can occupy without excluding directly through an act that would make the exclusion visible and transparent. At least that is how the indirect founding of relations between the processes in these small spaces might be understood. The argument of hardening can therefore be transferred to the question of the politics of surfaces:[67]

Surfaces regulate participation and possibilities of access as well as visibility.[68] They can be political because they order; they allocate what is perceptible and visible and thus classify these things socially.[69] What is important for site-specific social engineering is that, by means of surface design, attachments to certain places can be influenced. In the design of the environment, the choice of place always oscillates along the boundary between things and signs. This undermines the conscious perception of a practiced orthopedics of street furniture in the sense of learned and translatable techniques of the body.

---

64 See Foucault, *Society Must Be Defended*.
65 See Christa Kamleithner, "Was Architektur macht," *Arch + 217* (2014): 156–169.
66 See Frahm, "The Rules of Attraction," 94f.
67 Guided by the politics of things: "Politics stem from ordinary stuff. So the challenge for us is to articulate how politics work at the level of ontology ... Can things 'have politics'?" Woolgar and Neyland, *Mundane Governance*, 14.
68 See Leon Hempel, Susanne Krasmann, and Ulrich Bröckling (eds.), *Sichtbarkeitsregime: Überwachung, Sicherheit und Privatheit im 21. Jahrhundert* (Wiesbaden: VS, 2011): Special Issue 25 of *Leviathan*.
69 See generally Jacques Rancière, *Disagreement – Politics and Philosophy* (Minneapolis: University of Minnesota Press, 1998); Judith Butler, *Notes Toward a Performative Theory of Assembly* (Cambridge, MA: Harvard University Press, 2015).

While the small things – spikes and studs – mobilize, while they reject and divert every undesirable person, their strategy aims at the partial and temporary occupation of surfaces. The actual displacement always takes place on the surface, in what is visible, whether through built defense or attraction; this was already suggested by urban field research and observations at the beginning of the 1970s. Yet the filtering and sorting of people by means of attractive surfaces, the mutual attraction and integration of bodies, was intended to be more sustainable and successful than the demonstrative setting of sharply pointed mechanisms of exclusion, even if – or precisely because – they had a similar effect.

Hannah Zindel
Ballooning: Aeronautical Techniques from Montgolfier to Google

Since 2013, the technology company Google has sent hundreds of balloons into the stratosphere. As hovering radio towers, these are intended to provide an internet connection to the approximately 60 percent of the world's population who lack access to the World Wide Web. The balloons' nautical control system is based on algorithms that the tech giant extrapolates from weather data. Neural networks enable these aircraft, which usually cannot be steered horizontally, to move as planned by reading the conditions of the surrounding airspace. To navigate horizontally, the smart balloons move vertically toward targeted air streams. Taking this aeronautical technique as a starting point, this essay will focus on the historical interplay between aircraft knowledge and airspace knowledge. In order to take a critical look at Google's privatization of infrastructures and its economization of the stratosphere, I will concentrate below on two points of contact between balloon flight and meteorology that were forged around the year 1900: unmanned balloon flights and international simultaneous balloon launches. The hypothesis of this essay is that these two activities marked an epistemic turning point at which the ballistic stubbornness of balloons – their inability to be steered horizontally – was put to productive use. Balloons became established media of meteorology. At the same time, these lighter-than-air vessels were an essential impetus behind a transformation in meteorology from a qualitative and observational science into a quantitative and calculating one.

Even though balloons are still used today as meteorological instruments, there have been few historical studies of the interactions between aerostatic and atmospheric knowledge.[1] At most, historians of technology have regarded these

---

[1] Exceptions include Sabine Höhler, *Luftfahrtforschung und Luftfahrtmythos: Wissenschaftliche Ballonfahrt in Deutschland 1880–1910* (Frankfurt am Main: Campus, 2001); Robert Marc Friedmann, *Appropriating the Weather: Vilhelm Bjerknes and the Construction of a Modern Meteorology* (Ithaca: Cornell University Press, 1993); and Richard Holmes, *Falling Upwards: How We Took to the Air* (London: William Collins, 2013). In contrast to the historical-epistemological approach to balloons taken in my essay, the cultural geographer Derek P. McCormack has examined them from an ontological perspective. See his book *Atmospheric Things: On the Allure of Elemental Envelopment* (Durham: Duke University Press, 2018). Furthermore, in his essay "Elemental Infrastructures for Atmospheric Media" (2017), he has explicitly taken Project Loon as a point of reference

---

Translated by Valentine A. Pakis

Open Access. © 2020 Hannah Zindel, published by De Gruyter. This work is licensed under a Creative Commons Attribution 4.0 International License.
https://doi.org/10.1515/9783110647044-007

aircraft, which are filled with hot air or gas, as a failed intermediary stage within a history of technical progress. From the perspective of cultural techniques, however, the balloon is more than just a preliminary stage in the linearly conceived history of technological development – a stage that simply came before steerable blimps and airplanes happened to conquer the skies. It is more than just an actualization – in terms of the history of ideas – of the ancient dream of flying, and it is more than just a spectacular viewing platform or a decorative motif on wallpaper, dishes, and postcards. A research approach based on cultural techniques will enable me to examine the inconspicuous practices associated with balloon flight – techniques that converged at the intersection of widely various fields of knowledge such as observing and charting the surface of the earth, observing and collecting data about the atmosphere, and improving aircrafts and air travel itself.[2] Initially, these heterogeneous areas of application did not form a consistent epistemic field. From a cultural-technical perspective, however, it becomes clear that balloons have a different relation to the space that they pass through than trains or airplanes, which shoot across the landscape like projectiles.[3] Aerostats are not only media of that which moves them; they are also media of a type of writing that cannot be removed from the medium that bears the recording and inscribes itself into what has been recorded. The indexical relationship of the aircraft to its surrounding airspace is not only metonymic but reciprocal. Early balloon flight can thus be regarded as a site where concepts of environments and concepts of the relationship between bodies and their environments were repeatedly renegotiated.[4] These concepts seem to be central to today's efforts to measure, calculate, and simulate the atmosphere, and thus they are also central to our understanding of its privatization and economization, as the example of Google balloons will show.

---

in order to reflect on "atmospheric media" following Mark B.N. Hansen. In this approach, McCormack links Alexander Galloway's and Eugene Thacker's program of focusing on the "ambient," "environmental," "elemental aspects of networks" with approaches from infrastructure studies, in particular Lisa Parks and Nicole Starosielski's socio-technical understanding of "media infrastructures." This essay supplements this approach, which is more interested in ontological questions, with a perspective from cultural techniques and media history.

[2] See Hannah Zindel, *Ballons: Medien und Techniken früher Luftfahrten* (Paderborn: Fink, 2020, forthcoming).

[3] See Paul Virilio, "Fahrzeug," in *Fahren, fahren, fahren …*, trans. Ulrich Raulff (Berlin: Merve, 1978), 19–50, here 19.

[4] See Florian Huber and Christina Wessely, "Milieu: Zirkulationen und Transformationen eines Begriffs," in *Milieu: Umgebungen des Lebendigen in der Moderne*, ed. F.H. and C.W. (Paderborn: Fink, 2017), 7–17, here 13.

1 Unmanned Balloons

Fig. 1: One of Project Loon's internet balloons.

Internet Balloons

Google balloons are sent into regions where it is economically challenging to construct communication infrastructure or into regions where such infrastructure has been temporarily damaged by natural catastrophes. The radio towers in the air relocate these infrastructures into the sky. The first tests were conducted in New Zealand in 2013. In 2016, Indonesia became the first nation to access the internet via balloon on a permanent basis. And in October of 2017, Project Loon was granted a license to temporarily replace the communication infrastructure of Puerto Rico, which had been destroyed by Hurricane Maria, with internet balloons. At present, autolaunchers – large mobile scaffolds that automatically fill and launch balloons (Fig. 2) – are supposedly able to release a balloon into the stratosphere every 30 minutes. Every half hour, a new balloon can thus be added to the balloon network. In order to endure the extreme climatic conditions of the stratosphere, the Google balloons and their equipment have to be able to withstand temperatures of up to -90 degrees Celsius, wind speeds of up to 100 kilometers per hour, and intensive UV radiation.[5] Google balloons reach an altitude of 18 to 21 kilometers and are thus approximately twice as high in the air as passenger planes.

---

5 See https://loon.com/ (visited on March 30, 2019). Unless otherwise noted, this website is also the source of further information discussed in this section.

Fig. 2: Autolaunchers can purportedly add a new balloon to the hovering network every 30 minutes.

All of their equipment is highly energy efficient and powered by renewable energy. Solar panels serve as the energy source during the day, and they charge a battery to power the operations during the night. Each of the balloons transports a small box. The latter connects the balloons to one another via laser signals, links them to telecommunication networks via radio waves, and transmits high-speed internet from the stratosphere to smartphones and other LTE-compatible devices on the ground (Fig. 3).[6] If a balloon needs to be taken out of service, its helium is released through a valve, and the balloon falls back to earth with a parachute.

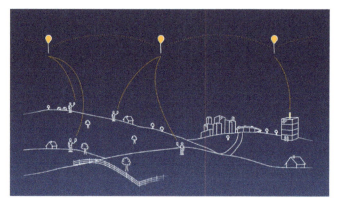

Fig. 3: Internet balloons are linked together through mobile radio towers and LTE-compatible devices.

---

[6] See Tom Simonite, "Project Loon: Billions of People Could Get Online for the First Time Thanks to Helium Balloons That Google Will Soon Send over Many Places Cell Towers Don't Reach," *Technology Review* 118, 2 (2015): 40–45.

So-called recovery teams then collect all the equipment that has floated down from the sky for the sake of reusing or recycling it.

The name Project Loon is a play on the words *balloon* and *loony*. At first, the project was conducted by Google X – the company's research division, which was also working on such things as contact lenses that can measure blood pressure, self-driving cars, and sunglasses equipped with microcomputers.[7] Google has said about the project that it will serve to make the world a better place by providing internet access to approximately 4.3 billion additional people. What Google touts as a grand humanitarian project is also, of course, a means for the company to access new customers, data, and markets. In order to make a critical assessment of Google's economization of the stratosphere and its privatization of infrastructures, it will be beneficial to take a historicizing and technically informed look at ballooning's techniques for measuring, calculating, and simulating the atmosphere.

From "ballons perdus" to "ballons-sondes"

In March of 1892, a journalist (and balloonist) conducted a series of experiments together with a scientist on the balcony of an apartment on Paris's Rue de Sébastopol. The two of them attempted to launch small unmanned balloons with self-recording instruments, including thermographs, hydrographs, and barographs. Gustave Hermite and Georges Besançon built their balloons for sake of studying the upper regions of the sky.[8] Manned balloons could only be used to investigate the atmosphere up to a certain altitude. At around eight or nine kilometers, the air would become too thin for the human actors in the basket, and even with an oxygen apparatus it was only possible to go a little bit higher. Thirty years earlier, for instance, the altitude at which James Glaisher had lost consciousness (8,800 meters) marked a spatial boundary that, in the accompanying research and travel report, had also become visible as an epistemic and aesthetic boundary: higher regions were inaccessible to the established recording methods of the time and were thus beyond the limits of existing knowledge.[9]

---

7 Since the beginning of 2018, Google X has been a subsidiary company called X, just as Google itself is now a subsidiary of the parent company Alphabet. However, because internet balloons are commonly referred to as "Google balloons," I will continue to use the term here.
8 On Hermite and Besançon's unmanned balloons, see Wilfrid de Fonvielle, *Les ballons-sondes de MM. Hermite et Besançon et les ascensions internationales* (Paris: Gauthier-Villars et fils, 1898).
9 See Hannah Zindel, "À ballon perdu: Forscherohnmacht an einem bewölkten Septembernachmittag 1862," in *Versteckt – Verirrt – Verschollen: Reisen und Nicht-Wissen*, ed. Irina Gradinari et al.

In the unmanned balloon flight attempted by the two Frenchmen in 1892, the problem of unconscious instrument readers on board was avoided by means of self-recording instruments. What was true of the epistemic subject, however, did not at all apply to the subject who more or less steered the balloon. It was nearly impossible to replace balloon captains as well. How were unmanned balloons supposed to land? How could they ever be found again? With no one on board, there would be no indications regarding the location of the balloon. Hermite and Besançon therefore devised a specific apparatus for landing and rediscovering their unmanned balloons (Fig. 4). They designed their aircraft in such a way that recording, landing, and rediscovery were partially automated. The instrument box with the self-recording devices, which hung from the balloon, was designed to sail back to earth with a parachute.

Ballon-sonde en papier pétrolé emportant le premier enregistreur.

Fig. 4: An unmanned paper balloon with self-recording instruments, 1892.

The balloon itself was made of rice paper, so that it would burst at a particular altitude.[10] The problem of ensuring that the balloons would fall from the sky was thus solved by using this fine paper. Yet how would it be possible to find them once they had landed? The two men experimenting on the balcony attached a bundle of postcards to each balloon; from this bundle, a single postcard would be released at regular intervals by means of a slow-burning wick. Those who found the cards were requested to write down the location of the discovery and return them to the address provided. The final postcard remained fixed to the balloon through its landing.

(Wiesbaden: Reichert, 2016), 347–366; and Zindel, "Clouds and Balloons: James Glaisher's Travels in the Air," in *Silver Linings: Clouds in Arts and Sciences*, ed. Dolly Jørgensen (Trondheim: Museumsforlaget, 2020), 27–37.

10 By using this paper instead of rubber-coated silk (which was more common at the time), the two tinkerers were in fact reverting to the material that had been used by the brothers and paper manufacturers Jacques-Étienne and Joseph Montgolfier during the early years of balloon flight in the late eighteenth century.

This semi-automated version of tracking balloons had various precursors.[11] Balloon travelers were followed by so-called pursuers on horseback, for instance. Postal techniques, too, had also been relied upon earlier. Moreover, carrier pigeons were brought along aboard balloons and returned one-by-one to the place of departure with short notes about the present time and location of the passengers. The last pigeon would usually be sent back from the location of a balloon's landing. In the event that passengers happened to know their location, they would send a request via this pigeon to be picked up there (along with the all the equipment, which was not exactly light) by a cart or a carriage (Fig. 5). The spatial logic of balloons and that of carrier pigeons thus complemented one another quite well when it came to issues of locating and addressing. Whereas balloons have a starting point A with n possible destinations, pigeons have n possible starting points and one destination B, their home pigeonhole.

Fig. 5: Loading heavy aircraft equipment.

Another postal technique was to drop prepared postcards. These, as in the case of Hermite and Besançon's balloons, would request that the discoverer of the card send it back to the given address with a note about the time and location of the

---

11 See Hannah Zindel, "Belagerung von Paris: 69 Freiballons, 381 Tauben und fast 11 Tonnen Post," in *Medien – Krieg – Raum*, ed. Lars Nowak (Munich: Fink, 2018), 141–160.

discovery. Later, balloonists would throw prepared telegrams overboard (Fig. 6). In addition to the delayed tracking of balloons through the mail and telegraphy, the simultaneous observation of traveling aerostats also became an established tracking method. During the early years of balloon travel, the Parisian Academy of Sciences had been unsuccessful in its attempts to record flight paths by stationing observers at certain places in advance. But it became common during the second half of the nineteenth century for local presses to publicize upcoming balloon flights, so that sightings could be reported in the newspapers.

Fig. 6: A route-tracking telegram by the balloonist and photographer Nadar.

These tricks discursivized and standardized a knowledge of airspace that factored in the uncontrollable nature of the balloon. Although it was unpredictable where a balloon would end up (on account of its inability to be steered horizontally), it was absolutely predictable that it would crash back down to earth at some point. Hermite and Besançon therefore made it their objective to turn the dropping of postcards into the standard form of communication for this type of air travel.

One term used during the early years of balloon flight reflects the negative connotations that had, to this point, been associated with the ballistic peculiarities of unmanned balloons: a balloon trip without passengers was also referred to as a *ballon perdu*, a "lost balloon."[12] In 1899, the meteorologist and balloon traveler Wilfrid de Fonvielle pointed out that, in order to market their unmanned balloons for an exploratory trip to the North Pole,[13] Hermite and Besançon had strategically replaced the term *ballon perdu* with the designation *ballons-sondes*. The negatively connotated *perdre* ("to lose," "to leak") was thus replaced with *sonder* ("to survey," "probe," "sound out"). This rebranding metaphorically condensed the shift that had been taking place in ballooning around the year 1900: a shift away from a project of disruptions and interruptions toward a project of flowing functionality.

Moreover, it also concerned the aesthetic question of charting and reproducing a trace – in the case of balloons, the question of documenting a moving object in three-dimensional space.[14] The methods for tracing and tracking balloons were all highly imprecise; the lag time between the event itself and its recording would lead to imagined arabesque or interrupted lines instead of calculated and continuous trajectories. The exploration of the upper regions of the sky was not only a matter of improving measurements of air temperature, humidity, and air pressure. In the nineteenth century, high-altitude flights were also about optimizing the techniques of addressing, tracking, and tracing.

From "ballons-sondes" to Radiosondes

At the turn of the twentieth century, unmanned balloons were no longer being used exclusively as so-called pilot balloons, which would be released in advance

---

12 See Dieter Zastrow, *Entstehung und Ausbildung des französischen Vokabulars der Luftfahrt mit Fahrzeugen "leichter als Luft" (Ballon, Luftschiff) von den Anfängen bis 1910* (Tübingen: Niemeyer, 1963), 39.
13 In the 1880s, the two inventors attempted to figure out how to execute the idea – first suggested in a publication by Delavile Decreux in 1863 – of making an excursion to the North Pole in a balloon. Although unmanned balloons were never used for polar explorations, Hermite and Besançon did prepare and train the scientist Salomon Auguste Andrée, who did in fact set out for the North Pole in 1897 in a balloon with two other passengers. See Wolfgang Struck, "Ingenjör Andrées luftfärd oder Die melancholischen Entdeckungen des Films," in *Literarische Entdeckungsreisen: Vorfahren – Nachfahrten – Revisionen*, ed. W.S. and Hansjörg Bay (Cologne: Böhlau, 2012), 29–52; and Kristina Kuhn and W.S., *Aus der Welt gefallen: Die Geographie der Verschollenen* (Paderborn: Fink, 2019).
14 It is rather symptomatic of balloon flight that the mathematician Leonard Euler happened to die in 1783 (shortly after the launch of the first Montgolfière) while trying to calculate the trajectory of a balloon. See Charles Coulston Gillispie, *Sciences and Polity in France: The Revolutionary and Napoleonic Years* (Princeton: Princeton University Press, 2004), 535.

of balloon launches in order to observe the approximate direction in which the manned balloon might travel. Despite the difficulty of determining their flight trajectory and their frequent disappearance, at the close of the nineteenth century they began to play an important role in studying the atmosphere, especially the upper regions of the sky. In 1896, Hermite and Besançon introduced their method at one of the first international meteorology conferences. Soon thereafter, the French meteorologist Léon-Philippe Teisserenc de Bort began to launch unmanned weather balloons systematically, and in 1902 he differentiated the atmosphere into the troposphere and the stratosphere.[15]

As of the beginning of the twentieth century, unmanned balloons became established meteorological instruments and were observed by means of theodolites, which functioned well in daylight but were difficult to use at night and during periods of bad weather even though lanterns would be held up beside them.[16] In meteorology, too, an improvement in measurement technology was tied to more sophisticated techniques of addressing, tracking, and tracing. Lag time between measuring and recording had to be minimized in order to justify meteorology's claims of being a prognostic science.[17] This required new developments in communications technology.

As before, the main application of unmanned balloons was to explore the climatically extreme stratosphere, which lay beyond the limits of manned balloon travel. In the 1860s, the French photographer and aeronaut Félix Nadar was still extolling balloons as instruments for observing the movement of soldiers, in which case the balloonist in the basket was supposed to deliver finished drawings to the field commander by means of a rope.[18] Now, however, the field of meteorology was attempting to use wires instead of rope and electrical signals instead of drawings. In 1843, for instance, the English physicist Charles Wheatstone had introduced the first "telemeteorograph." This was a device that made it possible to transmit data from thermometers, barometers, and hygrometers over multiple kilometers through a wire. In the same year, he presented a telemetering

---

[15] See Jean Mascart, "L'étude de la haute atmosphere et les travaux de Léon Teisserenc de Bort," *La Nature* 2080 (April 5, 1913): 296–300.
[16] See John L. Dubois et al., *The Invention and Development of the Radiosonde, with a Catalog of Upper-Atmospheric Telemetering Probes in the National Museum of American History, Smithsonian Institution* (Washington, DC: Smithsonian Institution Press, 2002), 26.
[17] See Matthias Heymann et al. (eds.), *Cultures of Prediction in Atmospheric and Climate Science: Epistemic and Cultural Shifts in Computer-Based Modeling and Simulation* (New York: Routledge, 2017).
[18] See Nadar, *When I Was a Photographer*, trans. Eduardo Cadava and Liana Theodoratou (Cambridge, MA: MIT Press, 2015 [1900]), 59.

thermometer. The latter, which was meant to be used with a tethered balloon, transmitted signals through two copper wires to a receiver on the ground.[19]

However, the conversion of analog weather data (noted on paper by self-recording instruments) into discrete data, which would be suitable for the transmissions of electromagnetic telegraphy, could only be accomplished if the discoverer of the balloon probe knew how to interpret the recordings. A central precondition for this further processing was thus the development of a digital meteorograph, so that it would no longer be necessary to convert analog measurements into digital signals. The latter was developed in 1896 by Luigi Cerebotini and Albert Silbermann, who patented their invention in the United States in 1900.[20] In the 1920s, this system would be used by Robert Bureau in his design of the first radiosonde. For it was not until 1929 that wireless communications technology had progressed to the point that it could also be used with balloons: from the *ballon-sonde* thus emerged the so-called *radiosonde* (Fig. 7). During the

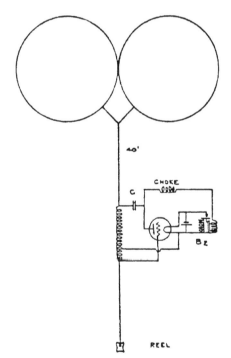

Fig. 7: William Blair's first radio-tracking transmitter, circa 1923–1924.

---

19 See Dubois et al., *The Invention and Development of the Radiosonde*, 15. This concept would not be tried again until 1917.
20 See Dubois et al., *The Invention and Development of the Radiosonde*, 18.

development of the radiosonde, the meteorologists of the 1920s were engaged first and foremost with the problem of making existing transmitter technologies function in the environment of balloons. This was a matter of adapting radio and transistor technology for free-floating balloons.[21]

During the 1920s, meteorology, balloon flights, and communications technology thus worked strategically together to improve military surveillance and economically relevant weather forecasting. A radiosonde was not only an unmanned balloon on which instruments could register data; it could also wirelessly communicate these data from a transmitter to the ground. Wireless tracking with radio signals made it possible to track balloon trajectories and thus, in addition, to measure the speed and direction of the wind and to determine wind vectors. After more than 140 years of experimenting on the techniques of tracking, tracing, and addressing in balloon flight, the fact that balloons could not be controlled was turned into a positive episteme.

2 Simultaneous Launches

Balloon Navigation from Kleist to Algorithms

One special feature of Google balloons is the fact that they can be navigated to move both vertically and horizontally. To achieve this, Google had to devise a technique of its own:

> In the stratosphere winds are stratified, and each layer of wind varies in speed and direction. To get balloons to where they need to go, Project Loon uses predictive models of the winds and decision-making algorithms to move each balloon up or down into a layer of wind blowing in the right direction. By moving with the wind, the balloons can be arranged to provide coverage where it's needed.[22]

---

[21] In 1921, Hugo Hergesell began to experiment with balloon-borne radio transmitters at the Lindenberg Observatory in Berlin. In France, Pierre Idrac and Robert Bureau, who were both employed by l'Office National Météorologique, launched free balloons with radio transmitters to an altitude of 14 kilometers in the stratosphere. Around the same time, William Blair, who was working at the Signal Corps Laboratory in Monmouth, New Jersey, managed to design and construct an oscillator for stratosphere balloons. As of 1928, these transmitters were flown extensively with clusters of four to six theodolite-tracked balloons (see Dubois et al., *The Invention and Development of the Radiosonde*, 29).

[22] Quoted from https://gnsec.wordpress.com/2017/07/12/technology-project-loon/ (visited on April 9, 2019). Vertical navigation functions by means of a built-in air pump that pumps air into or

The balloons have taken millions of kilometers of test flights and have collected weather data about these stratifications at various altitudes. On the basis of these data, software algorithms can now determine at which altitude the chances would be highest to locate an air stream in which a balloon would be able to remain for an extended time over a given area. The act of heading toward particular wind currents is controlled by software in one of Google's data centers, which incorporates wind forecasts from the U.S. National Oceanic and Atmospheric Administration in its simulations of stratospheric air streams. For these simulations, Google now relies on neural networks. Google balloons navigate by moving up or down into the appropriate current.[23] In this way, a balloon can remain for a longer period of time over a region that lacks widespread internet access. Knowledge about the stratosphere is measured in situ by sensors and is processed, stored, and retrieved at data centers that possess the necessary amount of computing power.

With the step from balloons to Loons, navigation has shifted from an unpredictable external force to a type of control that is realized by means of the wind. Control is not conceptualized and actualized on the basis of the moving object but rather on the basis of the space that surrounds it. Both approaches – the direct and indirect control of balloons – were also implemented in the nineteenth century.

Even in the early years of balloon travel, the desire for a controllable aircraft and controllable airspace necessitated the development of a new form of aerostatic and atmospheric knowledge. Numerous bold designs only ever existing on paper imagined the utopia of a technically exploitable airspace in which balloons could circulate as flying carriages or ships, and in which the terrestrial logic of space could be imposed on the sky.[24] At the same time, there also appeared another approach to controlling this aircraft by harnessing the logic of the airspace itself. This approach stems from Heinrich von Kleist, who wrote the following in 1810:

> The oilcloth manufacturer Mr. Claudius intends ... to travel into the air ... with a balloon and to steer it in a particular direction by means of a machine and thus independent of the wind. This undertaking seems strange, because the art of moving a balloon without any

---

removes air from a small balloon located within the large balloon and thus regulates the density of helium within the balloon as a whole. This technology enables the balloon to ascend or descend.
23 See Simonite, "Project Loon," 44.
24 See Elske Neidhardt-Jensen and Ernst Berninger, *Katalog der ballonhistorischen Sammlung Oberst von Brug: Fluglust – Fluges Beginnen – Fluges Fortgang* (Munich: Deutsches Museum, 1985), 59, 151, 161 (among many other examples in the book).

machinery – and in a quite easy and natural manner – has already been discovered. For, up in the sky, all possible currents (winds) overlap one another, and thus the aeronaut needs only to use vertical movements to locate the airstream that will lead him to his destination.[25]

According to Kleist, balloons can be controlled by intentionally harnessing the power of particular wind currents. Kleist's approach to solving the problem of navigability belonged to the domain of meteorology instead of mathematics; at least at the beginning of the nineteenth century, that is, this practice was more unpredictable and incalculable than not.[26] Over the course of the nineteenth century, however, meteorology transformed from a science that collected qualitative data into one that gathered quantitative data. In order to describe this transition, it will be beneficial to revisit a central scene in the history of balloon travel where meteorology was interwoven with mathematics and thus observation was integrated with calculation: the simultaneous balloon launches that took place around the year 1900.

Balloon Clusters and Wind Patterns

In 1896, a balloon was launched from the small Swiss town of Sion in the first attempt to cross the Alps. This Alpine journey occasioned the first European-wide simultaneous ascent – in Paris, St. Petersburg, Berlin, and Strasbourg. This simultaneous launch was organized by the Internationale aeronautische Kommission, which was also known as the Internationale Kommission für wissenschaftliche Luftfahrt.[27] As of this committee's second conference, which was held in 1900, simultaneous launches were systematically implemented as an observation program.[28] For, as one author remarked about the simultaneous launches that took place in 1902, a year in which 150 of them were undertaken in Europe and

---

25 Heinrich von Kleist, "Schreiben aus Berlin," in *Sämtliche Werke: Brandenburger (Berliner) Ausgabe*, ed. Roland Reuß and Peter Staengle (Frankfurt am Main: Stromfeld, 1888–2002), vol. 1, 65–66, l. 25–30; translation from the German by Valentine A. Pakis.
26 See Roland Borgards, "Experimentelle Aeronautik: Chemie, Meteorologie und Kleists Luftschiffkunst in den Berliner Abendblättern," *Kleist-Jahrbuch* (2005): 142–161.
27 See Höhler, *Luftfahrtforschung und Luftfahrtmythos*, 286–289. The idea for this simultaneous launch arose in September of 1896 in Paris at a meeting held by the directors of meteorological institutes from various countries in an effort to organize observations of the atmosphere across national borders. It was at this same conference that Hermite and Besançon introduced their unmanned balloons.
28 See Höhler, *Luftfahrtforschung und Luftfahrtmythos*.

the United States, "our highly inadequate understanding meteorological laws ... [needs] to be augmented by studies of the upper strata of the atmosphere."[29]

Data from the network of terrestrial weather stations were supplemented by data gathered in the upper regions of the sky. According to the meteorologist Julius Maurer's report about his balloon flight over the Alps, these supplemental data would make it possible to study not only local weather phenomena but also the global structure of weather conditions (Fig. 8).[30]

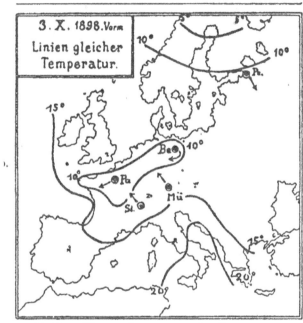

Fig. 8: The launch sites and wind directions of a European-wide simultaneous balloon ascent, 1898.

---

29 Alexander Supan, "Die internationalen Luftfahrten im Jahre 1902," *Petermann's geographische Mittheilungen* 50 (1904): 128–129, here 128; translation from the German by Valentine A. Pakis. A better understanding of the atmosphere was not only of interest for purposes of weather forecasting. It was rightly believed that weather forecasting would be of interest to agriculture as well as the military (see Friedman, *Appropriating the Weather*, 109). Commercial air travel would benefit from this knowledge later on.

30 Julius Maurer, "Die meteorologischen Ergebnisse," in Albert Heim et al., *Die Fahrt der "Wega" über Alpen und Jura am 3. Oktober 1898* (Basel: Schwabe, 1899), 82–125, here 119. In addition to Maurer, the others on board this flight were the mountain geologist Albert Heim and the balloonist and photographer Eduard Spelterini.

As historians of science have shown, the field of aerology played a significant role in the development of meteorology into a geographical and statistical science. Around the year 1900, meteorology transformed into a science devoted to "the quantitative recording of all environmental conditions."[31] The international collaboration of scientists on simultaneous balloon launches contributed to the establishment of atmospheric physics and to the formation of scientific guidelines for systematically measuring and charting the open air.[32] It was thus with the help of aerology that meteorology was able to transform from a science based on empirical observations into one grounded in theories and calculations.

Ballooning and Weather Forecasting

Collecting quantitative data is well and good, but what should be done with all the data? As of 1902, the data collected during the simultaneous balloon ascents flooded the desk of the chairman of the Internationale aeronautische Kommission in Strasbourg, Hugo Hergesell, in the form of standardized tables. There was so much new information, however, that at first no one was able to evaluate it.[33]

In 1904, the meteorologist Vilhelm Bjerknes therefore encouraged his colleagues to replace, on account of its inaccuracy and long delays, the graphical and synoptic evaluation of such data – as had been practiced with a chart, for instance, by the meteorologist Julius Maurer during his flight over the Alps – with a precise theoretical evaluation based on equations derived from the physical laws of the atmosphere.[34] This sort of weather forecasting would first be implemented in practice by the Englishman Lewis Fry Richardson, who published his results in his 1922 book *Weather Prediction by Numerical Process*, which likewise included a chart (Fig. 9).

Richardson proposed the following method: divide the earth's atmosphere into cubic regions and entrust each of the latter to someone proficient in mathematics. These researchers should take measurements of the temperature,

---

31 Höhler, *Luftfahrtforschung und Luftfahrtmythos*, 215.
32 See Höhler, *Luftfahrtforschung und Luftfahrtmythos*, 284–285.
33 In 1903, the gathered data was published in the *Veröffentlichungen der internationalen Kommission für wissenschaftliche Luftschiffahrt*, so that it could at least be made available to researchers.
34 See Friedman, *Appropriating the Weather*. In the absence of other ways to evaluate data, synoptic meteorology continued to be practiced into the 1970s. With this method, the movement of weather fronts could be represented in a graphical form. Maurer himself admitted the that, among other problems with this approach, the temperature measurements taken in balloons can be highly inaccurate (see Maurer, "Die meteorologischen Ergebnisse," 93, 104).

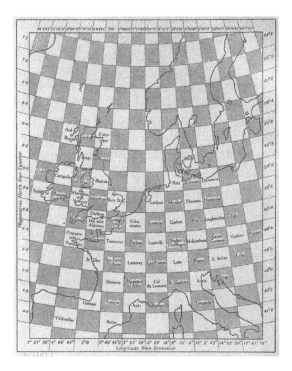

Fig. 9: Lewis Fry Richardson's map of Europe for numerical weather forecasting, 1922.

humidity, and pressure in their respective region and insert such data into formulas. The results should then be shared with those making the same calculations on adjacent regions. In order to test his formulas (which were based on physical laws) and the practicability of his method, Richardson worked through the necessary calculations for an example in 1917. His idea was to produce – long after the fact – a weather forecast for the day of May 20, 1910.

Because of the many simultaneous balloon launches undertaken over the course of the three previous days – known in English as "International Aerological Days" – there was, for May 20, a relative abundance of available data about the upper atmospheric regions, and this data set served as an especially good point of reference. After six weeks of making calculations, Richardson's results turned out to be false, but he had nevertheless demonstrated that numerical forecasting was fundamentally possible:

> In Ch. 9 will be found an arithmetical table showing the state of the atmosphere observed over middle Europe at 1910 May 20 d. 7 h. GMT [Greenwich Mean Time]. This region and instant were chosen because the observations form the most complete set known to me at

the time of writing ... Unfortunately this "forecast" is spoilt by errors in the initial data for winds. These errors appear to arise mainly from the irregular distribution of pilot balloon stations, and from their too small number.[35]

As Richardson remarks, his calculations were off the mark on account of certain errors in the data set pertaining to May 20, 1910, but he was confident that he had proven the viability of his approach. In order to speed up his method so that it could produce weather forecasts in real time (instead of being delayed by six weeks), he imagined a sort of forecasting factory (Fig. 10). His utopian vision of

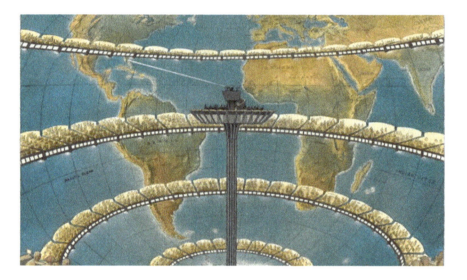

Fig. 10: Richardson's weather-forecasting factory (illustrated by François Schuiten).

a totally measurable and calculable world took the form of a large theater with 64,000 human computers, each responsible for performing calculations about a particular region of the sky and conveying this information to those working adjacently. This was of course impracticable, and forecasts of this sort would not be generated until the advent of the automated computer and its sufficient processing power. Today's forecasting models are all based on numerical methods, which are fed by new measurement data every six hours.[36] An increasing number of

---

[35] Lewis Fry Richardson, *Weather Prediction by Numerical Process* (London: Cambridge University Press, 1922), 2.
[36] See Gabriele Gramelsberger, *Computerexperimente: Zum Wandel der Wissenschaft im Zeitalter des Computers* (Bielefeld: transcript, 2010).

weather stations and ever-expanding computing power has admittedly enhanced the accuracy of more recent weather simulation, but the same problem of mathematical modeling that had created difficulties for Richardson still remains: minor deviations in input data can cause major deviations in output data.[37]

3 Concluding Remarks

Around the year 1900, knowledge about aircraft and knowledge about airspace coalesced into an epistemic turning point at which the ballistic stubbornness of balloons – their inability to be navigated horizontally – was used in a productive manner. Technology and practices for measuring, calculating, and simulating aerostatic and atmospheric dynamics were developed and established at several points of contact between balloon travel and meteorology. This essay has made clear, first, that unmanned balloons could be used as an instrument for tracking and recording balloon trajectories and wind vectors. Second, it has shown that simultaneous balloon launches could serve as a starting point for integrating data about local wind vectors and thereby identifying the structures of global weather conditions.

Examining these developments from a cultural-technical perspective makes it possible to describe Google's balloons in critical terms beyond the informational material provided by the tech giant itself and to illuminate the historically shifting references between travel space and traveled space, observing space and observed space, as well as calculating and calculated space. Because the sky could, from the outset, only be reached with machines, balloon flights served as a model for the epistemic specifics of the production of geographical knowledge in the nineteenth century: by means of particular technologies and practices, scientists constructed spaces, and as travelers found themselves subjected to the contingencies of these spaces.[38]

In the case of Google balloons, not only do the balloons become part of communication infrastructure; the stratosphere itself becomes infrastructure. Google balloons measure physical space and navigate by means of mathematical space. Google's stratosphere is not a thermodynamic space of random air currents and turbulence but rather a mathematically modeled and computer-simulated space; it is not unpredictable nature but rather a calculable and predictable environment.

---

37 See Gramelsberger, *Computerexperimente*.
38 See Johannes Fabian, *Im Tropenfieber: Wissenschaft und Wahn in der Erforschung Zentralafrikas* (Munich: C. H. Beck, 2001), 243–280.

This invites a historicizing media-theoretical and cultural-technical perspective from which it is possible to elucidate this "infrastructure-becoming" in light of its interwoven spatial, technical, and social aspects,[39] and from which it is also possible to describe collectives not only as metaphors but also as materialized and materializing agents that have become operative on a geopolitical level.[40] A cultural-technical perspective can make it possible to describe this "infrastructure-becoming" not only in light of its epistemic implications but also in light of its geopolitical implications. This raises not only the question of the extent to which balloon collectives are both abstract media of the space surrounding them and media of recording but also the question of how unmanned balloons and simultaneous ascents as such have formed, quite concretely, the ideal of a disruption-free airspace in which information can circulate without the distorting effects of noise.

With Project Loon – i.e., with the transition from balloons to Loons – these aircraft have become smart objects. The term "internet of things" implies that objects denoted as "smart" are "networked and context-dependent" and that they "operate on the basis of sensor data collected on a large scale."[41] They function independently, "in that they not only collect data but also predict future events or even make decisions on the basis of these data."[42] The aim of these twenty-first-century media is "to predicatively extrapolate, monitor, control, and ultimately economize (on the deepest level) movements, knowledge, and processes on the basis of sensor data."[43] In this light, one could maintain that Google's Project Loon economizes the stratosphere with its smart balloons and implements surveillance from it. This type of control via geographical and mathematical space is simultaneously a form of regulating sociotechnical space. Moreover, the borderless network of mobile, hovering radio towers is by no means stateless. Although governments can fill out a contact form on the project's homepage, legal regulations are a matter of negotiation or are undisclosed.

The alternative projects that have popped up to compete with Google's Project Loon are indicative of how lucrative it can be for large service providers

---

39 See Schabacher, "Medium Infrastruktur," 129.
40 See Lorenz Engell and Bernhard Siegert, "Einleitung," *Zeitschrift für Medienphilosophie und Kulturtechnikforschung* 3, 2 (2012): 5–13.
41 Florian Sprenger and Christoph Engemann, "Im Netz der Dinge: Zur Einleitung," in *Internet der Dinge: Über smarte Objekte, intelligente Umgebungen und die technische Durchdringung der Welt*, ed. Sprenger and Engemann (Bielefeld: transcript, 2015), 7–58, here 8.
42 Sprenger and Engemann, "Im Netz der Dinge," 8.
43 Sprenger and Engemann, "Im Netz der Dinge," 9. Here Sprenger and Engemann refer to Marc B.N. Hansen, *Feed-Forward: On the Future of Twenty-First-Century Media* (Chicago: University of Chicago Press, 2014), 4.

to make private mobile infrastructures available. Under the catchphrase "internet for everyone" – or, in Google's case, "Loon for everyone" – a competition has begun among infrastructures and transmission techniques. Companies such as Facebook and Tesla are likewise conducting experiments in the stratosphere with ultra-lightweight aircraft, solar-powered drones, and low-flying satellites. The internet for everyone implies a utopian vision of networking everyone and everything – and thus making everyone and everything calculable and predictable.

Texts/Bodies

Bernhard Siegert
Attached: The Object and the Collective

1 Cultural Techniques in 1983, 2000, 2020

As one of the reports written in the course of Friedrich Kittler's *Habilitation* procedure attests, in 1983 in the context of the humanities, the term "cultural techniques" carried the stigma of being unscientific. Kittler's *Aufschreibesysteme 1800/1900* (1985, translated as *Discourse Networks 1800/1900*) was regarded as belonging to a type of book that might be labeled "*kulturtechnisch*," which included other works such as Hans-Dieter Bahr's *Über den Umgang mit Maschinen* (1983, "On interacting with machines"), Jean Baudrillard's *L'Échange symbolique et la mort* (1976, *Symbolic Exchange and Death*), Oskar Negt and Alexander Kluge's *Geschichte und Eigensinn* (1981, *History and Obstinacy*), Wolfgang Schivelbusch's *Geschichte der künstlichen Helligkeit im 19. Jahrhundert* (1983, *Disenchanted Night: The Industrialization of Light in the Nineteenth Century*), or Jacques Derrida's *La carte postale* (1980, *The Post Card*). What these books have in common (if nothing else) is that they appear to suspend "the proven foundations of scientific knowledge."[1] Kittler himself had made use of the term "cultural techniques" at the end of the summer of 1983 in the preface he was pressured to write (which was later suppressed) for *Aufschreibesysteme 1800/1900*. "Even *écriture*, which has in the meantime become a hermeneutic slogan, does not use the term 'cultural techniques' to mean cultural techniques."[2] Describing reading and writing as cultural techniques meant, for Kittler (at that time), first, treating reading and writing as techniques for steering and programming people who could never be "mere individuals." Referring to Marcel Mauss's "techniques of the body,"[3] Kittler argues that an "approach concerned with cultural techniques" strips "even the most quotidian of practices of their apparent harmlessness ... since they suddenly became transparent as expressions of state or individual programs."[4] "Cultural tech-

---
[1] Wolfram Mauser, "Gutachten über die Habilitationsschrift von Friedrich A. Kittler: Aufschreibesysteme 1800/1900," *Zeitschrift für Medienwissenschaft* 6, 1 (2012): 169–197, here 169.
[2] Friedrich Kittler, " *Aufschreibesysteme 1800/1900*: Vorwort," *Zeitschrift für Medienwissenschaft* 6, 1 (2012): 117–126, here 125.
[3] Marcel Mauss, "Techniques of the Body," trans. Ben Brewster, *Economy and Society* 2, 1 (1973 [1934]): 70–88.
[4] Kittler, "Vorwort," 126.

Translated by Michael Thomas Taylor

𝟯 Open Access. © 2020 Bernhard Siegert, published by De Gruyter. This work is licensed under a Creative Commons Attribution 4.0 International License.
https://doi.org/10.1515/9783110647044-008

niques" was thus the term that allowed Kittler, in the early eighties, to formulate the scandalous idea that poetry was functionalized by the state into a control loop from 1800 onwards – for the training of education officials. Hence the concept of cultural techniques began to circulate, on the one hand, in the context of reinterpreting literary history as part of a history of data processing and, on the other hand, in order to brand the (as yet nameless) discourse of a media-theoretical and media-historical materialism with the stigma of being unscientific.

Together with the preface to *Discourse Networks*, the term "cultural technique" then disappeared from Kittler's conceptual repertoire for the next twenty years. "Technical media" took its place, only to again make way for cultural techniques with the founding of the Hermann von Helmholtz Centre for Cultural Techniques at the Humboldt-Universität zu Berlin. Instead of *Grammophone – Film – Typewriter* (1999 [1986]), the focus was now on "image – writing – number" (a somewhat strange triad that seems to imply numbers are not also writing). What does this double exchange of positions mean? Will "media theory" (at least in its Kittlerian vein) prove to have been but a brief interlude in the history of how research into cultural techniques was established as an academic field of study? Or was it necessary, after 1983, to replace the concept of cultural techniques with the concept of media in order to free the term "cultural techniques" from being deemed the epitome of all the subversive powers threatening the discursive sovereignty of the humanities? The latter seems more likely. This makes it all the more important to bear in mind that if, in 1983, the concept of cultural techniques was intended to decouple literature (or philosophy or art) from the individual (or from spirit or style) in order for literature to be described as a system of data processing, then today's research into cultural techniques must strictly de-individualize the practices on which it focuses, even if the concept of "discourse networks" has since become too narrow. Despite the general trend toward border crossing between media and social theory, this would require us to note crucial doubts about the use of the concept of cultural techniques in empirical media research, such that practices would again be viewed as originating in presuppositionless individuals.

Today, thirty-five years after the appearance of the term "cultural techniques" in the preface Kittler was compelled to write, a conference topic such as "Cultural Techniques of the Collective" makes it clear that the thrust of the concept and thus also its potential to cause controversy have changed and multiplied. The question of the cultural techniques of the collective would have to interrogate, first and foremost, formations of community with respect to those conditions that are not accessible to the community itself. These techniques would include, for example, the cultural technique of eating, inasmuch as eating, in contrast to food intake, engenders community. The question of which cultural programs determine our eating practices points to the cultural technique of sacrifice and this, in turn, points to the cultural technique of domesticating animals (because only the domesticated animal can be sacrificed).[5] An understanding of cultural techniques of the collective in the sense of cultural techniques of various collectives, on the other hand, would be a misunderstanding of the term cultural technique, insofar as such an understanding reconnects practices to a subject, to a collective subject, instead of to a program. The discourse of cultural techniques, however, has always aimed at hybrid collectives, that is, at collectives of subjects and objects or of quasi-subjects and quasi-objects. A view of the collective as a subject would therefore be antithetical to the basic concept of research into cultural techniques.

Cultural techniques of the collective are situated on much more primitive level than cultural techniques of reading and writing. They concern cultural-archaeological processes describing culture in layers far below the discourses of pedagogy, the university, and techniques of reading and writing. In contrast to 1983 and 2000, cultural techniques in 2020 are not only practices that create symbolic orders by preceding the concepts derived from them (such as image, text, number). On a level located below such complex practices as reading, writing, arithmetic, or drawing, etc., the concept of cultural techniques concerns the primary process of articulation as such. While it is undoubtedly true that reading, writing, arithmetic, and drawing are elementary cultural techniques, and that these cultural techniques consist of recursive operational chains that combine processes of hominization with higher media functions such as storage, transmission, and calculation, it is also indisputable that these practices rest, as semiotic practices, on a culturally primary process of articulation distinguishing signal and noise, message and medium, form and matter, communication and cacography, figure and ground.[6] With the second turn toward cultural techniques in German-language media theory that occurred

---

[5] See Walter Burkert, *Homo Necans: Interpretationen altgriechischer Opferriten und Mythen* (Berlin and New York: De Gruyter, 1972), 53–55; Thomas Macho, "Tier," in *Vom Menschen: Handbuch Historische Anthropologie*, ed. Christoph Wulf (Weinheim: Beltz, 1997), 73–75.

[6] Friedrich Kittler can be assumed to have understood media as cultural techniques, insofar as he considered it possible to analyze all individual media (from papyrus scrolls to the post office or up to the computer) as a more or less complex chain of the three functions of storage, transmission, and processing; see Friedrich Kittler, "Geschichte der Kommunikationsmedien," in *On Line: Kunst im Netz* (Graz: Steirische Kulturinitiative, 1993), 66–81; on this point, see also Sybille Krämer, "The Cultural Techniques of Time Axis Manipulation: On Friedrich Kittler's Conception of Media," *Theory, Culture and Society* 23, 7–8 (2006): 93–109. However, Kittler saw storage, transmission, and processing as media functions resting on a triad, postulated a priori of data, addresses, and commands, and he showed little interest in the processes of primary cultural techniques through which data, addresses, and commands were articulated in the first place.

at the end of the twentieth century, these "elementary cultural techniques" of reading, writing, arithmetic, and image making were joined by a series of further operations that might be called "primitive cultural techniques": drawing a line, filtering or sifting, separating land and sea from one another, measuring a field, erecting an enclosure, or sacrificing an animal. Primitive cultural techniques are used to process distinctions relevant to a concept of culture without reducing it to alphanumeric codes[7] (such as distinctions like those between inside and outside, pure and impure, human and animal, sacred and profane, etc.).[8] Whereas the production, processing, and deconstruction of the difference between male and female had been an essential characteristic of discursive systems, the concept of cultural techniques includes the articulation, processing, and deconstruction of distinctions that can hardly be located in the field of techniques for processing symbols.

Research into cultural techniques shifts the focus from ontological distinctions understood as naturally given to the problem of generating and stabilizing these distinctions to begin with. In short: the basic figure of this research is the enfolding of the ontic-ontological difference into the ontic. The being of beings falls under the rule of recursive, medial operational chains, where it is perpetually iterated, thematized, reflected, transformed, and re-mediated. Whereas media history examines what a given culture has been able to store, transmit, and process through the available media and codes, a history of cultural techniques is more concerned with the question of how distinctions that constitute symbolic orders (inside/outside, holy/profane, animal/human, sign/thing, analogue/digital) are recursively repeated, articulated, represented, and operationalized.

Only at this level of operative ontologies does the concept of cultural techniques gain its current strength. Geoffrey Winthrop-Young gets to the heart of this "philosophical" dimension of the theory of cultural techniques as follows:

> The term "cultural techniques" refers to operations that coalesce into entities which are subsequently viewed as the agents or sources running these operations. Procedural chains and connecting techniques give rise to notions and objects that are then endowed with essentialized identities. Underneath our *ontological* distinctions are constitutive, media-dependent *ontic* operations that need to be teased out by means of a techno-material deconstruction. To rephrase it in a more philosophical vein: the study of cultural techniques continues in a technologically more informed fashion a philosophical line of ontic-ontological question-

---

[7] See Sybille Krämer and Horst Bredekamp, "Kultur, Technik, Kulturtechnik: Wider die Diskursivierung der Kultur," in *Bild – Schrift – Zahl*, ed. S.K. and H.B. (Munich: Fink, 2003), 11–22.
[8] See Bernhard Siegert, *Cultural Techniques: Grids, Filters, Doors, and Other Articulations of the Real*, trans. Geoffrey Winthrop-Young (New York: Fordham University Press, 2015), 14.

ing opened up by Martin Heidegger. If German media theory in the Kittlerian vein focused on the materialities of communication, the study of cultural techniques takes aim at the materialities of ontologization.[9]

The ultimate aim of analyses undertaken by such research into cultural techniques is therefore a shift in the deconstruction of ontological categories to the level of technological materialism. The fact that the deconstruction of ontological distinctions takes place on the level of a (cultural) technical materialism marks an essential caveat against the enthusiasm with which collectivization, distribution, assembly, mixing, and posthumanization are currently in fashion everywhere in post-Actor Network Theories. From the point of view of the theory of cultural techniques, these methodological approaches are limited by the fact that they themselves often exhaust their possibilities in the conception of things as actions processed by networks of human and nonhuman actors. This approach may have an enormous potential to generate consensus across all kinds of disciplines inasmuch as it puts practical knowledge distributed in a network on an equal footing with subject-centered knowledge. The danger exists, however, that this merely replaces a notion of a self-present being with that of a self-present action. In not a few variations of post-ANT, such a romanticism of action can be found in an event in which subject and object, form and matter, passive and active, indistinguishably become one. Technology is transformed into a becoming that can no longer be questioned, except in esoteric formulations. The necessary diagnosis of the present – namely, that technology rules today as *physis* once did – threatens in posthumanist theories to conceptually dissolve into the self-presence of natural objects, into *entelecheia* and *ousia*, in a celebration of the life of things.[10]

---

[9] Geoffrey Winthrop-Young, "The *Culture* of Cultural Techniques: Conceptual Inertia and the Parasitic Materialities of Ontologization," *Cultural Politics* 10, 3 (2014): 376–388, here 387.

[10] This becomes clearest in Tim Ingold's contributions to the anthropology of technology, in which a nonhylemorphistic concept of technology remains restricted to craftsmanship, and "making" ultimately becomes a religion of life; see Tim Ingold, "Eight Themes in the Anthropology of Technology," *Social Analysis: The International Journal of Social and Cultural Practice* 41, 1 (March 1997): 106–138; Tim Ingold, "The Textility of Making," *Cambridge Journal of Economics* 34 (2010): 91–102. The clear-sightedness of media studies emerges, by contrast, precisely where the supposed actions of nature are revealed to be operations of technology. This does not mean, however, that Tim Ingold's texts have not delivered enormously helpful impulses to research into cultural techniques.

2 Quasi-objects: Dispersion, Compression, Attachments

Hybrid objects play a key role in the theory of cultural techniques. Hybrid objects are assemblages – which means they are couplings to themselves – that possess a specific agency due to the nature of this assemblage and the operations it makes possible. Bruno Latour already identified hybrid objects as quasi-objects in *Nous n'avons jamais été modernes* (1991, *We Have Never Been Modern*).[11] Quasi-objects refer to a becoming-collective in which objects and people always already possess and produce each other; the identity of subject and object, sign and thing, figure and ground, message and medium, is discovered only through recursively operative assemblages.

The quasi-object was first introduced by Michel Serres in his book on the parasite (1980). It is telling that the second cultural-techniques turn (after 2000) discovered Serres after Foucault, who was so central for *discourse networks* (*Aufschreibesysteme*) and their feedback control-systems of cultural techniques. In Serres's book on the parasite, the quasi-object is an object circulating between subjects (as in the children's game of "button button, who's got the button") that transforms the one to whom it attaches into a subject and all others involved in its circulation into a collective.[12] The theory of the quasi-object is therefore also a theory of the quasi-subject, since both penetrate each other by possessing each other. In his book on Rome published three years later, Serres uses the story of Romulus's *diasparagmos* that is told by Livius and Plutarch to reveal the mechanisms by which the collective emerges from the operations and transformations of the quasi-object.[13]

> After accomplishing these mortal deeds, Romulus was one day holding an assembly of the people on the Campus Martius near the Goat Swamp to review the army. Suddenly a storm arose with loud claps of thunder, enveloping him in a cloud so dense that it hid him from the view of the people. From then on Romulus was no longer on earth. The Roman people finally recovered from their panic when the turbulence was succeeded by a bright and sunny day. Seeing the king's throne empty, they readily believed the assertion of the senators who had been standing nearby that he had been snatched up on high by the storm. Nevertheless, they remained sorrowful and silent for some time, stricken with fear as if they

---

[11] See Bruno Latour, *We Have Never Been Modern*, trans. Catherine Porter (Cambridge, MA: Harvard University Press, 1993), 51.
[12] See Michel Serres, *The Parasite*, trans. Lawrence Schehr (Baltimore: Johns Hopkins University Press, 1982 [1980]), 224ff.
[13] Michel Serres, *Rome: Le livre des fondations* (Paris: Grasset, 1983), 126–127; *Rome: The First Book of Foundations*, trans. Randolph Burks (London: Bloomsbury Academic, 2015), 84.

had been orphaned. Then, on the initiative of a few, they all decided that Romulus should be hailed as a god, son of a god, king, and father of the city of Rome. With prayers they begged his favor, beseeching him to be willing and propitious toward the Roman people and to protect their descendants forever.[14]

The thunderstorm that hides the king from view appears *cum magno fragore*. *Fragor* means not only to crash but also to shatter, burst, break, or crack into pieces. This shattering, the fragment, and the fractal are etymologically closely related to noise. If one can no longer understand the voice of the other, the collective shatters. Noise swallows the voices of those who, as Plutarch reports, have scattered out of fear and who come back to the field of Mars only slowly, after peace and light have returned.[15] Dispersal, compression: a curve exists between noise and signal indicating the degree of statistical coupling between these individual elements. However, both Livy and Plutarch report an alternative version to this ascension:

> I suppose that there were some, even then, who privately claimed that the king had been torn into pieces by the hands of the senators. This rumor also spread, though in enigmatic terms.[16]

The second version explains the first. The religious aspect, the miracle of the ascension, is explained by an archaic ritual at the origin of the political that Serres calls *suffrage*. The alternative must accordingly not be read as an alternative but as a change of perspective that seeks to uncover the operation underlying the myth. Myth (the disappearance) stands opposed to history (*diasparagmos*), light to dark. The babble of voices, the *fragor*, comes together to become one voice: acclamation. And this is how Romulus becomes a God. The patricians surround him and tear him up; they then carry the fragments of his body, hidden under their togas, from the square. The *corps morcelé* (the *fragmented body*, with Lacan) of the king that is carried away under the robes of the senators becomes the founding object of Rome. Michel Serres reads this event ontologically: "There is no object without a collective, there is no human collective without an object. Rome constructs the object."[17] And even more so: "Voici le premier objet" ("Here is the first object").[18]

---

14 Livy, *The history of Rome* 1–5, vol. 1, trans., introduction, and notes Valerie M. Warrior (Indianapolis: Hackett, 2006), 25–26.
15 See Plutarch, *Lives* 1–11, vol. 1, trans., introduction, and notes Bernadotte Perrin (Cambridge: Harvard University Press, 1967), 176.
16 Livy, *The history of Rome*, 26. See also Plutarch, *Lives*, 175: "But some conjectured that the senators, convened in the temple of Vulcan, fell upon him and slew him, then cut his body in pieces, put each a portion into the folds of his robe, and so carried it away."
17 Serres, *Rome: Le livre des fondations*, 129; *Rome: The First Book of Foundations*, 86.
18 Serres, *Rome: Le livre des fondations*, 151; *Rome: The First Book of Foundations*, 100.

The first object is that which constitutes itself in the relationships of the group – as a game, a "fetish," or a commodity; conversely, the collective cannot form unless something circulates within it "que j'ai nommé quasi-objet" ("that ... I have called quasi-object").[19] These quasi-objects vary: sometimes they are body parts hidden in the fold of a toga; sometimes they are stones; sometimes they are voices. When stones converge, what results is a stoning; when voices are compressed, it is *suffrage*.[20] *Suffrage* in the anthropological and political sense is Serres's name for the operation leading to a substance, concept, or idea. There is no difference between what is stable in the mind, what is stable in the senses, and what is stable in phenomena. All these ontological regions are connected with each other by the operation of *suffrage*. For Serres, there is no being and no concept of being without operation. There is no substance without the operation of dismemberment, stoning, transformation, and coupling of fragments; substance is atomization and compression, but atomization and compression must also be traced back to concrete cultural techniques (such as those of sacrifice). Philosophy without any connection to the history of cultural techniques is and remains empty theorizing.

The origin of the collective, the founding act of Rome, is always a sacrifice; here, Serres follows René Girard's theory of the scapegoat. For the theory of the hybrid object, however, what is more interesting is how objects and individuals emerge from couplings: "I am a part of my group; I carry in my breast a part of the king's body".[21] This object "se transsubstantie" ("becomes transsubstantiated"): it becomes stone, becomes a bracelet, becomes a precious stone that patricians wear on their fingers.[22] The quasi-object is transformed, from thing to sign, but it always retains a reference to the object from which it originates; it always remains a fragment.

---

[19] Serres, *Rome: Le livre des fondations*, 128; *Rome: The First Book of Foundations*, 86.

[20] On this point, see Michael Cuntz, "Aufklärung über den Fetisch: Latours Konzept des *faitiche* und seine Verbindung zu Serres' Statuen," in *Gegenwart des Fetischs: Dingkonjunktur und Fetischbegriff in der Diskussion*, ed. Christine Blättler and Falko Schmieder (Vienna: Turia + Kant, 2014), 76: "What circulates in the collective that assembles together and thereby stabilizes itself are, first and foremost, the stones that serve as projectiles and the scraps of flesh from the sacrificial body that has been torn apart. This is the origin of the quasi-objects and this movement persists in the handling of fetishes and idols."

[21] Serres, *Rome: Le livre des fondations*, 150: "Je suis une part de mon groupe, je porte dans mon sein une part du corps du roi." *Rome: The First Book of Foundations*, 100.

[22] Serres, *Rome: Le livre des fondations*, 151; *Rome: The First Book of Foundations*, 100. Serres reads the story of Romulus's *diasparagmos* together with the story of Tarpeia, who betrayed the Romans to the Sabines and demanded, as a reward, what the Sabines were wearing on their left arm, namely, golden bracelets. After she had opened the city gate to the enemies and demanded her reward, the Sabines "stoned" her by throwing their bracelets and shields upon her.

Hybrid objects belonging to the family of quasi-objects can now be defined more precisely: they are by no means just any arbitrary things but special things that render recursive, that mark and thematize the transubstantiation of things into signs. Hybrid objects are not symbols in the sense of representing what is absent by something present; rather, hybrid objects embed things within signs.

Objects that embed the body parts of the saint in jewels (instead of representing them with jewels) are reliquaries. A secularized variant of reliquaries are the medallions that Elizabeth I gave to her most important counselors and defenders of the empire. Like miniature reliquaries, they were worn on the body, between one's shirt and one's skin, or they were held in the hand, kissed, and could be opened up. The locket known as the *Heneage Jewel* shows, on its front side, the portrait of Elizabeth in her formal imperial profile, surrounded by diamonds, rubies, and rock crystal from Burma and other parts of the colonial empire. The reverse side is hidden behind a cover, the outside of which depicts the ark of the Reformed church on a stormy sea and refers to Elizabeth's ecclesiastical authority as the *kybernetes* of the *Ecclesia Anglicana*. When the cover is opened, another portrait of Elizabeth appears showing her as a private person – a homage to her as a lady, as the "Astraea, Queen of Beauty" who is exalted in contemporary sonnets by John Davies. Her gaze is met inside the lid by the Tudor rose (Fig. 1). Insofar as the stones symbolized the empire's far-reaching trade relations, the hinge of the medallion connected the political-economic second body of the Empire, now defined also in terms of territory, with the portrait of the queen inside. The sources indicate that Elizabeth had a large number of such medallions, which she kept in a cabinet in her bedroom and showed to selected persons. All of them were wrapped in paper and given names by the queen. They are therefore quasi-objects whose owners directly participate in the intimate first body of the queen and, as a collective, guarantee the stability of English hegemony. The proximity to reliquaries representing the earliest examples of these miniatures follows from the fact that they were kept in ivory boxes covered with protective jewels.[23] These quasi-objects literally materialize the concept of attachments introduced by Antoine Hennion into Actor Network Theory, which he uses to describe the attachment of subjects to objects and vice versa.[24] These quasi-objects are literally attached to the quasi-subjects, and the quasi-subjects are, conversely, attached to the quasi-objects.

---

23 See John Murdoch et al. (eds.), *The English Miniature* (New Haven: Yale University Press, 1981), 73 and 76.

24 See Antoine Hennion, Sophie Maisonneuve, and Émilie Gomart, *Figures de l'amateur: Formes, objets, pratiques de l'amour de la musique aujourd'hui* (Paris: La documentation Française, 2000). It is in this sense that Ulrich Pfisterer also speaks of their "binding function" with regard to the medallions circulating in Rome during the Renaissance. See also Ulrich Pfist-

Fig. 1: Nicholas Hilliard, locket known as the Heneage Jewel (or Armada Jewel), ca. 1595, enameled gold, diamonds, Burmese rubies, rock crystal.

Left: front side: Portrait of Queen Elizabeth I; Second from left: reverse side: ark of the Anglican church in stormy sea (lid closed); third from left: half-opened lid; fourth from left: reverse side: portrait of Queen Elizabeth I (lid open 90°); right: portrait of Elizabeth I and Tudor Rose (lid fully opened).

What is essential is that the hybrid object's "mode of existence" is engendered by the operations of opening, closing, sewing, coupling, and folding. Signs and things are thus always already entangled by means of linking operations and are part of a continuum of operations that connects assemblages – and that also plugs in the subject attached to these structures.

---

erer, *Lysippus und seine Freunde: Liebesgaben und Gedächtnis im Rom der Renaissance* (Berlin: Akademie-Verlag, 2008).

Michael Cuntz
Monturen/montures: On Riding, Dressing, and Wearing. Nomadic Cultural Techniques and (the Marginalization) of Asian Clothing in Europe

In this text, I would like to link together three things: 1) elements of the history of the French word *monture* and its German loanword *Monturen*; 2) elements of the history of denying Asian influences through a long tradition of European exceptionalism, which I examine using the "example" of textiles and clothing; and 3) the resulting suppression of the Asian, nomadic, and interspecific origins of supposedly typical European garments that emerged from the domestication of horses on the Eurasian steppe. In doing so, I will attempt to bring together several traces and fragments of Eurasian processes for exchanging cultural techniques, living beings, and artifacts. In view of the geographical and temporal expansiveness of this macrohistory, my argument necessarily proceeds in leaps.

1 Words and Things

The question of the relation between words and things is no minor matter for an approach based in cultural techniques. It is thus significant that André-Georges Haudricourt, who will play a significant part in the second half of this essay, had a dual career – on the one hand, as an explorer of ensembles of technical objects, techniques of the body, and domestications, as well as their migration; and as a linguist, on the other. Yet this one side/other side is a separation that results from his appraisal by isolated professional communities, and that severs the indissoluble connection existing between words and things.[1] If one discredits, from the vantage point of an "enlightened" awareness of the complete arbitrariness of signs, the interest in words as objects of knowledge as cratylistic naïvety, that is, a naïvety that assumes a factual correspondence between things and words, then this in turn is based on a naive thinking of factuality, which believes that the

---
[1] See Michael Cuntz, "Kommentar zu André-Georges Haudricourts 'Technologie als Humanwissenschaft,'" *Zeitschrift für Medien- und Kulturforschung* 1, 1 (2010): 89–99.

Translated by Michael Thomas Taylor

∂ Open Access. © 2020 Michael Cuntz, published by De Gruyter. This work is licensed under a Creative Commons Attribution 4.0 International License.
https://doi.org/10.1515/9783110647044-009

mode of existence of the things in the world, or facts, are given independently of human manipulations.

It is precisely where there is talk of domestication, as will be the case in what follows, that the divide between nature and culture is always already overgrown by a web of connections between the two of them, which renders untenable the containment of existences fabricated in such a way within a purely human sphere of social constructions. If one assumes that there is no world "behind" our fabrications, our *faitiches*,[2] but rather a pluriverse[3] of worlds consisting of such fabrications, of worlds that overlap and exist in and through (discursive) exchange, then the critical uncoupling of words and things appears as the mutilation of a web in which actors, signs, materials, and milieus support and influence each other. But even staunch relativists would have to acknowledge that cultural techniques and, at least in part, their migratory movements have become sedimented in words. And even more so: that the naming and designating of actors, things, and operations is also a central cultural technique.

2 *Montures/Monturen*

I therefore begin this text by exploring a centuries-old process of migrations and shifts, between French and German, in the word pair "monture/Montur." The word "Montur" is no longer very common in German. It is most frequently used in the phrase "in voller Montur" ("in full kit"), where it refers to clothing and accessories, that is, to what one wears on one's body. It is used to denote a person who is optimally equipped, possibly even over-equipped. When used ironically in this sense, it means something like "Aufzug" (an "oufit," "getup," or "costume"). In some areas of southwestern Germany, such as on the Moselle River or in Saarland, *Montur* therefore also refers to "all of a woman's clothing" in the sense of a complete outfit with textile objects that are often seen as mere attire (*Rheinisches Wörterbuch*).[4]

On the other hand, according to *Duden*, it can also mean that one wears clothes for "a certain purpose," and hence perhaps for something other than an

---

[2] See Bruno Latour, *Sur le culte moderne des dieux faitiches, suivi de Iconoclash* (Paris: La Découverte, 2009).

[3] William James, *A Pluralistic Universe: Hibbert Lectures at Manchester College on the Present Situation in Philosophy* (New York: Longmans, Green & Co., 1909).

[4] For the other German-language dictionaries listed below, unless otherwise stated, as well as for *Meyers Großes Konversationslexikon*, quotes are from http://www.woerterbuchnetz.de (visited on July 21, 2019); translations by Michael Thomas Taylor.

everyday occasion.[5] This purpose is often work: *Monturen* are the liveries of servants or certain work clothes, for example, clothes worn by a mechanic.

But the reference to clothing is already a narrowing, because the word "Montur" – or as it was originally used in German, "Montierung" – indicates not only clothing but, analogously to the English word "outfit," also all the equipment or objects that accompanied an outfit, and specifically, the outfitting of soldiers. In addition, "Montur" designates, in both German and French, a setting in which something is fastened, such as precious stones.[6] The connection with "montieren" in the sense of a "Montage" is obvious: the soldier is thus a classic network actor capable of acting only in collaboration with the multitude of his objects. We thus find a soldier literally equipped, or *montiert*, in *Meyers Konversationslexikon* (published from 1905 to 1909):

> Montierung (in Österreich Montur, franz.), die Ausstattung des Soldaten, die ehedem jedem einzelnen oblag, bis mit Einführung der stehenden Heere der Staat es übernahm, die Truppen zu montieren, d. h. auszurüsten.

> Montierung" (in Austria, the French word, "Montur," is used), the outfitting of a soldier, which was formerly the responsibility of each individual, until the state took over the task, upon the introduction of the standing armies, of outfitting, i.e., equipping, the troops.

The verb "muntieren" can already be found in Middle High German (*Lexer*).[7] And the *Schweizer Idiotikon* contains the following example of usage from the year 1673: "Pferd, das mundiert ist zum Streit" ("a horse that has been equipped [mundiert] for battle").[8] Even if here it is, curiously, the horse itself that is equipped for battle instead of being the rider's equipment: this brings us closer to the starting point of this word history, because the first meaning of the French word "monture," for which we find examples going back to 1360 (*Petit Robert*), and from which all further word migrations with their semantic shifts are derived, is nothing other than mount, or riding animal, and, at least in the European context, the riding animal par excellence, the horse. To "montieren" something, or to put it together, i.e., *assembler*, is therefore derived from mounting something, and specifically, the mounting of an animal: The "monture" is that which is mounted; horse and rider, carrier and carried, together form an assemblage, and this basic constellation is repeated in the setting or the frame that carries something in or even on itself.

---

5 https://www.duden.de (visited on July 21, 2019).
6 https://www.duden.de/ (visited on July 21, 2019); *Le Nouveau Petit Robert: Dictionnaire alphabétique et anthologique de la langue française* (Paris: Le Robert, 1996).
7 Matthias Lexer, *Mittelhochdeutsches Handwörterbuch*, 3 vols. (Leipzig: Herzel, 1872–1878).
8 *Schweizerisches Idiotikon*, https://www.idiotikon.ch/ (visited on July 21, 2019).

In the migration of the word from French into German and between different contexts of use and social situations, a whole series of remarkable shifts occur. Not only the transition from the military sphere to the sphere of servants and work, up to the change in gender to women's clothing with its accompanying connotations of ridiculousness and frivolity; also the simultaneous reduction, to clothing and useless accessories, of what previously referred to useful and necessary equipment, which additionally, or primarily, included a domesticated living being. For the most radical revaluation already took place in the shift from French to German: from the elevation of the knight, who looks down from his mount upon those who are of low birth, to the humiliation of the soldier-servant who is likewise marked and equipped by his master.[9] Saying that these subalterns occupy the symbolic place of the horse would arguably mean underestimating the economic value of the horse.

Hence, in French, the meaning of the word already shifts from a *monture* for a riding animal to *monture* in the technical sense, or *montage*, which nevertheless does not appear in French until nearly four hundred years later, in the middle of the eighteenth century[10] – of course because *monter* contains the sense of "to erect, to set upright," and I would suggest that the conceptual proximity (which also exists, by the way, in the verb "dresser") is no coincidence. This is not only because the elevation of a person's point of view – or, to be more precise, the point where a person sits – to that of a mounted rider, and especially of a rider on horseback, has been the most important uprighting since humans began to walk on two legs, but also because horse and rider, as well as the articles of clothing and equipment (reins, bridles, tack, and saddles) and the gestures and techniques of the body required to ride an animal, likely form one of the most complex cultural-technical montages and assemblages of all. This assemblage is an interspecific history of technology, an emergence of interspecific network actors, that required a long period of preparation and, in its assembly, an alignment between

---

9 *Adelung*, "Montur," quoted from http://www.woerterbuchnetz.de; in *Adelung*, it is clear that the "common soldier" is nothing but a variant of the servant: both are equipped and outfitted by their master to be marked as property. On the meaning of liveries as a badge of servitude, or rather of servility, see also Thorstein Veblen, *The Theory of the Leisure Class* (Oxford: Oxford University Press, 2007 [1899]).

10 The *Petit Robert* gives 1740 as the year for the use of *monture* in the sense of a frame or setting. Whereas Nicot's *Thrésor de la langue française* from 1606 still presents *monture* exclusively as mount, the word "monture" appears in the fourth edition of the *Dictionnaire de l'Académie française* from 1762 in the sense of rack (of a cannon) or setting/frame, and hence expressly as assemblage: "L'assemblage des deux pièces d'une tabatière ou d'un étui, jointes l'une avec l'autre." The ARTFL project, *Dictionnaires d'autrefois*, http://portail.atilf.fr/dictionnaires/onelook.htm (visited on July 21, 2019).

humans and animals. Moreover, if one follows Jean-Pierre Digard (an expert on domestication),[11] and thus an approach established by André Leroi-Gourhan that is marked by attention to cultural techniques, this is also a history that was possible only via several detours: horse and rider constitute a *monture*, a montage, a small collective unit in a cultural-technical ensemble that was prepared over thousands of years.[12]

This interspecific *montur/montage* therefore forms the core for a reflection of the connection between things, persons, and other animate actors, as well as actions and habits, which is my aim in what follows. This consideration concerns the history of clothing that is completely familiar to us and supposedly typical for the West, and that takes us far away from today's Europe in space and time. It must also make headway in the face of a double resistance of disdain and contempt: European disdain for Asia that reaches back thousands of years, as well as the equally ancient contempt that settled peoples (who consider only themselves to be civilized) have toward nomads. At least the latter has a great deal to do with the question of the relationship between humans and animals, or with an anthropocentric world view. It is only by distancing oneself from such a condescending perspective – and this is something I will be tracing in what follows in relation to Asian textiles and garments – that it becomes clear how nomadism, domestication, and riding on the steppes of Asia are related to the development of techniques of the body for wearing clothing, as well as to the development of certain articles of clothing. This is not only a cultural issue, but also one of the connections between habits and riding habits.

3 Covers, Accessories, Spices: The Supplementary Topology of Exotica

It is therefore also necessary to speak about the attitude of European philosophers and scholars toward Asian textiles, because the puritanical and xenophobic affect aroused by exotica from Asia and the condemnation of their consumption as senseless waste or even a corruption of moral values can be traced back, in

---

[11] Jean-Pierre Digard, *Une histoire du cheval: Art, techniques et société* (Arles: Actes Sud, 2007).
[12] This is especially true if one assumes, with Digard, that horses were not domesticated to be ridden, but rather that riding horses arose much later; see Digard, *Une histoire du cheval*, 39.

Europe, at least to the branding of silk as the epitome of *luxuria*[13] in Rome.[14] It was apparently inconceivable that Europeans – and even more so, European men – might wear Asian textiles. Upon encountering the enthusiasm of rich Roman women for silk, Pliny the Elder was, not without good reason, concerned about Rome's stock of precious metals.[15] Seneca, by contrast, took moral offense at just how transparent and flowing this fabric was.[16] Tacitus reports that Roman men were forbidden by decree to dress in this "feminine" fabric.[17]

Pliny was consistent enough to also condemn amber or frankincense. Yet as Fitzpatrick clearly shows, Seneca's stoic invectives against Asian luxury goods served the same function as do the sermons held by today's defenders of national values and national economies.[18] They function as a hypocritical distraction from the more profane things that made up the lion's share of this trade, as well as from the facts of just who, exactly, was importing it at such a handsome profit: namely, the members of the Roman Senate (in its plutocratic constitution, which proved exemplary for later occidental empires), who accumulated their enormous fortunes primarily through trade with the East. Seneca had no trouble distinguishing between respectable wealth and offensive *luxuria*.[19] Which obviously meant not only: behave with dignity worthy of a Roman, but also wear real Roman clothing. For the money earned via long-distance trade, on the other hand, it was and remains true: *non olet*.

The textiles and garments that migrated from Asia to Europe are more than just exotic accessories, and the issue was not only the conspicuous consumption

---

13 Even more so than *luxus*, *luxuria* bears the connotations of moral and sexual depravation, a differentiation that persists in the different semantic values of French *luxe* (luxury) and *luxure* (lust). *Luxuria* is directly linked to *voluptas*.

14 To call Romans Europeans is a shortcut rooted in the fact that today's Europeans consider themselves the heirs of Greek and Roman culture. Of course, this equation between Greeks and Romans, on the one hand, and modern Europeans, on the other, must be deconstructed. Still, even if Romans like Pliny and Seneca were not Europeans: what they share with (many of) today's Europeans is their anti-Asian affect.

15 It can in fact be shown that Roman silver circulated in Asia on a large scale; see M. P. Fitzpatrick, "Provincializing Rome: The Indian Ocean Trade Network and Roman Imperialism," *Journal of World History* 22, 1 (2011): 27–54, here 50.

16 Peter Frankopan, *The Silk Roads: A New History of the World* (London and New York: Bloomsbury, 2015), 17f.

17 In a fashion that is equally absurd and consistent, this attitude is also transferred to riding: even the Greeks (Xenophon, for example) praise "bareback riding" and mock the riders of their Asian opponents as effeminate because they use saddles and blankets; see Digard, *Une histoire du cheval*, 68.

18 Valerie Hansen, *The Silk Road: A New History* (Oxford: Oxford University Press, 2015), 20.

19 See Fitzpatrick, "Provincializing Rome," 32.

of the ruling elite, even if this consumption was a strong driving force for both the peaceful and not so peaceful movement of coveted goods.[20] From the outset, it was important for subordinates to share in the profit from the flow of exotic goods, with the consequence that these goods became more common not only in the upper social stratum.[21]

Given the deep penetration of Asian goods into the West, the fact that it was exotic fabrics and garments, as well as spices, against which the Western imaginary constituted itself is perhaps anything but a coincidence in this context:[22] these goods stand metonymically for a view of "foreign" objects that, like covers or shells and spices, merely envelop and refine – or even supplement[23] – what is authentic, the genuine core, body, or food, and that can also be omitted if necessary, as for example in Renaissance France, where, long before "freedom fries," we find invectives against strongly seasoned food as an "un-French" corruption of national palates. But for millennia, thanks to economic-cultural exchange processes within long networks that would not have existed without *montures* like horses and camels, it is impossible to uncover, anywhere in Eurasia, the core of an "authentic" culture from beneath purely external additions. Instead, everywhere we encounter transcultural *Monturen* and assemblages.

4 Riding Pants and Centaurs: Palmyra – Sampul

What would Seneca and Pliny, both so offended by the transparent, unmanly fabric of silk, have said about the clothing worn by the inhabitants of Palmyra? In that city on the eastern edge of the Empire, which Rome had subdued a century before these philosophers lived, because the city was one of the gateways to trade with Asia, these two men would have seen many people dressed in a completely un-Roman fashion. Instead of wrapping lengths of wool or linen around their bodies and attaching them with clasps, as was customary in Greece and the

---

[20] Paul Treherne, "The Warrior's Beauty: The Masculine Body and Self-Identity in Bronze-Age Europe," *Journal of Historical Geography* 3, 1 (1995): 105–144.
[21] Barry Cunliffe, *By Steppe, Desert, and Ocean: The Birth of Eurasia* (Oxford: Oxford University Press 2015), 94f.
[22] See Stefan Halikowski Smith, "The Mystification of Spices in the Western Tradition," *European Review of History/Revue européenne d'histoire* 8, 2 (2001): 119–136.
[23] On the logic of the supplement, see Jacques Derrida, *De la grammatologie* (Paris: Minuit, 1967), 203–234.

Roman Empire, and with sandals bound to their feet, these people dressed in a "Persian" manner:[24] with boots and pants.[25]

This image of a "topsy-turvy world" in which the Persians appear to be the progenitors of our present-day clothing style rather than the Greeks (who would presumably consider us to be Asian barbarians because of our clothing), is more than a snapshot of history, for the form of many garments worn by Europeans and their descendants that we consider to be specifically European comes from Asia – as well as the way they function, meaning how clothing is put on and taken off, how it clings to the body and envelops it. Clothes, namely, cannot be separated from certain gestures, such as wrapping or covering up, or from cultural techniques. If one traces these gestures and techniques, it becomes clear that a person quite literally "carries" or "bears" clothing.

Furthermore, it appears that this "Persian" clothing is closely linked to the history of the horse's domestication – not primarily because the horse's skin (in the form of leather) or its coat (in the form of felt made of horsehair) would have provided the materials for this clothing, but because of the techniques of using horses as a means of transport, which has had consequences for the human body. I'll return to all of this later.

Still, it's not enough to travel to the Middle Eastern metropolis of Palmyra, because the Eurasian processes of exchange do not consist solely of encounters among diverse things and peoples at the margins of their respective territories and spheres of influence. On the contrary, early traces of fusioning can be found in far more remote places, which in turn prove, from a less Eurocentric perspective, to be central crossroads of the trade routes: Sampul, which today lies in Xinjiang province in western China but belonged, in the third century AD, to the kingdom of Khotan, which had been founded by steppe nomads. Graves from this time were found to contain a leg from a pair of woolen men's pants. (Fig. 1) This leg was in turn cut from a tapestry originating from the Parthian Empire. It shows the depiction of a centaur, a being from the Greek mythical world modeled, according to today's view, after nomadic equestrian peoples of the Asian steppe armed with bows. The tapestry probably originates from the Parthian Empire; the floral motifs on the

---

[24] Interestingly, the Roman Empire also invented the corresponding footwear for horses known as hipposandals; see, e.g., A. D. Fraser, "Recent Light on the Roman Horseshoe," *The Classical Journal* 29, 9 (1934): 689–691.

[25] Paul Veyne, *Palmyre: L'irremplaçable trésor* (Paris: Albin Michel, 2015), 12. This is also reflected in the depictions on the necropolis of Palmyra, which testify to the mixture of Greek and Parthian lifestyles: "The couches, drinking vessels, and hairstyles might be Hellenistic, but the long robes that clothe these figures are similar to those of the Parthians." See Xinru Liu, *The Silk Road in World History* (Oxford: Oxford University Press, 2010), 29.

collar of the warrior and the cloak of the centaur are of Central Asian origin, but the centaur motif spread deep into Asia, as far as Northwest China, and primarily where nomadic equestrian peoples lived.[26] If one follows the widely accepted thesis that the origin of the centaur myth lies in the perception of such Asian nomadic peoples on horseback by the early Greeks (who were themselves not familiar with horses and certainly did not ride them), then these riders were not only confronted with representations of this mythical image but apparently also willingly integrated them into their culture. Was this a form of self-reflexivity in relation to the fusion of rider and horse into a single entity, a mounted actor?

Fig. 1: Woolen Trouser Leg from Khotan.

The history of the horse's domestication leads us not only to Asia, but also away from the settled "high" cultures to the nomads of the steppes: first to the Scythians, then to the Turkic peoples and the Mongols, who were regarded for quite some time as being culturally inferior (otherwise the term "high culture" would not exist) and came to inspire respect (and fear) only when they conquered the urban realms of settled peoples, which itself gives evidence of the Mongols'

---

26 See Liu, 198ff.

superior military techniques and tactics.[27] In the representation and assessment of the contact between the nomads and the settled peoples they subjugated, cultural transfer is often conceived in only one direction, as a one-way street: It is in terms of the "wild" nomads who, after initially wreaking havoc, adopt the superior civilization. This attitude can be found even in some authors who are determined to correct the Eurocentric view of history by giving due consideration to Asia. In Jack Goody's *The Theft of History*, for example, the Mongols only appear as interrupters of trade between Europe and the Middle East;[28] beyond that, Goody mentions the Mongols only in noting that they adopted gunpowder from the Chinese.[29] For Goody, culture is always synonymous with urban culture. Things are similar with S. Frederick Starr, who argues that the Mongols were only able to maintain their rule by parasitically exploiting the abilities of others, whether in administration or science.[30] Unfortunately, Starr's rehabilitation of the urban cultures of Central Asia comes at the expense of these "culture-less" nomads – a perspective expressed in his work not least of all in a blatant Arabophobia. Barry Cunliffe makes a similar remark,[31] but the central claim of his book is that the Eurasian space was, for millennia, able to function only through permanent exchange between settled and nomadic cultures.

5 What's with the Hats? Hennins and *Panni Tartarici*

In his history of the Silk Road, Peter Frankopan also pushes back against this disparaging view in emphasizing not only the outstanding postal infrastructure of the Mongol Empire (which was of course based on horsemen) but also, as Marco Polo and Rustichello da Pisa had pointed out centuries earlier,[32] on the administrative and political skills of these rulers, which were based on tolerance

---

27 Jacques Paviot describes this shift of perception from fear to interest; see "England and the Mongols (c. 1260–1330)," *Journal of Royal Asiatic Society*, Series 3, 10, 3 (2000): 305–318.
28 Jack Goody, *The Theft of History* (Cambdrige, UK: Cambridge University Press, 2010 [2006]), 111 and 113; here, Goody writes of "Mongol invasions and other disturbances." The term "nomad(s)" is not found in the index to Goody's book.
29 Good, *Theft of History*, 103.
30 S. Frederick Starr, *Lost Enlightenment: Central Asia's Golden Age from the Arab Conquest to Tamerlane* (Princeton: Princeton University Press, 2013), 450f.
31 See Cunliffe, *By Steppe, Desert, and Ocean*, 424.
32 Marco Polo and Rustichello da Pisa, *Devisement ou Description du Monde* (1298).

(including religious tolerance) and open-mindedness.[33] The question is thus justified of whether these nomads were also more intellectually agile than many of their settled neighbors.

Frankopan furthermore points out how the Mongol empire, founded by nomads, influenced both high culture and everyday life, after its expansion in the thirteenth century, not only in China but also in India and Europe.[34] While he concludes that the Mongols had an influence on China's culinary habits, when he turns to Europe he notes the influence of Mongol clothing on fashion. For example, he argues that the hennin, which European women wore in the fourteenth and fifteenth centuries, is directly inspired by Mongol head-coverings. (Fig. 2) In addition, the emblem of the Order of the Garter, known above all for its

Fig. 2: Women wearing Hennins (right), a man wearing a vest (left), which did not exist in Europe before the very late Middle Ages (Jean Froissart, *Chroniques*, 3d book, page from an illuminated manuscript [detail], J. Paul Getty Museum).

---

33 Peter Frankopan, *The Silk Roads: A New History of the World* (London: Bloomsbury, 2015), 177ff.
34 Frankopan, *The Silk Roads*, 180f.

motto "honi soit qui mal y pense" (the old French literally means: "shame on him who thinks badly of it"), was made of "Tatar cloth."[35]

Indeed, from the thirteenth century onwards, what was known as *panni tartarici* – silk interwoven with gold or silver threads, a kind of brocade fabric made in Central Asia – enjoyed great popularity throughout Europe as prestigious luxury textiles.[36] These fabrics were extremely popular with the Mongols (called Tartars by the Europeans); and the Mongols were doubly responsible for their spread in Europe: on the one hand, because the Europeans wanted to imitate the Mongol's clothing style; and on the other, because trade between Europe and Asia flourished due to the safety of the routes in the Mongol Empire (and the elimination of the Arab middlemen, which went hand in hand with an increasing tendency to generally view the Mongols as allies against the "Saracens"). The *panni tartarici* were thus available in significant quantities.

The *panni tartarici* and hennins – or conical headdresses, as the latter were also called – even belong together, because these headdresses were often decorated with "goldcloth" or "drap d'or," as Gustaaf Schlegel (a renowned Dutch sinologist who, among other things, researched the emergence of gunpowder and engaged with geographical reports in Chinese historiography) wrote at the end of the nineteenth century.[37] Schlegel also explains that the hennins can be traced back to the sixth century AD, where they are found among the Hepthalites (belonging to the Huns) and were common among the Kyrgyz and Kazakhs up until his own day (and such hats are still worn today in Asia). This headdress of these Asian nomads thus has a long tradition.

The image of European women in the fifteenth century was also shaped by the hennin. That said, one would probably hesitate to ascribe to *panni tartarici* and hennins any fundamental significance for how Europeans dress. Yet the way in which their significance is marginalized is striking. In the one paragraph that Frankopan dedicates to both items, he speaks of "new fashions," describing the Mongol style as "modish" and the hennin as "fashion accessory." Hence here, again, this argumentation operates according to the logic of spices in relation to food, namely as a mere accessory (would Frankopan also call the bowler an

---

[35] Two decades after the foundation of the order, Andrea de Bonaiuto still makes this connection when he depicts, in his 1368 frescoes for Santa Maria Novella in Florence, a Mongol in conversation with an English knight of the Order of the Garter; see Paviot, "England and the Mongols," 318.

[36] See, for instance, Geertje Gerhold, in *Berichte und Forschung aus dem Domstift Brandenburg*, vol. 2 (2009): 187–204, here 189f.

[37] Gustaaf Schlegel, "Hennins or Conical Lady's Hats in Asia, China and Europe," *T'oung Pao* 3, 4 (1892): 422–429, here 424.

accessory?) that is connected to the ephemerality of fashion and thus marked as a passing fancy.

And even as late as the year 2000, Jacques Paviot's portrayal of Asian clothing still conveys such an impression of marginality for Europeans. His text on the Mongol craze that swept England in the thirteenth and fourteenth centuries offers only a concluding note about clothing; at a jousting tournament held on September 21, 1331, King Edward III and his entourage dressed (up) like Mongols with capes made of camel hair, among other things. It is precisely the highlighted singularity of this event that makes it appear as a kind of carnival or primal scene of Oriental exoticism in Europe. And this is no coincidence, because the fact that Mongol clothing, in Paviot's argument, slips into secondary importance is itself a symptom of the "decay" of England's relations with the Mongol empire. It symbolizes the fading away of a temporary fascination with the Mongols that was a mere episode leaving no lasting political or cultural traces. Instead of cultural imprinting, we are apparently dealing here with a momentary infection, like a cold that must be waited out.

Although it is more discreet and implicit, an attitude prevails here that is even more blunt in Schlegel's assessment of the hennin a century earlier. Even if Schlegel's aim is to prove that trade relations with Asia were already quite close in the Middle Ages, he does not get much from hennin. Schlegel, whose view of top hats was apparently so overly habitualized that he could not reflect on their shape, constantly mocks the monstrous height of this headpiece; he quotes the phrase "curious head-dress" from Viollet-Le-Duc, and even goes beyond him with phrases like "extravagant headdress," "ridiculous fashion," "foolish fashion," and "strange and absurd fashion."[38] Schlegel furthermore mentions it was via Mongol rule that this headdress fashion also reached the Chinese court, where it dominated from the eleventh to the fourteenth centuries, despite decrees against it. The hennin arrived in Europe via rich Flemish merchants. Schlegel notes that European contemporaries had already reacted to the garment with mockery and abuse; among other things, he mentions a priest whose sermons railed against the cone-shaped headdress. Schlegel closes with the words:

> We have already proved in a former article, that the ridiculous fashion of Crinoline, with the reappearance of which we are now again threatened, was likewise introduced from Asia, as also the still more ridiculous fashion of the Mouches or patches-of-beauty, which the European ladies pasted upon their faces in the 18th century.[39]

---

38 Schlegel, *Hennins*, 424, 425, 426, 428.
39 Schlegel, *Hennins*, 429.

Of course, it is no coincidence that the notorious moralist Schlegel only targets the transgressions of women's fashion brought in from Asia. Whether the opponent of the hennin is Seneca, the fifteenth-century priest who attempted in vain to exorcise the devilish horned hennin, or the erudite scholar Schlegel: the issue is always controlling female madness that falls victim to Asian decadence, and thereby, above all, defending male reason and masculinity threatened by Asian effeminacy, and thus ultimately defending the decency of Europe.

Perhaps the writer Julien Gracq had the better intuition in comparing the hennin to the horn of a unicorn, the animal that stood for purity and could be tamed by virgins.[40] Perhaps the hennin was appropriated precisely as a symbol of purity in the Christian context, an insight that escaped Schlegel's phalloeurocentrism, which felt threatened by high women's headdresses.

However, as the pants and boots in Palmyra in fact show, the influence of Asian nomadic peoples on European clothing can be limited to neither women nor fashion.

6 The Nomadic Genealogy of Pants and Boots: *Monturen* in Equestrian Cultures

This directly or indirectly has to do with the domestication of the horse. It is not the mere fact of this domestication that is crucial here, but its development and the changing practices and techniques in which these animals were involved. Although it is difficult to accurately reconstruct the origins of its domestication, it can only have taken place in the milieu of the steppe, where horses still survived after intensive hunting by prehistoric humans. Even if it is possible that this process was multicentric, the thesis accepted today is that it took place in the fifth millennium BC in the region of the Pontic-Caspian steppe,[41] with localizations varying between present-day Ukraine[42] and present-day Kazakhstan.[43]

---

[40] "*hennins qui tanguent raidement à chaque mouvement du cou comme la corne de la licorne*" (Gracq), quoted from http://encyclopedie_universelle.fracademic.com/47471/hennin (visited on July 22, 2019).
[41] David W. Anthony, *The Horse, the Wheel, and Language: How Bronze-Age Riders from the Eurasian Steppes Shaped the Modern World* (Princeton: Princeton University Press, 2007), 134ff.
[42] Jean-Pierre Digard, *Une histoire du cheval*, 42.
[43] Cunliffe, *By Steppe, Desert, and Ocean*, 77f.

If one follows Jean-Pierre Digard, the development occurred in stages similar to the domestication of the reindeer: at first the horse served only as food, which has remained the case on the steppes of Asia. In a next step, a millennium later, the horse was used as a pack animal and then as a draft horse, and only in a final step did the technique of riding develop (the riding saddle is congruently derived from the Mongol pack saddle).[44] Developments in military technology correspond to this process: the invention and use of chariots long preceded the development of cavalry.[45] The use of a complex vehicle, which also required the invention of the wheel and the axle, was thus a necessary detour, as it were, on the way to a supposedly closer, immediate connection between people, animals, and weapons technology. If one follows Digard, this would be evidence of the effort and process of cultural co-evolution represented in the assembling together of horse and rider.[46]

The natural milieu also played its part in this development: Riding horses originated in mountain regions where it was not possible to use chariots, unlike on the plains of Mesopotamia or in northern India, for example.[47]

Digard makes clear how close, even symbiotic, this connection is in equestrian cultures, in contrast to cultures in which an exclusively equestrian culture develops among the elites, as was the case with knights in the European Middle Ages: in nomadic equestrian cultures, social reality is completely permeated by the horse, which is food, a religious point of reference, and a means of trans-

---

44 Digard, *Une histoire du cheval*, 45ff.
45 Similarly, the differentiation between settled and nomadic people only arises around 900 BC with the Scythians; there are no Asian nomads without horses.
46 Of course, scholars from Anglo-American anthropology, such as Cunliffe and especially Anthony, have a completely different opinion: For them, domestication and riding emerge together. Neither, however, are scholars of cultural techniques – unlike Digard, who quotes Mauss, Leroi-Gourhan, and Haudricourt, featuring the former prominently in his *Introduction*. From this perspective on domestication, concerned with cultural techniques and techniques of the body, the supposedly obvious equation of "man + horse = riding" appears to be the result of protracted processes of mutual transformation requiring complex technical, habitual, and symbolic measures of reconstruction. Asserting that this result is the beginning of this process, as Anthony does, is a *retrofitting* on the basis of an outcome. Or, to incorporate Latour's distinction between ready-made science and science in the making: it asserts ready-made interspecific culture instead of interspecific cultural techniques in the making. See Bruno Latour, *Science in Action: How to Follow Scientists and Engineers through Society* (Cambridge, MA: Harvard University Press, 1987).
47 Digard, *Une histoire du cheval*, 49 and 50, as well as 68. A plain is also necessary in order for it to be possible to gain enough speed with heavy war chariots. Of course, this applies for heavy cavalry, as well.

port. Everybody learns to ride from a young age, which becomes the only form of movement, even for the shortest distances. Running is absolutely unusual. For this riding culture, the principle of economy applies, i.e., the development of simple, manageable, and pragmatic riding techniques.[48] It is only logical that the outfit, or clothing, is attuned to such an interspecific technique of the body, which shapes the rider's habitus deep into the apparatus of their muscles. Hence, without the cultural technique of riding and the peoples who would have developed it, there would be no pants (Fig. 3).[49]

Fig. 3: Reconstruction of the dress of a Saka chieftain, 4th to 3rd century BC (the Saka were a group of nomadic peoples living in the steppe between the Caspian See and the Pamir).

It is therefore no coincidence that the people in Palmyra who did not dress like Romans were Persians, or more precisely Parthians, members of the great empire that existed on the territory of today's Northern Iran from the third century BC to the third century AD. The Parthians were a settled subtribe of the Scythians, a confederation of Indo-European tribes who developed riding, and who invented

---

[48] Digard, *Une histoire du cheval*, 193.
[49] Digard, *Une histoire du cheval*, 76.

not only the saddle but also pants and the forging of iron.[50] Accordingly, the heart of the Parthian army consisted of mounted archers and armored riders, the cataphracts.[51] This garment then not only spread to Europe from Iran: via the Xiongnu, a nomadic people who rode horses and waged war against the Chinese for centuries, pants also reached the Middle Kingdom.[52] The extent to which pants were tied to riding is also manifest ex negativo in China: around 300 BC, the introduction of mounted archers by the Chinese Emperor almost failed because of the refusal to exchange the long skirt, a sign of social prestige, for the pants that were necessary for riding.[53]

The pants leg simultaneously enables close contact between the human leg and the horse's body, while also protecting the leg and allowing great freedom of movement without fabric fluttering around. Another device was nevertheless needed to attach the feet, and thus also to steer the horse directly: it is well known that the stirrup is one of the most important developments in riding technology, and since the work of Lynn White, its role in the development of European warfare and the caste of knights from the eight century AD onward has also been known and extensively debated.[54] Stirrups, however, were also developed in Asia, in a long process of development that began a millennium earlier.

The first fixed metal stirrups were made in Central Asia in the first century AD, presumably in the Kushan Empire (whose territory covered present-day Afghanistan, Pakistan and northern India). Hence the stirrup not only made possible

---

[50] Hermann von Wissmann, "Badw. II: The History of the Origin of Nomadism in its geographical Aspect," in *The Encyclopedia of Islam*, vol. 1, ed. Hermann v. Wissmann and Friedrich Kussmaul (Leiden: E.J. Brill, 1960), 874–880, 878f.
[51] Cunliffe, *By Steppe, Desert and Ocean*, 258f. Rome later integrated Parthian armored riders into its own army, which represents another common variant of technology transfer: the buying up of know-how. On the cataphracts, see also Digard, *Une histoire du cheval*, 70f. These were forerunners of the European knights, even without stirrups. They instead utilized other forms of fusing rider and horse through shared armor, which also served to stabilize the rider's position. That is to say: we are dealing here with a form of technical concretion in the sense of Gilbert Simondon because several elements, such as the quiver that holds arrows and stabilizes the rider, are multifunctional. One might therefore ask: is it perhaps the interspecific *assemblage* that produces the first concrete technical forms? A further factor is the breeding and selection of particularly robust horses, which can, in the first place, *carry* and *bear* this form of waging warfare, fought through physical impact, as well as this armor. On the notion of "concretion," see Gilbert Simondon, *Du mode d'existence des objets techniques* (Paris: Aubier, 1958), 19ff.
[52] Von Wissmann, "Badw.," 879.
[53] Digard, *Une histoire du cheval*, 77.
[54] Lynn White, Jr., *Medieval Technology and Social Change* (Oxford: Oxford University Press, 1964), 1ff.

European knights, but, earlier, the Scythian bow riders from whom the figure of the centaur emerged as we know it from Greek mythology.

It was thus a shrewd costume that Edward III and his courtly entourage had chosen with their Mongol outfits, for without the invention of the stirrup in the Asian steppe, the jousting tournament they held on that September day of 1331 would not have been possible at all.

Of course, warfare had been waged on horseback before; the cataphracts are an obvious example. Bows and arrows had also been used. But the specific tactics of the harcèlement, or the lightning-fast attack and retreat combined with the use of curved composite bows with great penetrative power and reach, required riders firmly supported by stirrups. Only such robust metal stirrups gave knights (in Europe) and archers (on the Asian steppe) the stability in the saddle (even if the cataphracts offer an earlier history of this stabilization) necessary for the development of these dreaded and efficient techniques for waging war.

The human foot, however, is not stable enough to withstand this strain without injury (especially as the first stirrups were extremely narrow bars), which means that riding with stirrups requires riders to wear sturdy footwear, and a nomadic life in the saddle requires riders to wear boots. Stirrups and boots, initially made of hard felt, thus belong together; they are necessarily part of the same technical ensemble or, as Didier Gazagnadou writes, of the same technical-military-political structure.

> Il s'agit de penser le problème de l'invention et de l'emprunt des étriers en tant qu'agencement entre le technologique, le militaire et le politique, et donc de mettre en relation étriers, selles, pantalon, bottes, organisation sociale nomade, et de considérer le rapport de ce type de société à l'espace, au cheval et à la guerre.[55]

> It is a question of thinking about the problem of the invention and borrowing of stirrups as an arrangement between the technological, the military, and the political, and thus of relating stirrups, saddles, pants, boots, nomadic social organization, and considering the relationship of this type of society to space, the horse, and war.

The war technology of the nomads on the Asian steppe emerges from a political and social system of organization based entirely on the domestication of the

---

[55] Didier Gazagnadou, "Les étriers: Contribution à l'étude de leur diffusion de l'Asie vers le monde," *Techniques & Culture* 37 (2001): 4; http://tc.revues.org/266 (visited on August 16, 2017); my translation. Hence, this contradicts Lynn White and all his genealogies, based on the inability to conceive of other, advanced civilizations (India and China) as the inventors of something as important as stirrups. And: *aspandak*, the Persian word for stirrup, means "attache à cheval," which reminds us that we are dealing here with a history of attachments between species, and hence of quite specific technical attachments, namely *montage*.

horse: a social system of organization that exists between species, and a culture of riding that is not reserved for an elite but in which everyone participates. This culture distinguishes itself (just like that of the nomadic Arabs, who were early adopters of the stirrups) by agility, archery, and the harcèlement, and this technical ensemble also includes metallurgy (since functional stirrups must be made of metal) as well as the manufacture of boots. This footwear, common today in Europe, also came to us from Asia, from Turko-Mongol nomadic peoples. And the migration of these elements of the ensemble reveals in turn: such elements can be separated from their original ensemble at any time. Even if pants, stirrups, and boots belong together: they were exported to societies that were settled and in which riding was an elite culture. And, of course, pants and boots could eventually become completely detached from the cultural technique of riding.

7 Techniques for Wearing Clothing and Carrying Loads

But such observations cannot only be made for footwear and legwear (which, in any case, developed late in Europe). In a text from 1948 entitled "Relations entre gestes habituels, forme de vêtements et manière de porter des charges" (Relationship between habitual gestures, the shape of clothing and the manner of carrying loads), another Mauss disciple, André-Georges Haudricourt (who often stands in the shadow of Leroi-Gourhan),[56] noted that the Europeans of antiquity knew only two types of clothing: a type of tunic wrapped around the body, and a type of poncho, originally with an opening that was pulled over the head (as we do today with T-shirts or sweaters).[57] One would also have to include in this category the tabard, which knights wore over their armor.[58] Many of these garments

---

[56] Yet see also Erhard Schüttpelz, "Die medienanthropologische Kehre der Kulturtechniken," *Archiv für Mediengeschichte*, 6 (2006): 87–110, and Cuntz, "Kommentar zu André-Georges Haudricourt."

[57] André Georges Haudricourt, "Relations entre gestes habituels, forme de vêtements et manière de porter des charges," in A.G.H., *La technologie science humaine: Recherches d'histoire et d'ethnologie des techniques* (Paris: Editions de la Maison des sciences de l'homme, 1987), 171–182, here 175.

[58] Hence the word *tabard* has *nothing* to do with the garment known in German as a *Tappert* (Duden: "a cloak-like coverlet from the fourteenth and fifteenth centuries worn mostly by men"), despite the obvious derivation of the German word from this source; the *Tappert* corresponds, instead, to the *houppelande*.

were therefore loosely fitted to the body and did not even require sewing,[59] which means that cutting the fabric to size was either not practiced or at least played a minor role.[60]

Garments such as jackets, waistcoats, coats, or shirts that are open at the front and can be closed with buttons or the like were, however, unknown. As the first Western garment of this type, Haudricourt identifies the *houppelande*, or in German, *Tappert*, which first appears in the thirteenth century[61] and was replaced in the fifteenth and sixteenth centuries by the *Schaube* (a gown or robe with sleeves), which, unlike the *Tappert*, is worn open like a vest. (Fig. 2) Haudricourt traces the French and Spanish words for vest, "gilet" and "jaleco," back to the Turkish word "yelek" (in contemporary Turkish, this is still the word for vest), and thus to a nomadic origin. The process of introducing such garments into Europe therefore took place between the thirteenth and sixteenth centuries, with the beginning of the Mongol "fashions," and at the same time of the *panni tartarici* and the hennins: if one follows Haudricourt's chronology and genealogy, their marginalization thus conceals simultaneous processes of the Asianization of European clothing that carry a far greater significance.[62]

Haudricourt then connects the development of clothing with the development of techniques for carrying. He distinguishes between active carrying, in which the arm muscles are constantly tensed and placed under stress, and passive carrying, which is more economical because it can be integrated into the movements of the muscles by holding an upright posture and walking.[63] Like Marcel Mauss, Haudricourt begins with habitual gestures and thus techniques of the body, and with the assumption that there is a development from active carrying to passive, energy-saving carrying. He attests a dominance, in Europe in the 1940s, of active carrying and thus a backwardness in comparison to other parts of the world, including Asia; for passive carrying, he distinguishes several possibilities, each of which produced different objects and fastening mechanisms.[64]

---

59 Veyne, *Palmyre*, 12.
60 Haudricourt, "Relation entre gestes habituels," 177.
61 Confirmed by the *Petit Robert. Houppelande* dates to the fourteenth century; its precursor form *hopelande* dates back to 1280.
62 Haudricourt, "Relation entre gestes habituels," 176.
63 Haudricourt, "Relation entre gestes habituels," 173.
64 The fact some things never change is shown by the malicious attack of the television personality Piers Morgan on Daniel Craig, for carrying his baby in a papoose, as an "#emasculatedBond"; see "Piers Morgan mocks Daniel Craig for carrying baby," https://www.bbc.com/news/uk-45873664 (visited on July 22, 2019). Sonja Haller: "Piers Morgan ridiculed James Bond actor Daniel Craig for wearing a baby carrier ... in 2018," https://eu.usatoday.com/story/life/allthemoms/2018/10/16/piers-morgan-mocks-daniel-craigs-manhood-wearing-baby-carrier/1658291002/

Monturen/montures: On Riding, Dressing, and Wearing. Nomadic Cultural Techniques ― **161**

Regarding gestures and movements not as natural, but as a culturally developed reservoir of techniques of the body, learned in the same way as the phonology of individual languages, also means regarding supposedly banal sequences of movement, such as carrying, to be the result of long developmental processes. In this view, certain cultures have only certain forms of active and passive carrying, which in turn form the physical habitus and thus imprint themselves on and implant themselves into the body. There is thus a fundamental difference between attaching a carried load to the head, or with straps crossed over the chest.

Another, separate form of attachment is that with parallel shoulder straps, as is the case with backpacks. This requires a sequence of movements that Haudricourt does not at all take for granted, but which he sees as the result of a longer learning process relating to techniques of the body: of putting the straps on one after the other. Haudricourt then notices that the same, comparatively complex, sequence of movements is carried out when one slips on a shirt or jacket, which brings him to the conclusion that the manner of carrying loads and wearing clothes, the gestures, and corresponding objects emerged at the same time. Indeed, the backpack or winegrower's vat, which is also fastened with straps, arrived in Europe at about the same time as the jacket.[65]

As the first known representation in Europe of a garment of this type, he names an illumination in a manuscript from the turn of the tenth to the eleventh centuries. It depicts a Bulgarian warrior.[66] As one learns from the archaeologist Barry Cunliffe, the Bulgarians were originally an Asian nomadic tribe from the Pontic-Caspian steppe that was driven out to the Balkans in the seventh century by the Khasars, a Turkic tribe, but had previously likely adopted the clothing of those who displaced them, or of the ethnic groups related to them.[67] According to Haudricourt, the first wearers of garments adapted to the body, i.e., requiring the fabrics to be cut to size, that are open at the front and pulled on as one puts on a backpack, by slipping into the straps, were Turkic peoples and the Mongols, who also introduced these garments in China. Anyone today who wears pants,

---

(visited on July 22, 2019). The white, Eurocentric virility-fetishist Morgan finds it unmanly that Craig makes use of an Indian technique of passive carrying instead of stupidly wasting his energy by actively carrying his baby in his arms – like the Greeks, who preferred to ride their crotch raw without a saddle (bareback) rather than use unmanly Asian saddles. Those harping on this old tune, however, definitely do not seem to become much smarter over the centuries.

65 Haudricourt, "Relations entre gestes habituels," 178. It can be proven that in Asia, too, this spread occurred simultaneously (see 179f).
66 Haudricourt, "Relations entre gestes habituels," 177.
67 Cunliffe, *By Steppe, Desert and Ocean*, 324f.

sturdy shoes, a blouse, a sweater, a shirt, or suits is wearing clothing developed by nomads of the Asian steppe: *Monturen* made of bodies, fabrics, gestures, and environments.

8 Interspecific *Montagen/montures* instead of Anthropocentric Projections

Even if Haudricourt does not identify, in "Entre gestes habituels, forme de vêtements et manière de porter des charges," any direct causality between the domestication of animals and the emergence of this form of carrying loads and clothing, he still does give a brief but relevant reference in his text. Nomadic peoples in the far north of Asia use reindeer and dogs as motors to pull loads transported on sledges. These animals are harnessed with shoulder straps crossed over their chest.

As discussed above, more than fifty years later Didier Gazagnadou pointed out for the assemblage or outfit of pants/saddle/boots/stirrups that the entire social organization of the nomads and their relationship to space must be taken into account.[68] Given this fact, a logical conclusion can be drawn from Haudricourt's entire ethnobiological approach, with its attention to energies, gestures, and habits: it is probably no coincidence that in a nomadic society whose mobility implies the permanent transport of loads, the most efficient and, as Haudricourt puts it, "passive" forms of carrying loads are formed, or are spread and adapted from this origin.

Haudricourt's deliberations must be supplemented by the idea that, in addition to the perfect use of one's own body, the delegation of carrying (including in the transformation of movement into pulling), i.e., having something be carried (or pulled), is an efficient form of passive carrying: *make do* instead of doing it yourself. What is equally obvious is the connection between locomotion and warfare on horseback and an optimization of clothing that requires hides, skins, or fabrics to be cut in order to achieve a better fit and thus the greatest possible freedom of movement, as we have already observed for pants. This becomes necessary in a climate in which, unlike in India, for example, wearing clothing is essential for survival.

---

[68] Gazagnadou, "Les étriers."

Especially Haudricourt's works on animal motors (above all in mills)[69] and the resulting forms of movement of animals and machines – and (to a certain extent) any confrontation with domestication processes that examines practices and gestures – demonstrate how problematic it is to generalize the anthropocentric-hylomorphistic organ projection thesis, formulated by the pseudo-technical philosopher Ernst Kapp,[70] beyond the purely phantasmatic dimension that made it so attractive for Freud and his discourse about "a kind of prosthetic God,"[71] or McLuhan's model of narcissism and narcosis.[72] What actually applies to the ear, i.e., the reproduction of the functional mechanism of the perceptual organ in technical objects such as loudspeakers, cannot be transferred to other organs and certainly not to motor processes.[73] Rather, it proves to be a projection of the human or perhaps only the European desire to be the origin and starting point of everything.

But in riding, it is not simply that the legs of human beings are projected over the stirrup into the horse and its legs; rather, *both* horse and rider must habitually incorporate the body of the other living creature until the muscular and perceptive apparatus of the human being changes. It is not only that the horse learns various gaits or to move with a carriage, wagon, or cart; riding also requires the activation of different human muscle groups than those used in running. Horses have learned to perceive and interpret the smallest muscle movements of human beings, and the view of the world from the horse's back is different than on foot, especially when it comes to perceiving and mastering distances. To these are added gestures and movements performed on another living being, such as grooming or putting on a harness. Dealing with the animal's body requires exploring the animal's characteristics and anatomy; it requires objectifying and analyzing what can be done with this familiar yet alien body, what this body can be made to do, what it can bear, carry, and endure. Hence, what would it mean if

---

69 André-Georges Haudricourt, "Les moteurs animés en agriculture," in A G H., *La technologie science humaine*, 157–167, see also Haudricourt, *Des gestes aux techniques: Essai sur les techniques dans les sociétés pré-machinistes*, ed. Jean-François Bert (Paris: Maison des sciences de l'homme, 2010), 117–126. On this question, see also Schüttpelz, "Die medienanthropologische Kehre," 87–110.
70 Ernst Kapp, *Grundlinien einer Philosophie der Technik* (Braunschweig: G. Westermann, 1877).
71 Sigmund Freud, *Civilization and its Discontents*, trans. James Strachey, in *The Standard Edition of the Complete Psychological Works of Sigmund Freud*, vol. XXI (1927–1931) (London: Vintage, 2001), 91–92.
72 Marshall McLuhan, *Understanding Media: The Extensions of Man* (London and New York: Routledge, 2001 [1964]), 45–52.
73 For the analogy between ears and soundboxes, see, for example, Michel Serres, "Caisses," in M.S., *Les cinq sens* (Paris: Grasset, 1985).

the development of a new gesture of putting something on and pulling it over had taken a detour via the practice of attaching a device for harnessing draft animals, or at least that there was an interspecific exchange between the attachment of loads and vehicles to human and animal bodies?[74]

Haudricourt, who always also includes the history of naming in his considerations, notes that the French word for a bridle, "bretelle," was not used until the thirteenth century. The word derives from the Old High German "brettil" or "brittil," which itself refers to nothing other than the bridle of the horse.[75] The word was therefore apparently used first for riding or draft animals and only thereafter for vessels or containers attached to the human body and, finally, for the clothing itself.[76] Moreover, the bridle, too, was in all probability invented by Mongols.[77] In French, this relationship can still be seen in the similarity of the words "bretelle" and "bride," both of which derive from "brittil."

Domesticated animals, above all horses, were therefore not only indispensable as transport media in overcoming the distances that made the migration of objects throughout Eurasia possible. In addition, their domestication and the techniques of riding and pulling have themselves produced objects, namely garments, that spread from the steppes and high plateaus of Asia to China and Europe, and from Europe to the whole world. The reality of these Eurasian migration processes of cultural techniques and things, without which neither our current technical civilization nor the global economic networks that have existed for millennia would have emerged, is not a phantasmatic projection or externalization of the human being and its organs into its environment, from the inside out, as every anthropocentric hylomorphism would have it. Rather, it is the formation of interspecific and milieu-dependent *Monturen* or assemblages composed of animals, human beings, and gestures.

---

[74] In *L'utilisation de l'énergie naturelle*, Haudricourt nevertheless identifies (one must concede) direct transmissions from humans to animals in northern Asia and America: "certain dog or reindeer teams that appear to be directly inherited from the human mode of pulling" (186). In any case, however, there is central technical problem to be solved: how to efficiently fasten something to the body of an animal?
[75] Haudricourt, "Relations entre gestes habituels," 178.
[76] *Althochdeutsches Wörterbuch des Wörterbuchnetzes*, http://awb.saw-leipzig.de/cgi/WBNetz/wbgui_py?sigle=AWBode=Gliederungemid=AB02951#XAB02951 (visited on August 16, 2017).
[77] See Goody, *Theft of History*, 188.

Jörg Paulus
Self-Imprints of Nature

1 An "Unpacked Event"

At what was billed as an "Unpacked Event" on February 20, 2019, a well-known South Korean electronics group presented its latest generation of smartphone models.[1] Interest centered especially on a mobile phone called Fold that opens up like a book. At the event, images of the device were officially revealed to the world (although they had previously been circulating) that showed, on each half of the display, a pair of butterfly wings in intense colors (orange, green, and blue), with the area of transition in the middle, the place of the virtual insect body, remaining empty. During everyday use, this is the critical area for potential damage. In addition to other technical details, press reports highlighted the fact that a fingerprint sensor was attached to the side of the phone – right where the thumb usually rests when using a mobile device. Yet both this making-appear of pairs of related butterfly wings upon folding or hinged objects and the function of places preformatted to be accessed or touched by fingertips can be found much earlier in interlinked assemblages of cultural techniques, as the following essay shows.[2]

2 Packages

At some point, the three slipcases in the format of 16.7 × 10.7 × 2.9 cm (Fig. 1) covered with dark green patterned paper must have looked more or less identical. In the meantime, however, they have become easy to tell apart even if one disregards the inscriptions applied with black ink. The slipcase no. 1 has a discol-

---

[1] Details had already been leaked shortly before; see https://www.indiatoday.in/technology/news/story/samsung-galaxy-fold-toldable-phone-leaks-out-in-press-renders-ahead-of-unpacked-event-1460330-2019-02-20 (visited on February 22, 2019); illustrations can be found, for example, at https://www.nytimes.com/2019/02/20/technology/personaltech/samsung-galaxy-s10.html?action=clickodule=Discoverygtype=Homepage (visited on February 22, 2019).

[2] Regarding the specific form of this making-appear of folding or hinged objects, see Helga Lutz, "Medien des Entbergens: Falt- und Klappoperationen in der altniederländischen Kunst des späten 14. und frühen 15. Jahrhunderts," *Archiv für Mediengeschichte* 10 (2010): *Renaissancen*: 27–46, here 32.

Translated by Michael Thomas Taylor

Open Access. © 2020 Jörg Paulus, published by De Gruyter. This work is licensed under a Creative Commons Attribution 4.0 International License.
https://doi.org/10.1515/9783110647044-010

Fig. 1: Three labeled slipcases – in German, Schuber: (1) Papillones, (2) Bombices et Noctu[ae], (3) Spinges, Geometridae, Tortices, Pyralides, Tinides, Allucides (presumably around 1850, private collection).

oration in the upper right corner caused by moisture, which extends across the narrow right side and affects the entire left third of the back. On the lower right side of the front, a thick spot of ink rests like a black bird amid the foliage of the dark-green paper. The other two slipcases, one of which (no. 2) was repaired with adhesive tape, have lost their bottoms, exposing the lower edges to increased abrasion. Slipcase no. 3 is the only one that contains an inner case, made of colored paper; this inner case is visible at the upper edge, where a semicircular notch has been created to make it easier to pull out the contents.

The individual changes and losses refer, ex negativo, to the basic function of all slipcases: they are "protective containers" made by machine or (as in this case) manually by scratching, folding, and/or assembling, intended to protect the contents from damage caused by dust, moisture, impact, tearing, and climatic fluctuations.[3] Yet from the perspective of linguistic history, the German word *Schuber* (sg./pl.) has multiple references. In the archival world, it can mean a bound or unbound heap or pile (the latter corresponding to the word *Schober* in the sense of "layered pile of grain"),[4] but it can also denote a container or a cover

---

[3] "Schuber," in *Lexikon des gesamten Buchwesens*, vol. 7, ed. Severin Corsten, Stephan Füssel, and Günther Pflug (Stuttgart: Hiersemann, 2007), 634.

[4] Friedrich Kluge, "Schober," in *Etymologisches Wörterbuch der deutschen Sprache* (Strassburg: K.J. Trübner, 1910), 411.

for documents or bundles of files.⁵ The word points etymologically to practices of *schieben*, or pushing, whereas the English "slipcase," by contrast, more likely originates from the slipping/gliding of the contents it is meant to protect. Beyond these etymological and historical conditions, systematic implications arise from the cultural-technical practices that converge in the object *Schuber*: *Schuber* are operational assemblages that aim to stabilize what they hold and thereby, through their respective formats, realize dimensional templates.

But how do these three "slipcases" relate to what they hold as something capable of slipping out? Symbolically, the inscriptions on the front sides (which is what makes these sides the front to begin with) refer to contents from the order of butterflies or moths, the *Lepidoptera* (named thus because of their characteristic wing scales). The insertions are made available, in cultural-technical terms, through the semicircular notch I mentioned above, which is positioned at the top in the middle of the front and back sides of the *Schober*: paper restorers call it a *Griffmulde* or *Greifausschnitt* (a notch or cutout for grasping or gripping) or just an *Eingriff* (slit, intrusion).⁶ Their symmetrical attachment to the slipcase has an affordance function adapted to the physiology of the human hand, designed especially for the thumb and the index or middle finger, which are most adept at grasping the *Schober* of paper and pulling it out.⁷

3 Unpacking

At first glance, what slips out of the slipcase seems to be simply an assemblage of cultural techniques: a depiction of butterflies on sheets of slightly rough paper. In part, they have been given inscriptions in pencil or ink (in the case of the inscriptions in pencil, the genus is shown at the top left, the species name at the bottom

---

5 Heinrich Otto Meisner, *Archivalienkunde vom 16. Jahrhundert bis 1918* (Göttingen: Vandenhoeck & Ruprecht, 1969), 49.
6 Translator's note: this use of the word *Eingriff* is unusual. An *Eingriff* usually refers to "an (unjustified or unauthorized) intervention," a surgical operation, or the fly on a pair of pants (*Duden*); like *Griffmulde* and *Greifausschnitt*, it is based on the German word *greifen*, to grasp, grip, or grab.
7 See Richard Fox, Diamantis Panagiotopoulos, and Christina Tsouparopoulou, "Affordanz," contibution to SFB 933 "Materiale Textkulturen": https://www.degruyter.com/downloadpdf/books/9783110371291/9783110371291.63/9783110371291.63.pdf (visited on February 22, 2019); the exhibition catalogue *Das bewegte Buch*, which assembles a wealth of materials into forms and formats of collecting, touches upon this arrangement of cultural techniques in passing in relation to Lavater's experiments with loose-leaf books; see *Marbacher Magazin* 150/151/152, ed. Heike Gfrereis (Deutsche Schillergesellschaft: 2015), 30, and related illustrations.

Fig. 2: Sheet 112r from slipcase 2. **Fig. 3:** Sheet 75r from slipcase 1.

right, Fig. 2, while the categorizations in ink are found beneath the illustrations, Fig. 3). Often the inscriptions are missing completely.

The flat, two-dimensional plainness of the arrangements is crossed by a (usually vertical) folding of the sheets, which runs exactly down the middle between the feelers and the wings of the butterflies. The two-dimensionality of the sheets, already eliminated by folding, finally disappears completely on the microlevel, revealed when the objects are placed under a microscope or stereomicroscope that shows the spatial structure of the scales as it actually exists, lying on the wings like roof tiles. The insect bodies lying between the wings, including the eyes and antennae, remain flat: they have indeed been painted by watercolor. On about one fifth of the sheets, these "artistic" additions are missing and only the wings are present. In these cases, specifically, the dried residue of a previously more or less liquid substance can be seen, left behind on the paper as a darkened area (Fig. 4).

The objects collected in the three slipcases were created using what is known as the *Naturselbstdruckverfahren* (nature self-printing or self-imprinting) as it was essentially developed – with earlier precursors – in the sixteenth and seventeenth centuries, to then be used throughout the eighteenth and nineteenth centuries, and as it is still being used today, in various modified forms, to document plants, animals, and other natural objects. Nevertheless, the practices denoted by this name (other names in German include *Naturdruck*, *Physiotypie*, and *Autoplastik*)

Fig. 4: Slipcase 1, sheet 31r (detail).

are quite diverse and the tools used to carry them out are quite varied. By far the majority of documented attempts concern self-imprints of plants, as we already see in early evidence from the thirteenth and fourteenth centuries.[8] The case I am discussing here concerns a variant that, strangely enough, only plays a secondary role, if any, in more recent scholarly depictions of *Naturselbstdruck*, although it has sometimes also been described as *the* technique of *Naturselbstdruck* in the true and narrower sense, in which "natural objects or parts thereof are directly used to make an impression, as carriers of color," compared with the more

---

[8] Previous work to be noted includes Peter Heilmann, "Über den Naturselbstdruck und seine Anwendung," in *Die Sache selbst*, ed. Silke Opitz and Gerhard Wiesenfeldt (Weimar: Bauhaus-Universitäts-Verlag, 2002), 100–110; Amin Geus, *Natur im Druck: Eine Ausstellung zur Geschichte und Technik des Naturselbstdrucks* (Marburg an der Lahn: Basilisken-Presse, 1995), especially the chapter by Armin Geus, "Natur im Druck: Geschichte und Technik des Naturselbstdrucks," 9–27; Gianenrico Bernasconi, "The Nature Self-Print," in *Objects in Transition: An Exhibition at the Max Planck Institute for the History of Science, Berlin, August 16–September 2, 2007*, catalogue to the exhibition ed. Gianenrico Bernasconi, Anna Maerker, and Susanne Pickert (Berlin: Max Planck Institute for the History of Science, 2007), 14–23; Lorraine Daston and Peter Galison, *Objectivity* (New York: Zone Books, 2010), 105–133; "Exkurs: Der Naturselbstdruck," in *Die Buchkultur im 19. Jahrhundert*, vol. 1, ed. Walter Wilkes, Frieder Schmidt, and Eva-Maria Hanebutt-Benz (Hamburg: Maximilian-Gesellschaft, 2016), 235–242; Friedrich Weltzien, *Fleck – Das Bild der Selbsttätigkeit: Justinus Kerner und die Klecksografie als experimentelle Bildpraxis zwischen Ästhetik und Naturwissenschaft* (Göttingen: Wallstein, 2011), 303–309; and Simon Weber-Unger, *Naturselbstdrucke: Dem Originale identisch gleich* (Vienna: Album, 2014).

general technique of "nature printing."[9] For this reason, with this technique it is also impossible to make several, in principle nearly identical, prints from one source element, as happens in the case of plants prepared with printing ink or the use of various lead or copper form transfer media on which "natural" objects can leave a trace.[10] Descriptions of the technology can be found above all where the objects involved call for it, for example, in a 2016 contribution to a collection ensemble from the holdings of the Letter-Stiftung:

> In contrast to the more common natural self-printing of objects such as leaves or feathers, which are inked and then printed, the natural self-printing of butterfly wings is based solely on the transfer of the colored wing scales onto paper that has been previously prepared with a thin, adhesive layer. The exoskeleton and veins of the wing can then be carefully removed while the scales remain stuck to the surface.[11]

A relevant historical source from 1889 describing the specific process for the production of natural self-prints with butterflies is also mentioned here. In that source, one reads:

> *Dr. Vanhöffen presented a collection of butterfly prints and talked about the process of preserving butterflies as a natural self-print.* One takes a small sheet of white paper folded in the middle and coats one side of it in a thin layer of highly viscous white gum arabic. Then, one folds the paper together, spreading the gum onto the other half to form an equally large area and rubs the gum very thinly and evenly on both sides. The wings of one side of a butterfly are then placed with a pincette just as the butterfly usually holds them in flight, whereupon both halves of the paper are folded up again and glued together by gentle pressure. Gradually, the wings are pressed together a little harder until the gum is almost dry and then the paper is unfolded. This succeeds easily, since the gum dries earlier where the wings form a dry intermediate layer than it does in the surrounding area, where two damp spots are touching. After removing the wing exoskeleton, one washes off the excess adhesive with a brush and water, dries the sheet between tissue paper, and completes the print by painting the butterfly body with watercolors. This results in a natural self-print that shows the upper side of the butterfly wings on the right and the underside on the left, or vice versa. If the aim is to make an impression of the top of the whole butterfly, however, one must place the four

---

9 Geus, *Natur im Druck*, 24.
10 Examples of this can be found in the publications listed above in note 8.
11 Collection cabinet with 510 natural prints of butterfly wing pairs, held by the Letter-Stiftung, new acquisitions, available at http://www.letter-stiftung.de/index.php/leser-83/items/schmetterlings-kaestchen.html (visited on February 15, 2019). I owe my finding of this trace to Andrea Hübener (Braunschweig); I would also like to thank Viola Richter from the Museum für Naturkunde, Leibniz-Institut für Evolutions- und Biodiversitätsforschung (Berlin) for numerous hints and insights.

wings in the correct position on a sheet treated with gum and cover it with another piece of gum-coated paper.[12]

It remains noteworthy that even if the practice described here represents, in a certain sense, an opposite procedure to the other processes described as "natural self-printing" (in one case, the material is colored; in the other, it is preserved with all its colors and made to adhere by the transparent gum arabic), it is nevertheless usually grouped together with them in historical and systematic terms. The use of the transmission medium is simply understood as another variant[13] of accepting the affordance (or "invitation") of nature to be represented. But it is precisely the relationship between an "invitation" and a reply in terms of cultural techniques that differs significantly in the case of butterflies and their transformation into natural self-print ensembles, compared to what happens in the case of other techniques. The difference lies not only in the additional spatial dimension that is gained but also in the interconnected cultural-technical, abstract spatial prerequisites and consequences of the associated operations.

This concerns, first of all, the coupling of the operations: the simple folding of the paper generally couples each pair of the four wings that every butterfly has, making them into a double: the sheet thereby follows the movement the butterfly employs to fly, in which all four wings are moved simultaneously – perhaps giving the impression that the insect has only two wings. The coupling of cultural techniques is thus, in a certain sense, preformatted as a natural technology. At the same time, however, the printing process leads to what we might call a distortion of the natural format and thus also of the reality effect linked to it, which makes us believe that we *see* the shape of a butterfly (even without the addition of a body – as was also the case with the folding mobile phone). However, there are no such things as butterflies with wings whose upper and lower sides are simultaneously visible, and because *both* pairs of wings have been removed from *one* original pair of wings, they cannot be transformed into a natural state, as it were, by operations in two or three-dimensional space. The procedure of removing both pairs of wings of a butterfly at the same time that was mentioned in the last sentence of the description cited above avoids this disfiguration but only by abandoning the integrity of clear scientific illustration: the front and back sides

---

[12] Meeting of May 2, 1889, in *Schriften der Physikalisch-Ökonomischen Gesellschaft zu Königsberg* 30 (1889): 24–25, here 24 (accessible online at https://www.biodiversitylibrary.org/item/51236#page/110/mode/2up, visited on February 15, 2019); in the description provided by the Letter-Stiftung (see above, note 11) the speaker is listed as Ernst Vanhöffen (1858–1918).

[13] Weltzien, for example, writes of a "rubber-like substance" that made it possible to refine the previous technology; Weltzien, *Fleck – Das Bild der Selbsttätigkeit*, 308.

of each pair of wings (in the first method) can represent the color and shape patterns of the whole wing apparatus, whereas the view of only the front sides of four wings (in the second method) can never represent the whole. In both cases, however, the transfer of the scales to the paper is accompanied by a further shift: the top side of the scales that are pressed onto and adhere to the paper always faces the paper base, so that the viewer of the finished preparation can only perceive the backs of the scales of the upper and lower wings (which normally remain invisible inside the wing). This usually goes unnoticed given that butterfly scales are colored on all sides. Only the blue color, Vanhöffen notes, poses "some difficulty," especially in the case of the blue male butterflies of the genus *Lycaena*: these butterflies have shorter brown scales under the blue scales, which cover the blue scales in a natural self-print. The impression of the male therefore shows the simple brown color of the female, which lacks the blue scales.[14]

The butterflies represented in natural self-printing by means of only one pair of wings are thus assemblages that undergo recto/verso transpositions on several levels. First, they adapt to the ground on which they are held (namely, the sheet of paper) in the format of folding; second, they separate themselves from the ground (in that the recto and verso dyad of a double wing appears here juxtaposed on the recto of the paper); and third, they present themselves with a view of the scales on the back that is hidden in nature.

As a transition area between the two sides of the folded geometric construction, the free space between the pairs of wings at first remains ontographically uncoded. If bodies are inserted by means of watercolor, the structure becomes something like an artistic trompe-l'oeil, although, for the reasons given above, the effect goes beyond the Euclidean scheme of reference that a first glance might suggest.[15] If these bodies are lacking, however, then the constructedness of the arrangement comes to the fore, often accentuated by the traces of gum arabic left behind in such self-prints as the medium enabling this operational disfiguration. Yet in the case of this collection, as a rule only those specimens with painted bodies were deemed worthy of being labeled, whereas the wings that remain by themselves are almost always unlabeled. The trace of the pencil

---

14 Meeting of May 2, 1889, in *Schriften der Physikalisch-Ökonomischen Gesellschaft zu Königsberg*, 30 (1889): 25.

15 Indeed, the arrangement could be understood as a superposition of different spaces that are "enacted" in the first place by the underlying operational chains (according to Helga Lutz and Bernhard Siegert, with reference to John Law, in an essay on collapsible and foldable pictorial objects); see Lutz and Siegert, "In der *Mixed Zone:* Klapp- und faltbare Bildobjekte als Operatoren hybrider Realitäten," in *Klappeffekte: Faltbare Bildträger in der Vormoderne*, ed. David Ganz and Marius Rimmele (Berlin: Dietrich Reimer, 2016), 109–139, here 112.

thus becomes the countersignature of a union of the hybrid thing-space with the drawing space; since the view of the front and back sides of the wings would be quite sufficient to determine the species, here the hybrid body is to be understood as the symbolic medium that, at the interface between the material and the conceptual dimensions of the object, activates the space of meaning in the first place.[16]

4 Affiliations

The butterflies are redimensioned as epistemic objects by nature's self-printing, that is, they are given a manageable and transportable form[17] while remaining comparatively fragile, from a conservationist point of view: they rely on stabilizing media such as slipcases, cassettes and – on another level of nesting – protective cabinets,[18] while indirect natural self-prints might also end up in locations more exposed to physical wear and tear, such as book pages. The article from Königsberg further highlights corresponding advantages and possible applications of self-printing in areas of science, pedagogy, and art.[19] The referential chains of cultural technologies into which the ensemble of the collection is integrated, however, extend beyond such pragmatic classifications and similarly also connect to many pragmatic contexts.[20] For the subsequent history of natural self-printing in the nineteenth and twentieth centuries, the criminological practices of fingerprinting and the history of early photography are the best-known affiliations.[21] As single specimens and in the special form of the butterfly natural

---

[16] See Lutz and Siegert, "In der *Mixed Zone*," 138–139; and Hans-Jörg Rheinberger, "Zettelwirtschaft," in *Schreiben als Kulturtechnik*, ed. Sandro Zanetti (Berlin: Suhrkamp, 2012), 441–452, here 442.
[17] See Rheinberger, "Zettelwirtschaft," 443, with reference to Bruno Latour's text, which has by now acquired the status of a classic: Bruno Latour, "Visualisation and Cognition: Drawing Things Together," *Knowledge and Society: Studies in the Sociology of Culture and Present* 6 (1986), 1–40.
[18] See Anke te Heesen and Anette Michels (eds.), *Auf/zu: Der Schrank in den Wissenschaften*, exhibition catalogue (Berlin: De Gruyter, 2007).
[19] Meeting of May 2, 1889, in *Schriften der Physikalisch-Ökonomischen Gesellschaft zu Königsberg*, 30 (1889): 25.
[20] Here, my concept of the "referential chain" follows Latour's concept in a generalized form as explicated in Bruno Latour, "Selbstportrait als Philosoph," http://www.bruno-latour.fr/sites/default/files/downloads/114-UNSELD-PREIS-DE.pdf (visited February 28, 2019).
[21] On the systematic connection between natural self-prints and dactylograms, see Geus, *Natur im Druck*, 20 and 27; on natural self-prints and photography, see Rolf H. Krauss, "Fotografie und

self-print, they are, in accordance with the historical circumstances of the collection, only occasionally involved in such affiliations. For example, only those sheets of the collection are exposed to fingerprints that belong to the collective of holdings that were handed down in slipcases without a protective case (slipcases 1 and 2).

Overall, however, many connections remain to be clarified. It is true that the nearly 350 specimens in the collection almost seem to be objects directly demonstrating the method that Ernst Vanhöffen presented to the physical-economic society in Königsberg and then explained in its journal. Since the origin of the present collection itself is unknown and no relevant notes can be identified in the three convolutes, one can only speculate as to whether there was a real historical and local connection to the surroundings of Ernst Vanhöffen (1858–1918). Incidentally, just a few weeks after his lecture in Königsberg in July 1889, Vanhöffen set out to take part in the first expedition to the Atlantic dedicated specifically to the exploration of plankton, and in later years he traveled nearly the entire globe on further expeditions.

Conversely, the individual butterflies whose wings were added to the collection – surrendering those specific body parts while also revealing their existence as a species – came from relatively distant regions and biotopes, but they are essentially European, not exotic, species. Just as the gum arabic forms a kind of halo around each individual preparation, the butterflies can be embedded as a collective in a kind of virtual cloud of their origin. As Merck's *Warenlexikon* of 1884 emphasizes, the gum arabic itself ultimately came not from Arabia but from Africa and would, according to the editors, therefore be "better called African gum, because Africa is the true home of the thorny acacias or mimosas that exude the material."[22]

Yet it is not so much the origins that matter as the routes these techniques traveled. By the early nineteenth century, the technique of natural self-printing had probably already spread to Japan, where it prominently remained in use.[23] It would need to be clarified whether prints or natural self-printed preparations were also produced in Korea. The same holds for a possible connection of the latest technological designs of mobile devices to the technology of natural self-printing of butterfly wings, which is comparatively marginal in the context of naturally reproductive processes. In the future, if the hopes of the manufac-

---

Auers Naturselbstdruck: Kontaktbilder als Medien einer mechanischen Objektivität," *Fotogeschichte* 121 (2011): 13–24.
22 Art. "gum arabicum," in *Merck's Warenlexikon* (1884), http://www.retrobibliothek.de/retrobib/seite.html?id=45534#Gummi arabicum (visited on February 15, 2019).
23 Peter Heilmann, "Über den Naturselbstdruck und seine Anwendung," 107–108.

turers are fulfilled, their optically "adjusted" shape enriched with a brilliant blue will form a new branch in the family tree of hybrid operational chains, with the natural self-printing process of the fingerprint having been transferred into digital coding.

Jürgen Martschukat
Identifying, Categorizing, and Stigmatizing Fat Bodies

1 The Malleable Meaning of Body Shapes

As I was flipping through the pages of Sander Gilman's book *Obesity: The Biography*, my curiosity was piqued by the reproduction of a chronophotograph and its caption "Eadweard Muybridge, 'A Gargantuan Woman Walking.' Collotype (1887) .... (*Wellcome Collection*)."[1]

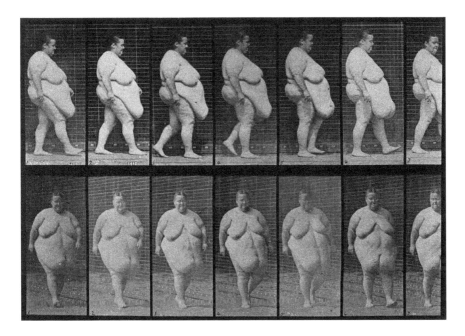

Fig. 1: Chronophotograph by Eadweard Muybridge, 1887, Wellcome Library.

---

[1] Sander L. Gilman, *Obesity: The Biography* (Oxford: Oxford University Press, 2010), 108; this chapter builds on Jürgen Martschukat, "Identifying, Categorizing, and Stigmatizing Fat Bodies," *Food, Fatness and Fitness: Critical Perspectives*, January 1 (2018), http://foodfatnessfitness.com/2018/01/01/identifying-categorizing-stigmatizing-fat-bodies/ (visited on March 24, 2019).

∂ Open Access. © 2020 Jürgen Martschukat, published by De Gruyter. This work is licensed under a Creative Commons Attribution 4.0 International License.
https://doi.org/10.1515/9783110647044-011

Identifying the walking woman as "gargantuan" struck me as awkward. It led me to trace the history and meaning of this image and its caption, which – as it turns out – tells a lot about how fat bodies have been identified, categorized, and stigmatized in modern history and how this changed over time.

The woman shown in this chronophotograph performs the body technique of walking.[2] Yet whereas existing research on body techniques tends to focus on the acquisition and habitualization of certain body techniques, this essay explores how walking bodies are identified, categorized, and related to certain body shapes and types and how various other cultural techniques, such as photographing, filing, and classifying, have endowed this woman's body with meaning. They tell us what she does and what she is, or rather what she is made to be in the eyes of the observer. Furthermore, this chapter argues that how images and their depictions are categorized, how they are arranged into a certain order – which then again produces a certain "truth" about human beings and their bodies – changes significantly over time. The history of this chronophotograph shows how bodies are embedded in power relations, which revolve around the administration of bodies, life and populations, and are shaped by cultural techniques.[3]

2 *Animal Locomotion*

From 1884 to 1887, Eadweard Muybridge shot 781 chronophotographs at the University of Pennsylvania. He titled the collection *Animal Locomotion*, which serves as a visual anthropology of body techniques while also very much representing the modern desire to observe, fractionalize, identify, and classify.[4] We encounter a seemingly endless series of images of people throwing a punch, sitting down, turning away, and walking, walking, and walking – a focal practice of

---

[2] The seminal article is Marcel Mauss, "The Notion of Body Techniques," *Sociology and Psychology: Essays by Marcel Mauss*, trans. Ben Brewster (London: Routledge and Kegan Paul, 1979), 95–123. See also Erhard Schüttpelz, "Körpertechniken," *Zeitschrift für Medien- und Kulturforschung* 1, 1 (2010): *Kulturtechnik*: 101–120.
[3] Netzwerk Körper (ed.), *What Can a Body Do? Praktiken und Figurationen des Körpers in den Kulturwissenschaften* (Frankfurt am Main: Campus, 2012); Michel Foucault, *The Birth of Biopolitics: Lectures at the Collège De France 1978–79*, trans. Graham Burchell (London: Palgrave, 2008).
[4] Eadweard Muybridge, *Animal Locomotion: An Electro-Photographic Investigation of Consecutive Phases of Animal Movements: Prospectus and Catalogue of Plates* (Philadelphia: J.B. Lippincott Company, 1887); Andreas Mayer, *Wissenschaft vom Gehen: Die Erforschung der Bewegung im 19. Jahrhundert* (Frankfurt am Main: Fischer, 2013), 195.

chronophotographic research.[5] Muybridge's images display many bodies almost or completely naked so as to make them better objects of observation and analysis. In June 1885, he also began to put the bodies in front of a grid pattern, which is familiar from anthropological photography.

Muybridge's work stands at a specific moment in modern history. It connects the body and society, revealing the extraordinary momentum of their ties and the body's significance in the intense longing for renewal in America in the progressive era. The late nineteenth and early twentieth centuries have been described as an age of the body in American history, when a biopolitical rationality took the enhancement of the body as its subject.[6] For example, modern industrialization demanded a laboring body, whose productivity became a focal point of the contemporary sociocultural configuration. The individual body's productivity was seen as something that could be increased by making it part of a complex wheelwork of men and machines in the modern factory. Making a factory run smoothly and successfully required knowing the body's mode of operation, which was fractionalized into the smallest possible units and recomposed with other bodies and machines into a single complex operation. Muybridge's work very much interacted with the emerging contemporary concept of the "scientific management" of work processes and of society as a whole, as his chronophotography rendered bodies comprehensible, controllable, and usable.[7]

Furthermore, at this particular moment in history, ability and fitness were considered more crucial for the advancement of societies than ever. Next to productivity, fitness became a key term of the period, which praised competition and its powerful dynamics as key principle of societies and diagnosed the "struggle for existence" between individuals, nations, and even races as a major driving force of mankind and its history.[8] This endowed the body with utmost significance for

---

[5] Schüttpelz, "Körpertechniken," 102; Michel Frizot, "Der menschliche Gang und der kinematographische Algorythmus," in *Diskurse der Fotografie: Fotokritik am Ende des fotografischen Zeitalters*, ed. Herta Wolf (Frankfurt am Main: Suhrkamp, 2003), 456–478, here 456.
[6] TJ Jackson Lears, *Rebirth of a Nation: The Making of Modern America: 1877–1920* (New York: Harper Perennial, 2009); Jürgen Martschukat, "'The Necessity for Better Bodies to Perpetuate Our Institutions, Insure a Higher Development of the Individual, and Advance the Conditions of the Race': Physical Culture and the Shaping of the Self in Late Nineteenth and Early Twentieth Century America," *Journal of Historical Sociology* 24, 4 (2011): 472–493; Foucault, *Birth of Biopolitics*.
[7] David E. Nye, *America's Assembly Line* (Cambridge, MA: MIT Press, 2013); the classic for the European context is Anson Rabinbach, *The Human Motor: Energy, Fatigue and the Origins of Modernity* (Berkeley: University of California Press, 1992).
[8] Jürgen Martschukat, "The Age of Fitness: The Power of Ability in Recent American History," *Rethinking History* 23, 2 (2019): 157–174. A classic is Richard Hofstadter, *Social Darwinism in American Thought, 1860–1915* (Boston: Beacon Press, 1992 [1944]).

the success and even the survival of societies, groups, nations, or "races." This importance of the body made another fact thoroughly irritating: that, at the same time, many Americans fathomed themselves to be living in increasingly disembodied modern times. Members of the white urban middle-class, in particular, experienced a manic fear of developing weakened neurasthenic bodies. White men countered their fear of neurasthenia by efforts to strengthen their bodies through exercise that were at times obsessive and demanded precise observation, measurement, instruction, optimization, categorization, and differentiation of their bodies.[9]

Thus, it is no coincidence that Philadelphia's white middle class was most interested in Muybridge's work and supported and sponsored him financially. In the original catalogue from 1887, the first 514 plates of *Animal Locomotion* show men and women, mostly naked, performing some kind of movement or work effort. Male models were either sports students at the University of Pennsylvania or local craftsmen. Women were picked from all social classes; their age and size were most important to Muybridge, rather than a particular craft. In the original catalogue, Muybridge arranged the plates according to types of body techniques (such as "walking" or "arising from the ground"), and the arrangement of the images in the book was gendered in a surprisingly stereotypical manner: whereas both women and men walk, the exercising and working body is male, while the female body is graceful, caring, and often depicted with children. The erotic connotation of many images is unmissable.[10]

The final 219 plates of the catalogue show different animals, which are of no further interest here. However, placed between humans and animals are 27 plates in a section titled "abnormal movements." These photos include a boy without legs, a hysteric woman with spastic paralysis, and several more, with the models recruited from the poorhouse of the city of Philadelphia and set apart from the rest of the book as 'the other.' The disabled body appears as lower class and genderless, not fully human and only partly able to perform the techniques of the body. In progressive America, the disabled body encountered increasing attention from both a sensationalist public and the medical profession.[11]

---

9 Tom Lutz, *American Nervousness, 1903: An Anecdotal History* (Ithaca, NY: Cornell University Press, 1991); John F. Kasson, *Houdini, Tarzan, and the Perfect Man: The White Male Body and the Challenge of Modernity in America* (New York: Hill & Wang, 2002).
10 Shawn Michelle Smith, *At the Edge of Sight: Photography and the Unseen* (Durham, NC: Duke University Press, 2013), 75–98; Elspeth H. Brown, "Racialising the Virile Body: Eadweard Muybridge's Locomotion Studies 1883–1887," *Gender & History* 17, 3 (November 2005): 627–656.
11 Kim E. Nielsen, *A Disability History of the United States* (Boston: Beacon Press, 2012), 88–91.

Most interestingly, in this catalogue from 1887, the fat woman is hard to find. She models on two plates, which are neither part of the section on "abnormal movements" nor given any other special place in the book. She also did not play a role in the medical research on abnormal bodies that was conducted in conjunction with Muybridge's project.[12] Instead, she is integrated among the first 514 plates according to the technique of the body that she is performing on the respective plate. The first plate, no. 19, is one of 58 plates depicting people walking, simply titled "walking, commencing to turn around." We do not encounter her again until plate 268, in the section on lying down and standing up. In the original, this plate is simply entitled "Arising from the ground."[13]

Muybridge sold his plates in packages of one hundred to subscribers, who were interested in his work and wealthy enough to afford buying it. Yet, in the original catalogue, the fat model was not presented to customers as particularly noteworthy or as belonging to a specific category, neither by the presentation of her body nor by her placement in the collection. Very little information is provided about her. Bracketed between technical and other explanations, it indicates only that she is model no. 20, unmarried, and weighs 340 pounds. It is somewhat difficult to trace her by going through the long list of 31 pages. No special description draws the reader's attention to her body, no special categorization makes her appear different from normal, no othering of her body takes place that would define other bodies as "normal." In Muybridge's original catalogue, she can only be identified by her number, and she simply belongs to the people walking or rising from the ground.[14]

When Muybridge worked on *Animal Locomotion*, an age of fitness and competition was just beginning to emerge in America, and body fat was only beginning to acquire the connotation of being problematic. Throughout the preceding nineteenth century, fat persons – and in particular extremely fat persons – had performed as spectacles in fairs and traveling circuses, and later in vaudeville theaters and dime museums. Fat persons served as depictions of "human

---

[12] Marta Braun and Elizabeth Whitcombe, "Marey, Muybridge and Londe: The Photography of Pathological Locomotion," *History of Photography* 23 (1999): 218–224. Muybridge cooperated for this section with the neurologist Frances Derkum. See Douglas J. Lanska, "The Dercum-Muybridge Collaboration and the Study of Pathologic Gaits Using Sequential Photography," *Journal of the History of the Neurosciences* 25, 1 (2016): 23–38; Francis X. Dercum, "A Study of some Normal and Abnormal Movements Photographed by Muybridge," in *Animal Locomotion: The Muybridge Work at the University of Pennsylvania: The Method and the Result*, ed. University of Pennsylvania, foreword by William Pepper (Philadelphia: J.B. Lippincott, 1888), 103–133.
[13] Muybridge, *Animal Locomotion*, Catalogue, ii, xiii.
[14] Muybridge, *Animal Locomotion*, 13.

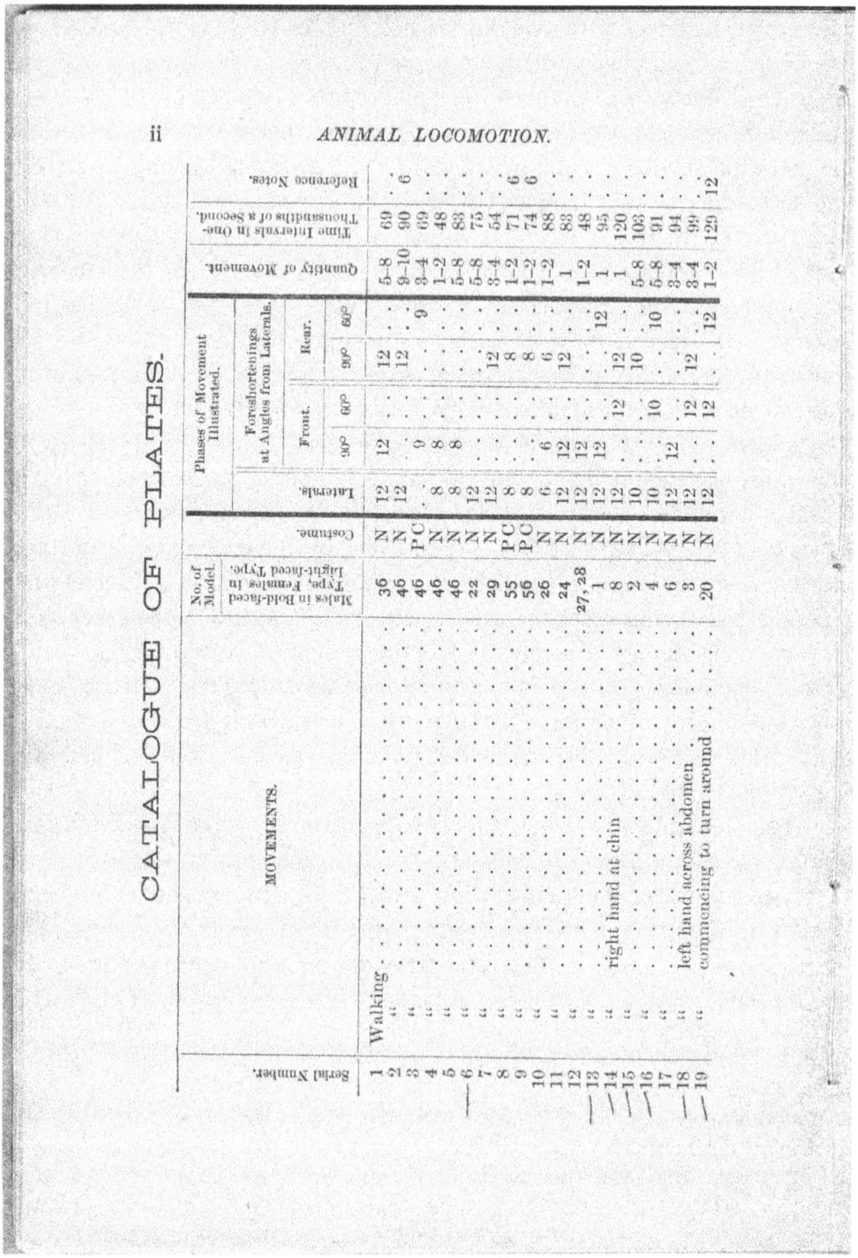

ii ANIMAL LOCOMOTION.

CATALOGUE OF PLATES.

Serial Number.	MOVEMENTS.	No. of Model. Males in Bold-faced Type, Females in Light-faced Type.	Costume.	Phases of Movement Illustrated. Laterals.	Foreshortenings at Angles from Laterals. Front. 90°	Front. 60°	Rear. 90°	Rear. 60°	Quantity of Movement.	Time Intervals in One-Thousandths of a Second.	Reference Notes.
1	Walking	36	N	12	12	.	12	.	5–8	69	.
2	"	45	N	12	.	.	12	.	9–10	90	6
3	"	46	PC	.	9	.	.	9	3–4	69	.
4	"	46	N	8	8	.	.	.	1–2	48	.
5	"	46	N	8	8	.	.	.	5–8	83	.
6	"	22	N	12	.	.	.	.	5–8	75	.
7	"	29	N	12	.	.	12	.	3–4	54	.
8	"	55	PC	8	.	.	8	.	1–2	71	.
9	"	56	PC	8	.	.	8	.	1–2	74	6
10	"	26	N	6	6	.	6	.	1–2	88	6
11	"	24	N	12	12	.	12	.	1	83	.
12	"	27, 28	N	12	12	.	.	.	1–2	48	.
13	"	1	N	12	12	.	.	12	1	95	.
14	" right hand at chin	8	N	12	.	12	12	.	1	120	.
15	"	2	N	10	.	.	10	.	5–8	103	.
16	"	4	N	10	.	10	.	10	5–8	91	.
17	"	6	N	12	12	.	.	.	3–4	94	.
18	" left hand across abdomen	3	N	12	.	12	12	.	3–4	99	.
19	" commencing to turn around	20	N	12	.	12	.	12	1–2	129	12

Fig. 2: Page from Muybridge, *Animal Locomotion* (1887), showing how difficult it is to identify our walking woman.

grotesquery," in particular, when they were women.[15] However, in general, until the late nineteenth century the meaning of body fat as sign of success was almost undisputed, and being fat rather than skinny counted as healthy. It was only in the age of fitness and of the "struggle for survival" that fat began to be perceived as an indicator of limited physical and mental ability and as a medical problem. The fat person was conceived of as lower on the evolutionary ladder and therefore also as a collective social problem. Thus, at the time when Muybridge carried out his research at the University of Pennsylvania, body fat began to change from a sign of success to an indicator of individual and collective degeneration, with the potential to marginalize those people who did not conform to the norm. In the following decades, slowly but surely and in areas ranging from everyday life to the sciences, being fat was tied to laziness, passivity, and the inability to cope with the demand for constant movement in modern life and to take good care of oneself. Body fat was also tied to femininity: even women's body cells were seen as more passive than male ones. Fat was said to resemble dead tissue, with a flabby body indicating a weak mind, as well as moral and biopolitical failure. The tone was set for the twentieth century.[16]

3 "A Gargantuan Woman Walking"?

After the Great Depression and World War II, the fear of fat began to gain new momentum as wealth and even abundance returned in the 1950s, at least for the growing middle class in American society. After the 1970s, this fear became stronger, and in the age of neoliberalism, when fitness and having and showing

---

15 Amy E. Farrell, *Fat Shame: Stigma and the Fat Body in American Culture* (New York: New York University Press, 2011), 32.
16 Hillel Schwartz, *Never Satisfied: A Cultural History of Diets, Fantasies and Fat* (London: Collier Macmillan, 1986); Peter Stearns, *Fat History: Bodies and Beauties in the Modern West* (New York: New York University Press, 2002), 6-9; Nina Mackert, "'I want to be a fat man / and with the fat men stand': U.S.-Amerikanische Fat Men's Clubs und die Bedeutungen von Körperfett in den Dekaden um 1900," *Body Politics* 3 (2014): 215–243, http://bodypolitics.de/de/wp-content/uploads/2015/07/ Heft_3_10_mackert _fat_End.pdf (visited on March 24, 2019); Nina Mackert, "Women Are Cooler: On the Feminization of Fat," *Food, Fatness and Fitness: Critical Perspectives* (February 1, 2018), http://foodfatnessfitness.com/2018/02/01/women-cooler-feminization-fat/ (visited on March 24, 2019). For a most recently published historical overview see Christopher E. Forth, *Fat: A Cultural History of the Stuff of Life* (London: Reaktion, 2019).

a foresighted relation to one's body became more important than ever, the fear of fat developed into an obsession.[17]

Fittingly, the caption that characterizes Muybridge's model no. 20 as "gargantuan" and first peaked my curiosity is a late twentieth-century invention. Yet, as I will now argue, the history of this caption is even more complex, and it reveals the twists, ramifications, and pitfalls of the meaning of fat bodies in modern history.

In 1996, this plate was given a new caption by an art historian of the Wellcome Library in London, most likely when the library introduced electronic catalogues and search systems for its impressive collection in the history of medicine and the body, which also contains hundreds of plates from Muybridge's project.[18] Obviously, different from the late nineteenth-century catalogue, in the late twentieth century a fat body had to be marked by key words so that it could be easily identified and traced by researchers and search engines. Still, "gargantuan" is an awkward adjective and rarely used. It refers to big and grotesque bodies, and dictionaries provide "enormous" in connection with "appetite" to explain its meaning. Etymologically, the word can be traced back to the 1530s and François Rabelais' novels on Pantagruel – specifically to Pantagruel's father Gargantua, a giant, guzzling, farting glutton.[19] Today, the use of the word to describe Muybridge's images conveys a message of physical and moral transgression. Furthermore, calling the woman "gargantuan" is – most likely unintentionally – fat shaming by an art historian of the Wellcome Library, who might have a special interest in early modern French culture, or who may have just read Mikhail Bakhtin's book on *Rabelais and His World*.[20]

Thus, it is only fairly recently that model no. 20 was described as "gargantuan," and this act of describing needs to be seen as embedded in the evolving history of body fat and its changing meanings. Nevertheless: the way the caption was put together and presented by the Wellcome collection gives the impression that the image and caption had belonged together since Eadweard Muybridge took this picture in his laboratory at the University of Pennsylvania. The caption makes it sound as if model no. 20 had always been described as "gargantuan." Its historicity is obscured, as well as the historicity of the perception of fat bodies. Furthermore, through internet platforms such as Wikimedia and Pinterest, this

---

[17] Wendy Brown, *Undoing the Demos: Neoliberalism's Stealth Resolution* (New York: Zone Books, 2015); Martschukat, "The Age of Fitness"; and *Das Zeitalter der Fitness: Wie der Körper zum Zeichen von Erfolg und Leistung wurde* (Frankfurt am Main: Fischer, 2019).
[18] Information by the Wellcome Library to the author in an email from November 4, 2017.
[19] François Rabelais, *Gargantua and Pantagruel* (London: Penguin Classics, 2006).
[20] Mikhail Bakhtin, *Rabelais and His World*, trans. Helene Iswolsky (Cambridge, MA: MIT Press, 1968).

combination of image and caption has spread across the globe, finding its way into publications for academic and lay audiences.[21]

The story continues. In December 2009 fat activist Charlotte Cooper published a blog post on Muybridge's model no. 20. In its combination with the description as "gargantuan," Cooper writes, the images reduced the woman to her naked body; she was deprived of her humanity, just like the objects of colonial photographers. Described in an academic register (history of art and literature, anthropology), her othering is endowed with the tinge of objectivity and eternity rather than being a specifically modern phenomenon. It is an ironic twist to this story that Cooper only found the image because of its tagging.[22]

William Schupbach, curator of the Wellcome Collection, read and reacted to Cooper's comments in another blog post, critically reflecting upon Muybridge, fat, images, cataloguing, and its techniques. Nevertheless, in the early twenty-first century, not categorizing this woman's body seemed impossible to the Wellcome Collection. Today, fat bodies seem to demand classification, and an electronic catalogue obviously needs to serve its users' desire to search for images of fat bodies by describing and classifying its items as precisely as possible. Yet "gargantuan," Schupbach wrote, would not make any sense, and "why drag in Rabelais anyway?"[23]

4 From "Gargantuan" to "Obese"

William Schupbach replaced "gargantuan" with "obese," which meant throwing the baby out with the bathwater. As a medical term, obese expresses the normativity and politics of health, and it depicts fat bodies as in need of treatment. Furthermore, today, health and body shape are considered to be dependent on lifestyle and individual decisions, which are judged as right or wrong, good or bad. Choosing wisely, eating right, and taking good care of one's body contribute

---

[21] See for instance https://commons.wikimedia.org/wiki/Category:Obese_women#/media/File:A_gargantuan_woman_walking._Collotype_after_Eadweard_Wellcome_V0048623.jpg (visited on March 24, 2019).

[22] Charlotte Cooper, "Archival Images of Fat People, Pathology and Medicalisation," *Obesity Timebomb*, (December 16, 2009), http://obesitytimebomb.blogspot.de/2009/12/medical-history-obesity-pathology-and.html (visited on March 24, 2019).

[23] William Schupbach, "Obesity and Personality," *Wellcome Library Blog*, (December 22, 2009), http://blog.wellcomelibrary.org/2009/12/obesity-and-personality/ (visited on March 24, 2019).

substantially to being recognized as a productive member of society.[24] Therefore, in the twenty-first century, "obesity" is even more moralizing and politically charged than the early modern "gargantuan." In the sixteenth century, Rabelais' text on Gargantua was humorous, ironic, and ambivalent, in contrast to the moralizing use of "obesity" today. The obesity discourse claims absolute authority when it comes to fat, with being fat presented as proof of misconduct and being unproductive, as pathological and abnormal.[25]

Let us take a final look at model no. 20. In a new 2010 coffee-table edition of *Animal Locomotion*, she is described as "a young woman, grotesquely obese, [who] is made to squat in front of Muybridge's camera to help represent the genre 'medical abnormities.'" Even though "'medical abnormities'" is marked as a quote, this category does not exist in Muybridge's original. And in an article by Muybridge's co-researcher, the neurologist Francis X. Dercum, on "Normal and Abnormal Movements," model no. 20 plays no role at all; she is not even mentioned in the text. Additionally, the editors of this new luxury volume rearranged the order of things. The plates with our fat model are now placed at the very beginning of the section on "abnormal movements," in the medicalized freak show of living creatures who are depicted as unable to move properly. From 1887 to 2010, the fat woman changed from having just an ordinary body, maybe sparking some curiosity in others, into being the pathological prototype of the unhealthy body. Cultural techniques of identifying, categorizing, and tagging have substantially contributed to this transformation.[26]

---

24 Charlotte Biltekoff, *Eating Right in America: The Cultural Politics of Food and Health* (Durham, NC: Duke University Press, 2013); Nina Mackert and Jürgen Martschukat, "Introduction: Fat Agency," *Body Politics* 7 (2015): 13–24, http://bodypolitics.de/de/wp-content/uploads/2016/01/Heft_5_01_Mackert_Martschukat_Intro_End-1.pdf (visited on March 24, 2019).
25 Bakhtin, *Rabelais and His World*; Jennifer A. Lee and Cat J. Pausé, "Stigma in Practice: Barriers to Health for Fat Women," *Frontiers in Psychology* 7 (2016), https://www.frontiersin.org/articles/10.3389/fpsyg.2016.02063/full (visited on November 15, 2018). Criticizing the normativity of the health discourse does not mean one is against health; Jonathan Metzl and Anna Kirkland (eds.), *Against Health: How Health Became the New Morality* (New York: New York University Press, 2010).
26 Hans Christian *Adam, Eadweard Muybridge: The Human and Animal Locomotion Photographs* (Cologne: Taschen, 2010), 32, 564f.; Dercum, "A Study of Some Normal and Abnormal Movements."

Kathrin Fehringer
Techniques of the Body and Storytelling: From Marcel Mauss to César Aira

> Indeed, Mauss's famous lecture is indispensable
> for an expanded understanding of cultural techniques.
> Geoffrey Winthrop-Young (2013)[1]

1 "Excusez-moi si je vous raconte" ("Les Techniques du Corps," 1934)

To speak of cultural techniques means to speak of techniques of the body: it has almost become a reflex to refer to Marcel Mauss's lecture "Les Techniques du Corps" (1934) when it comes to determining what cultural techniques can be. "Les Techniques du Corps" has been widely summarized and more closely considered, yet almost exclusively in the context of (German) media studies.[2] Hence it's also hardly surprising that the most popular – in the literal sense of the word – passage[3] from "Les Techniques du Corps" is the often-quoted anecdote about the cinema that Erhard Schüttpelz, in his essay "Körpertechniken" (2010), reads as a

---

[1] Geoffrey Winthrop-Young, "Cultural Techniques: Preliminary Remarks," *Theory Culture & Society* 30, 6 (2013): 3–19, here 10.
[2] Lorenz Engell and Bernhard Siegert, "Editorial: Kulturtechnik," *Zeitschrift für Medien- und Kulturforschung* 1, 1 (2010): *Kulturtechnik*: 5–9; Harun Maye, "Was ist eine Kulturtechnik?" *Zeitschrift für Medien- und Kulturforschung* 1, 1 (2010): *Kulturtechnik*: 121–135; Erhard Schüttpelz, "Die medienanthropologische Kehre der Kulturtechniken," in *Archiv für Mediengeschichte* 6 (2006): *Kulturgeschichte als Mediengeschichte (oder vice versa?)*: 87–110; E.S., "Körpertechniken," *Zeitschrift für Medien- und Kulturforschung* 1, 1 (2010): *Kulturtechnik*: 101–120; Geoffrey Winthrop-Young, "Cultural Techniques: Preliminary Remarks," *Theory Culture & Society* 30, 6 (2013): 3–19. Another classic reference text on the topic of cultural techniques alludes to Mauss's "Techniques du Corps" without, however, explicitly naming them or citing them as a source: Sybille Krämer and Horst Bredekamp, "Kultur, Technik, Kulturtechnik: Wider die Diskursivierung der Kultur," in *Bild – Schrift – Zahl*, ed. S.K. and H.B. (Munich: Fink, 2003), 11–22.
[3] Namely, in the words of François de Pitaval, "celèbre et intéressant": "famous and interesting – famous, because [it is] interesting," as Nicolas Pethes defines "popular." See Nicolas Pethes, "Vom Einzelfall zur Menschheit: Die Fallgeschichte als Medium der Wissenspopulari-

---

Translated by Michael Thomas Taylor

Open Access. © 2020 Kathrin Fehringer, published by De Gruyter. This work is licensed under a Creative Commons Attribution 4.0 International License.
https://doi.org/10.1515/9783110647044-012

prerequisite for Mauss's discovery of techniques of the body (and which leads to Schüttpelz's diagnosis that "techniques of the body are media").[4] Mauss tells his anecdote as follows:

> Une sorte de rélévation me vint à l'hôpital. J'étais malade à New York. Je me demandais où j'avais déjà vu des demoiselles marchant comme mes infirmières. J'avais le temps d'y réfléchir. Je trouvai enfin que c'était au cinéma. Revenu en France, je remarquai, surtout à Paris, la fréquence de cette démarche ; les jeunes filles étaient Françaises et elles marchaient aussi de cette façon. En fait, les modes de marche américaine, grâce au cinéma, commençaient à arriver chez nous. C'était une idée que je pouvais généraliser.
> ("Les Techniques du Corps," 368)

> A kind of revelation came to me in hospital. I was ill in New York. I wondered where previously I had seen girls walking as my nurses walked. I had the time to think about it. At last I realised that it was at the cinema. Returning to France, I noticed how common this gait was, especially in Paris; the girls were French and they too were walking in this way. In fact, American walking fashions had begun to arrive over here, thanks to the cinema. This was an idea I could generalise. ("Techniques of the Body," 72)[5]

Up to now, mentioning techniques of the body (and especially of Schüttpelz's essay with the same title) seems to automatically trigger a discussion in the terms of media studies. But this is currently hindering any other fundamentally different view of Mauss's lecture.[6] I would thus like to separate the discussion from arguments made in the context of media studies by examining "Les Techniques du Corps" in a different way: What new findings might result from reading the text from the perspective of literary studies?[7] Two problems need to be pursued more closely:

---

sierung zwischen Recht, Medizin und Literatur," in *Popularisierung und Popularität*, ed. Gereon Blaseo et al. (Cologne: DuMont, 2005), 63–92, here 63.
4 Schüttpelz, "Körpertechniken," 114.
5 Marcel Mauss, "Les Techniques du Corps," in M.M., *Sociologie et Anthropologie,* introduction by Claude Lévi-Strauss (Paris: Presses Universitaires de France, 1985 [1950]), 365–388. Cited hereafter as TdC. Translated as "Techniques of the Body" by Ben Brewster, *Economy and Society* 2, 1 (1973): 70–88. Cited hereafter as ToB. In the following I will refer to both the original French text and the English translation when details are crucial.
6 Schüttpelz himself pointed to what he saw as a kind of dud effect with Techniques du Corps: "Something appears to be hindering the concept of techniques of the body – not in its effect, but in its development"; Schüttpelz, "Körpertechniken," 108.
7 To date, there have been no considerations of "Les Techniques du Corps" from a literary-critical perspective. The most recent study stems, once again, from the field of media studies: a twenty-page chapter titled "Körpertechniken: Rhythmen von Mensch und Maschine," in Timo Kaerlein, *Smartphones als digitale Nahkörpertechnologien: Zur Kybernetisierung des Alltags* (Bielefeld: transcript, 2018), 175–196.

First, it should be noted that until now, Marcel Mauss's lecture has been read unquestioningly as a text, despite the fact that it consistently exhibits its orality. In addition to addressing his audience a number of times, Mauss even analyzes the scene of his own speaking as an example of the symbolics of techniques of the body:

> Je suis en conférencier avec vous; vous le voyez à ma posture assise et à ma voix, et vous m'écoutez assis et en silence. Nous avons un ensemble d'attitudes permises ou non, naturelles ou non. (TdC 372)

> I am a lecturer for you [literally: I am with you as a lecturer]; you can tell [see] it from my sitting posture and my voice, and you are listening to me seated and in silence. We have a set of permissible or impermissible, natural or unnatural attitudes. (ToB 76)

Mauss's lecture, as a recounting of privately experienced events (as in the anecdote about the cinema) is not only a public act of constituting meaning. Rather, it opens up public space for a collective scholarly practice when Mauss, as he himself emphasizes, is seen and heard by the scholars of the Société de Psychologie to whom he is speaking.[8] Such a space is opened up textually, too, when Mauss's lecture is printed in 1935 in a scholarly journal and reprinted in 1950 in an edited volume (an issue to which I will return later).

Second, the question arises regarding just which occasions for stories (that are obviously components of a scholarly practice), exactly, one is dealing with in "Les Techniques du Corps." The question regarding the status of what I would like to call *storytelling* in the following has already been raised in discussions in media studies, namely in connection with the terms *exemple* and *anecdote* that Mauss himself uses. These terms have been dismissed as rhetorical ornaments, as elements of an entertaining and charming style of presentation, while also being labeled with keywords that are highly interesting from the perspective of literary studies, and which are worth examining more closely: a list, an inventor's story, report to an academy.

What Harun Maye aptly observed in writing about the "list of examples" in "Les Techniques du Corps"[9] is that occasions for storytelling can be systematically[10] sought out and also found in the text, analogous to Mauss's own dogma "example and order, that's the principle" (TdC 384; ToB 85).[11] For these things

---

[8] This could be pursued further with Hannah Arendt, *Vita active oder Vom tätigen Leben* (Munich: Piper, 2002); *The Human Condition* (Chicago: University Press of Chicago, 1958).
[9] Harun Maye, "Was ist eine Kulturtechnik," 122.
[10] This could be followed further elsewhere; see Jack Goody, "What's in a list?" in Goody: *The Domestication of the Savage Mind* (Cambridge and New York: Cambridge University Press, 1977), 338–395.
[11] For this reason, the anecdote as an object of literary studies will not be the main focus of what follows.

function as analytical instruments of innovative research, namely of that "descriptive ethnology" (TdC 365; ToB 70) initially identified by Mauss, which he not only opposes to but places above the theoretical sciences.[12] This is innovative inasmuch as Marcel Mauss is considered not only the founder of French ethnology but also the inventor of social and cultural anthropology.

In situating "Les Techniques du Corps" in the interdisciplinary context of their emergence and the rhetoric on which they are based, I am thus also responding in what follows to the label of the "funny inventor's story" ("possierliche Erfindergeschichte") that Erhardt Schüttpelz applied to Mauss's lecture, and under which "Les Techniques du Corps" has been consistently taken up in research into cultural techniques.[13]

> To put it more precisely, it [Marcel Mauss's "successful inventor's report"] is a travesty and parody of an inventor's report, a *report to an academy* in the sense of Kafka, a series of variety numbers for mutual amusement. This striking textual arrangement comes from Marcel Mauss's social anthropology ...: According to Mauss, if one study's techniques of the body by studying one's own body, one encounters trained animals, and one encounters the first trained animal, the human being.[14]

It is from Mauss's use of the word *dressage* to describe techniques of the body that Schüttpelz here derives his thesis of a striking textual arrangement, the narrative use of which he describes as variety numbers: he compares Mauss's *lecture* with the fictional lecture setting in Kafka's story (written in 1917), in which a first-person narrator who had appeared "on the great variety stages of the civilized world" reports to the "Gentlemen, esteemed academicians!" about his previous "apedom."[15]

I do not think that Mauss's language bears such Kafkaesque, "apish" traits, something to be described as the travesty of an inventor's report; Schüttpelz himself here identifies the scientific, anthropological argument that Mauss has in mind with his numerous examples and anecdotes. Mauss's language follows a logic of vividly comparing the body with a machine, on the one hand, and a trained animal, on the other, but not of equating them (French: *dresser*, "to train";

---

12 "I recognize the higher certainty of the descriptive sciences compared to the theoretical sciences (in the case of very complex phenomena) even when I practice a theoretical science." Mauss cited from Stephan Moebius, *Marcel Mauss* (Konstanz: UVK, 2006), 9.
13 Schüttpelz, "Körpertechniken," 103.
14 Schüttpelz, "Körpertechniken," 105–106.
15 Franz Kafka, "A Report to an Academy," in F.K., *Metamorphosis and Other Stories*, trans. Michael Hofmann (London: Penguin, 2007), 225–236, here 225.

"to set up"; "to erect"; TdC 375).¹⁶ "Les Techniques du Corps" is therefore precisely not a Kafkaesque report; nor is it basically "funny," since Mauss relies on narrative as a scientific instrument of analysis.

Hence, in my opinion, the relating of examples and anecdotes in "Les Techniques du Corps" must also be distinguished from an *elocutio (ornatus)* in the sense of the general stylistic and rhetorical shaping of speech. Rather, it is a form of storytelling, namely a way of representing one (or more) occurrences. *Storytelling* is the act that generates an oral (or written) discourse that tells of such an event – for example, an epiphanic moment in a New York hospital.¹⁷

The aim of what follows is to analyze this storytelling and these discourses. In the sense of discovering what already exists, I understand *storytelling* in "Les Techniques du Corps" to be fundamentally part of an *inventio of* greater significance, which is bound to a scientific practice that I will trace, as it were.

Marcel Mauss himself had a love for language and its sciences. "To be a good sociologist," the 26-year old wrote in 1898, "one must be a good philologist." The study of comparative philology was a prerequisite for Mauss's interdisciplinary research interests, ensuring that he regularly read texts in the Greek or Hebrew original, in the way of a humanistic scholar.¹⁸ Hence his lecture of 1934 begins with a commentary on his own serial narration and a related gesture of apology.

---

16 The passage reads as follows: "Les techniques du corps peuvent se classer par rapport à leur rendement, par rapport aux résultats de dressage. Le dressage, comme le montage d'une machine, est la recherche, l'acquisition d'un rendement. Ici c'est un rendement humain. Ces techniques sont donc les normes humaines du dressage humain. Ces procédés que nous appliquons aux animaux, les hommes se les sont volontairement appliqués à eux-mêmes et à leurs enfants. ... Je pourrais par conséquent les comparer [les techniques] dans une certaine mesure, elles-mêmes et leur transmission, à des dressages, les ranger par ordre d'efficacité" TdC 374–375 (my emphasis); "The techniques of the body can be classified according to their efficiency, i.e. according to the results of training [résultats de dressage]. Training [le dressage], like the assembly [le montage] of a machine, is the search for, the acquisition of an efficiency. Here it is a human efficiency. These techniques are thus human norms of human training [dressage humain]. These procedures that we apply to animals men voluntarily apply to themselves and to their children. ... As a result I could to a certain extent compare these techniques, them and their transmission, to training systems [à des dressages], and rank them in the order of their effec tivenes" (ToB 77–78; my emphasis).
17 For an introduction, see Matias Martinez and Michael Scheffel, *Einführung in die Erzähltheorie* (Munich: Beck, 2005 [1999], sixth edition), here 29–30, and Albrecht Koschorke, *Wahrheit und Erfindung: Grundzüge einer Allgemeinen Erzähltheorie* (Frankfurt am Main: Fischer, 2012); additionally, see the relevant works on narration and narrating by Gérard Genette.
18 Mauss quoted from Moebius, *Marcel Mauss*, 112–114. Mauss's humanistic and philosophical education, which provided him mastery of more than a dozen dead and living languages (including ancient Greek and Sanskrit), founded his reputation among students: "Mauss sait tout!" ("Mauss knows everything!"); Moebius, *Marcel Mauss*, 7.

Both have probably gone unnoticed up till now because the translation into German is inaccurate: the French *raconter* means "to tell (stories)," but it has evidently always been understood in the sense of *rapporter* ("to tell" as in "to report"), which is how it has been translated into English and German.

> Excusez-moi si, pour former devant vous cette notion de techniques du corps, <u>je vous raconte</u> à quelles occasions j'ai pu poser clairement le problème général. Ce fut une série de démarches consciemment et inconsciemment faites. (TdC 366; my emphasis)
>
> Forgive me if, in order to give this notion of techniques of the body shape for you, <u>I tell you</u> about the occasions on which I pursued this general problem and how I managed to pose it clearly. It was a series of steps consciously and unconsciously taken. (ToB 71)

At the same time, Mauss speaks of a chain of insights connected with a series of personally experienced events that he subsequently narrates (sometimes in more detail, and sometimes less) and analytically sets forth as a "series" of conscious and unconscious steps of thought. In this regard, Mauss's lecture (which was presumably read aloud) could thus be described as a kind of autohistoriography of his own scientific practice, which produced a new discipline, clarifying Schüttpelz's label of an "inventor's report," the subject of which is often Marcel Mauss's own body.

In connection with the term *raconter*, Mauss himself uses the terms *exemple* and *décrire* ("example/to describe") or *anecdote* and *raconter* ("anecdote/to tell stories") as synonyms, or he merges or overlaps them in his usage:

> Je dis bien *les* techniques du corps parce qu'on peut faire la théorie de *la* technique du corps à partir d'une étude, d'une exposition, d'une <u>description</u> pure et simple *des* techniques du corps. J'entends par ce mot les façons dont les hommes, société par société, d'une façon traditionnelle, savent se servir de leur corps. En tous cas, il faut précéder du concret à l'abstrait, et non pas inversement. / Je veux vous faire part de ce que je crois être une des parties de mon enseignement <u>qui ne se retrouve pas ailleurs</u>, que je répète dans un cours d'Ethnologie descriptive ... / ... <u>Excusez-moi si</u>, pour former devant vous cette notion de technique du corps, <u>je vous raconte</u> à quelles occasions j'ai poursuivi et comment j'ai pu poser clairement le problème général. Ce fut une série de démarches consciemment et inconsciemment faites. (TdC 365–366; italics in the original; my emphasis underlining)
>
> I deliberately say techniques of the body in the plural because it is possible to produce a theory of *the* technique of the body in the singular on the basis of a study, an exposition, a <u>description</u> pure and simple of techniques of the body in the plural. By this expression I mean the ways in which from society to society men know how to use their bodies. In any case, it is essential to move from the concrete to the abstract and not the other way round. / I want to convey to you what I believe is one of the parts of my teaching <u>which is not to be found elsewhere</u>, that I have rehearsed in a course of lectures on descriptive ethnology ... / <u>Forgive me if,</u> in order to give this notion of techniques of the body shape for you, <u>I tell you</u> about the occasions on which I pursued this general problem and how I managed to pose it clearly. It was a series of steps consciously and unconsciously taken. (ToB 70–71)

The numerous concret facts function as case studies and thus as constitutive components of an *inventio* in the sense of a discovery of arguments, designed as an experiment: On the one hand, the piling up of the *exemples* and *anecdotes* represents a problem-solving strategy, or method, that still takes place in narration, and which is therefore processual and investigative. And on the other hand, these reported and narrated case studies serve as supporting evidence for the solutions already discovered, which Mauss provides as concise definitions; one of these sentences here forms the prelude to the lecture and has in the meantime become a commonplace in research into cultural techniques ("By this expression I mean the ways in which from society to society men know how to use their bodies").

In this context, I would like to propose that "Les Techniques du Corps" be understood as a *popular case history*, a text genre that has, in recent years, become a focus of work in literary studies. Case histories are characterized by the fact that they constitutively depend on storytelling (and thereby take recourse to other stories) to describe phenomena whose observation precedes the creation and institutionalization of new scientific disciplines.[19] This is true, for example, of the development of modern psychology (in the eighteenth and nineteenth centuries) and, with Mauss, obviously for the emergence of the new disciplines of ethnology and anthropology following the First World War.

However, the concept of case history not only provides a central view of the role of narrative in Mauss's "inventor's story." In particular, it also enables a retrospective view of the individual case – largely lacking in current debates about cultural technologies, and the express concern of this volume – that Mauss takes to be the foundation for any theory; and from here, it enables an innovative view of literary texts. This is why I will finally turn, with Mauss, toward a literary text that stages techniques of the body as bodily injury and, by doing so, to the primary cultural techniques of drawing and writing: *Un episodio en la vida del pintor viajero* ("An Episode in the Life of the Traveling Painter") by the Argentine writer César Aira (2000).[20]

The novel, which claims to be an episode in its title, is a case history insofar as it relates to a serious riding accident and its consequences. It takes place in the emptiness of the Argentine pampas, which motivated the painter and geographer

---

[19] For an exemplary treatment of this topic, see Nicolas Pethes, "Vom Einzelfall zur Menschheit," in *Das Beispiel: Epistemologie des Exemplarischen*, ed. N.P., Jens Ruchatz, and Stefan Willer (Berlin: Kadmos, 2007).

[20] César Aira, *Un episodio en la vida del pintor viajero* (Mexico City: Era, 2001; first edition 2000 by Rosario, Buenos Aires, Argentina), translated by Chris Andrews as *An Episode in the Life of a Landscape Painter*, in *Three Novels by César Aira*, preface by Roberto Bolaño (London: Penguin, 2015; 141–230).

Moritz Rugendas to set out on a mission of scientific documentation and exploration during the 1830s, prompted by Alexander von Humboldt. The novel is structured as a fictionalization of private letters written by Rugendas, who did in fact live in Augsburg.

The text is amenable to description with classical methodological categorizations and concepts from literary studies (for example, the notion of *écriture automatique*) and is therefore well suited for making visible the interdisciplinary possibilities of a perspective concerned with cultural techniques. My aim is to make clear how a perspective concerned with (cultural) techniques of the body can, as a method, enrich literary analysis. Conversely, I would like to try to show that, and to what extent, literatures can partake in determining techniques of the body and hence cultural techniques.

This will bring me back to the discussion in media studies, namely, to one of the great, controversial and unresolved questions of current research into cultural techniques: whether "techniques of the body can be subsumed under cultural techniques or, conversely, cultural techniques can be derived from the techniques of the body," as Harun Maye puts it.[21] By considering Mauss's speech and Aira's text, this question can also be posed in a different way.

2 Backgrounds: Textual Space and Interdisciplinarity

Mauss gave his lecture on May 17, 1934, to the Société de Psychologie in Paris, an association of worldwide importance founded by Pierre Janet in 1901 that Mauss himself chaired in 1924. As part of his efforts to expand and scientifically consolidate the school of his uncle and teacher Émile Durkheim after the First World War, Mauss initiated a new dialogue between sociology and psychology.[22] In his lecture he explicitly addresses the research positions of some psychologists, such as George Dumas (TdC 369; ToB 73), the former chairman of the Société de Psychologie and co-editor of its periodical, in which Mauss's speech also appeared.

In his lecture, Mauss advances the strong thesis that there are no natural postures (TdC 368; ToB 72) or modes of action (TdC 371; ToB 75) of the body: He argues these were all "not simply a product of some purely individual, almost completely

---

21 Maye, "Was ist eine Kulturtechnik," 122.
22 This went hand in hand with the development of ethnology; see Moebius, *Marcel Mauss*, 9. Durkheim's exclusion of psychology is thus expressly undone by Mauss (see 92). On the consultation of the Durkheim School, see 51–64.

psychical arrangements and mechanisms" (TdC 368; ToB 72) but rather social and learned. The same applied to perception, he argues, which he sees as equally socially conditioned, and evidently also to the representations of the trained body, as Mauss shows with the example of the imitation of fashionable gaits that his nurses copied from cinema actresses.[23]

Mauss claims he needed considerations from three disciplines ("triple viewpoint," TdC 369; ToB 73) – sociology, psychology, and biology – to develop his method, namely "the ways in which from society to society men know how to use their bodies" (TdC 365; ToB 70) to make the body something that can be described and systematized. He argues that the social element (*habitus*, "collective and individual practical reason," TdC 369; ToB 73) and with it the question of education, fashions, and prestige, were of primary significance for the development of techniques of the body. In second place, however, he places the psychological element ("psychological *momentum*," TdC 371; ToB 75, which he examines using the example of magical action, formulae, and objects) and the anatomical-physiological element (overcoming "biological resistance," TdC 371; ToB 75, using examples of words and magical objects), in which imitation and talent or faculty played a major role (TdC 369; ToB 73).

In a culmination of his argument, Mauss follows by speaking of "elements of the art of using the human body" (TdC 369; ToB 73; my emphasis) and thus no longer of "ways." He finally grasps this art of acting by using the concept of technology as individual and collective habits (*habitus*) stemming from "practical reason."[24] Mauss understands this art of using the body in the ancient Greek sense as *techné*, i.e., in the sense of an embodied practical knowledge of how to approach and execute something.

In emphasizing their theorization as a "série d'actes montées" ("series of assembled actions"; TdC 372; ToB 76), i.e.: as a chain of operations, these techniques of the body themselves recall their own proven method of analysis as "a series of steps consciously and unconsciously taken."

---

23 Hence "Les Techniques du Corps" is among the first contributions of a sociology (if not a pre-Foucauldian archaeology) of the body; see Moebius, *Marcel Mauss*, 77. It is the body as a "social formation" that first guides how Mauss perceives our body as a "physical formation," as the social anthropologist Mary Douglas wrote about Mauss in 1970. Mary Douglas, *Natural Symbols: Explorations in Cosmology* (London: Routledge, 1996 [1970]). On the summary of "Les Techniques du Corps" and their classification in Mauss's thinking, see Moebius, *Marcel Mauss*, 77–78 and 105.

24 Already in 1901, Mauss specified these collective habits as the proper central object of sociological research, which he understood as "intersections of social practices that vary according to society and tradition and can change over time." In "Les Techniques du Corps" he chooses the Latin term *habitus* to describe these practices; see Moebius, *Marcel Mauss*, 77.

As a text, although quite obviously unchanged in its original orality, Mauss's lecture appeared one year later in 1935 in the periodical of the Société de Psychologie (Society of Psychology), the *Journal de Psychologie Normale et Pathologique* ("Journal of normal and pathological psychology"), published from 1904 to 1915 and 1920 to 1940 (i.e. pausing during the First World War, and until the beginning of World War II). This journal usually refers to the minutes of the Société meetings held several times a year, while also printing the lectures given at these meetings. The Société de Psychologie always only forms a subschapter in the journal's structure. Above all, it offered a platform for new, interdisciplinary approaches to research, including from cultural philosophy.[25] One special volume *Psychologie du Language* ("Psychology of language," 1933), for instance, lists texts by Karl Bühler and Ernst Cassirer in its section on "language theory."[26]

"Les Techniques du Corps" are therefore also located textually in an interdisciplinary context – here, *between* existing disciplines; this makes it possible to recognize the discursive effort with which Mauss tries to link his theses to the scholarly horizon of an event that is mainly attended by psychologists. In understanding Mauss's lecture as a case history, however, I advocate for the strong thesis of locating it on the border of unexplored realms, namely, the creation of a new scholarly discipline, ethnology and ethnography ("Ethnologie descriptive").

In 1950, the year of Mauss's death, an anthology is published in his honor in which "Les Techniques du Corps" is printed for the second time, and which literally bears the signature of the book's ethnological approach, even if it is not reflected in the book's title: *Sociologie et Anthropologie* ("Sociology and Anthropology"): A foreword by Claude Lévi-Strauss, the last pupil of Mauss (and an ethnologist) is intended to help the volume – and thus Mauss's research – reach a larger, international audience and hence to fulfill a desideratum of sociologists and ethnographers, as well as students of these disciplines who had until then been obliged to go looking for his texts with great effort (TdC VII).[27] In this anthology of seven texts, containing first and foremost

---

25 Marcel Mauss, "Les Techniques du Corps," *Journal de Psychologie Normale et Pathologique* 32, ed. Pierre Janet and George Dumas (Paris: Presses Universitaires de Paris, 1935): 271–293. https://gallica.bnf.fr/ark:/12148/bpt6k9656899d/f7.item (visited on May 22, 2019).
26 Ernst Cassirer, "Le langage et la construction du monde des objets," and Karl Bühler, "L'onomatopée et la fonction représentative du language," *Journal de Psychologie Normale et Pathologique* 33 (1933): *Psychologie du Langage*: 8–44 and 101–119.
27 The introduction by Lévi-Strauss to the work of Marcel Mauss was so famous and controversial that it had a considerable influence on the reception of Mauss as an "ingenious, yet ultimately unsystematic predecessor of structuralism." See Stephan Moebius and Frithjof Nungesser, "La réception de Mauss en langue allemande," *Trivium: Revue franco-allemande de sciences*

the "Théorie générale de la magie" ("general theory of magic") and the famous "Essai sur le don" ("essay on the gift"), "Les Techniques du Corps" occupies the sixth place.

"Les Techniques du Corps" is a by-product of his lectures in the context of his institutionally founded "descriptive ethnology" to which he explicitly refers at the beginning of his lecture (TdC 365; ToB 70).[28] Mauss had founded the Institut d'Éthnologie de l'Université de Paris (Institute of Ethnology of the University of Paris) with an École Pratique (or practical school) in 1925, which, at the time of his speech, had already been extremely successful for ten years in training medical doctors, colonial officials, missionaries, and ethnographers.[29]

In his home country, Marcel Mauss exerted an enormous influence on the humanities. Throughout his life, he worked to ensure cooperation between disciplines.[30] This might be the reason why "Les Techniques du Corps" is so convincing as a key text in current discussions about cultural techniques. Lévi-Strauss' foreword has ensured that Mauss is perceived as a highly gifted man who[31] "gets bogged down in a thousand details," as René König, one of the first German Mauss specialists, critically writes, which is why König also called in 1978 for Mauss's "detailed analyses" to be valued as "particularly instructive."[32]

Scholars repeatedly refer to the fragmentary character of Mauss's works (including the "Techniques du Corps"), which Mauss himself titled "essays," "drafts," or "sketches."[33] His ideal *is* to work in fragmentary forms and comes from his declared research interest in wanting to examine and understand "l'homme total," "the 'total man'" (TdC 369; ToB 73). In this sense, there is no

---

humaines et sociales 17 (2014): 1–9, here 5–6. https://journals.openedition.org/trivium/4828 (visited on March 25, 2019). On this point, see also Moebius, *Marcel Mauss*, 130–131.

28 "I want to convey to you what I believe is one of the parts of my teaching which is not to be found elsewhere, that I have rehearsed in a course of lectures on descriptive ethnology" (TdC 365; ToB 70).

29 Yet, Mauss was himself never a traveler; see Moebius, *Marcel Mauss*, 9-10.

30 As the first holder of a chair of social sciences in France, Durkheim founded the journal *L'Annee sociologique*, which served as a kind of laboratory for interdisciplinary collaboration in the service of Durkheim's and Mauss's ideal of collective collaboration. On this point, see also Moebius, *Marcel Mauss*, especially 43–47.

31 Claude Lévi-Strauss, "Introduction à l'Œuvre de Marcel Mauss," in *Sociologie et Anthropologie*, ed. C.L.-S. (Paris: Presses Universitaires de France, 1950), IX–LII; see also Moebius, *Marcel Mauss*, 129–136.

32 René König quoted from Moebius, *Marcel Mauss*, 129; on René König as researcher who worked on Mauss since the 1930s, see Moebius and Nungesser, "La réception de Mauss en langue allemande."

33 Moebius, *Marcel Mauss*, 129.

"pure" discipline for Mauss. The fragmentary character of his works reflects an understanding of research, as Mauss explains, prefacing his remarks[34]:

> En tous cas, il faut procéder du concret à l'abstrait, et non pas inversement./ ... Or, l'inconnu se trouve aux frontières des sciences là où les professeurs "se mangent entre eux," comme dit Goethe (je dis mange, mais Goethe n'est pas si poli). ... Il y a toujours un moment où la science de certains faits n'étant pas encore réduite en concepts, ces faits n'étant pas même groupés organiquement, on plante sur ces masses de faits le jalon d'ignorance: "Divers". C'est là qu'il faut pénétrer. (TdC 365)

> In any case, it is essential to move from the concrete to the abstract and not the other way round. ... Now the unknown is found at the frontiers of the sciences, where the professors are at each other's throats, as Goethe put it (though Goethe was not so polite). ... There is always a moment when, the science of certain facts not being yet reduced into concepts, the facts not even being organically grouped together, these masses of facts receive that posting of ignorance: "Miscellaneous." This is where we have to penetrate. (ToB 70)

Mauss's selection of *faits divers*, or "miscellaneous,"[35] represents the nucleus of his storytelling. At the boundary of research where Mauss quotes Goethe with a wink, while actually marking the emergence of a new scholarly discipline, Mauss establishes himself as a *observateur* ("narrating observer"): Mauss's method of "Ethnologie descriptive" (TdC 365; ToB 70) requires (initially) taking recourse to storytelling. For instance, the scholarly and thus factual speech intended to serve *description* ("descriptive") is in fact a more or less adventurous, narrative quest for knowledge (What are techniques of the body?) and its order (How are techniques of the body to be systematized?). They are adventurous to the extent that the numerous, often extremely personal *exemples* and *anecdotes* tell especially of Mauss's time as a soldier in the First World War and amateur athlete (swimming, mountaineering, boxing); they tell, for example, of sleeping on piles of stones in the field or on a ledge overlooking an abyss in the mountains, and in doing so they testify to a desire for (serial) storytelling that, from the perspective of literary studies, is worth taking seriously despite, or precisely because, of its entertainment value.

---

34 He thus also describes the approach of a current interest of research into cultural techniques that this volume would like to represent, and that revolves around working on concrete materials in defining the concept of "cultural techniques."

35 The French expression "fait divers" refers to a marginal newspaper article. Such an article, for instance, respectively served nineteenth-century authors Gustave Flaubert and Theodor Fontane, to give two examples, as the nucleus for their novels *Madame Bovary* and *Effi Briest*. See also Roland Barthes, "Structure du fait divers," in R.B., *Œuvres complètes*, vol. 1 (Paris: Seuil, 1995), 1306–1319, and Nicolas Pethes' essay in this volume.

I therefore propose understanding "Les Techniques du Corps" as an autohistoriographic case history: case histories (from the Latin *casus*, a medical or legal case) are inevitably disciplinary because they precede the development of new disciplines; they are characterized by narrative modes that elevate empirical observations to a narrative principle. To describe Mauss's use of storytelling at the boundary of a new discipline – and of what is perhaps an equally novel narrative strategy that transforms itself, in its serial character, to become like its object (and not, for example, the other way around) – one could argue, following Nicolas Pethes, that Mauss's attempt at such a survey of a new field of knowledge shows that, "as a case history … ["Les Techniques du Corps"] is not a subsequent or secondary addition to a theorem that is already known as such but rather the sole illustrative material for its justification."[36]

3 Mauss's Body: Occasions for Storytelling

The wit of Mauss's lecture lies above all in the fact that Mauss presents himself as his own *casus*. The seriousness required to appear before a scholarly audience is ensured by the great temporal distance to his youthful self, which the speaker – by then as aged as he was famous – regards with a certain self-irony. This is probably why Erhard Schüttpelz describes Mauss's "inventor's story" as "funny," speaking of "variety numbers for mutual amusement." Thus, we might expect that Mauss's language would generate a *popularizing effect*, possibly resulting from its proximity to the historical genre of case history.

Case histories do not serve to "popularize already existing knowledge but [are] rather a popular medium within an [existing] scientific system, which is always accessed when it is necessary to measure new fields of knowledge."[37] Casuistic modes of storytelling do not draw from the "inventory" of "literature and science [Wissenschaft]" but constitute this inventory in the first place.[38] Case histories deal with exemplary cases, which are to be understood as exemplary narratives, and which both document and enable comparisons. As with Mauss, they seek out "the relationship between individual observation and general law."[39]

---

[36] Pethes, "Vom Einzelfall zur Menschheit," 75.
[37] Pethes, "Vom Einzelfall zur Menschheit," 73.
[38] Pethes, "Vom Einzelfall zur Menschheit," 78. Pethes' arguments concern texts from the late eighteenth century from the field of early psychology.
[39] Pethes, "Vom Einzelfall zur Menschheit," 69–71.

Such an approach can already be observed structurally in "Les Techniques du Corps," in which the narrative part takes up about a quarter of the lecture, because Mauss argues in four steps (I–IV) in order to move from the individual case to the theory or abstract, which is then tested once again in its application to the individual case. His lecture thus follows a systematic procedure, the key words of which Mauss himself names: *raconter* (I, "to tell stories"), *énumérer* (II, III, "to list"), *considérer généralement* (IV, "to judge").

Read in this way, the anecdote about the cinema in New York serves Mauss solely as a *popular* prelude to his remarks, because it remains isolated as an anecdote. Mauss draws above all from a great stock of personal memories as a soldier and young athlete (at the time of his lecture he was over sixty, during the war he was in his forties) or, as here, from his school days:

> Exemple: je crois pouvoir reconnaître aussi une jeune fille qui a été élevée au couvent. Elle marche, généralement, les poings fermés. Et je me souviens encore de mon professeur de troisième m'interpellant: "Espèce d'animal, tu vas tout le temps tes grandes mains ouvertes! "Donc il existe également une éducation de la marche. (TdC 368; my emphasis)
>
> For example: I think I can also recognise a girl who has been raised in a convent. In general she will walk with her fists closed. And I can still remember my third-form teacher shouting at me: "Idiot [You brute] ! why do you walk around the whole time with your hands flapping wide open?" [literally: "You piece of an animal always walking with those big hands open!"] Thus there exists an education in walking, too. (ToB 72)

Just before this passage, Mauss introduces the topic with personal memories of his swimming lessons and the First World War. He relies upon a horizon of experience that he shares with his audience – namely, scholars of his age who had served as soldiers as he did – in order to diagnose a paradigm shift in techniques of the body, using swimming as an example. He does so by comparing the human body with a machine, the steamboat. Physical techniques, as habitualized techniques, are thus seen as subject to the logic of processual stabilization and destabilisation.

> [N]otre génération, ici, a assisté à un changement complet de technique: nous avons vu remplacer par les différentes sortes de *crawl* la nage à brasse et à tête hors de l'eau. De plus, on a perdu l'usage d'avaler de l'eau et de la cracher. Car les nageurs se considéraient, de mon temps, comme des espèces de bateaux à vapeur. C'était stupide, mais enfin je fais encore ce geste: je ne peux pas me débarrasser de ma technique.
> ... Dans le même temps [pendant la guerre] j'ai eu bien des occasions de m'apercevoir des différences d'une armée à l'autre. Une anecdote à propos de la *marche*. Vous savez tous que l'infanterie britannique marche à un pas différent du nôtre: différent de fréquence, d'une autre longueur. ... Pendant près de six mois, dans les rues de Bailleul ..., je vis souvent le spectacle suivant: le régiment avait conservé sa marche anglaise et il la rythmait à la

Techniques of the Body and Storytelling: From Marcel Mauss to César Aira — **201**

française. Il avait même en tête de sa clique un petit adjudant de chasseurs à pied français qui savait faire tourner le clairon et qui sonnait les marches mieux que ses hommes. Le malheureux régiment de grands Anglais ne pouvait pas défiler. Tout était discordant de sa marche. Quand il essayait de marcher au pas, c'était la musique qui ne marquait pas le pas. Si bien que le régiment de Worcester fut obligé de supprimer ses sonneries françaises.

(TdC 366–367; italics in the original; my emphasis)

[H]ere our generation has witnessed a complete change in technique: we have seen the breast-stroke with the head out of the water replaced by the different sorts of *crawl*. Moreover, the habit of swallowing water and spitting it out again has gone. In my day swimmers thought of themselves as a kind of steam-boat. It was stupid, but in fact I still do this: I cannot get rid of my technique.

... In the same period I had many opportunities to note the differences between the various armies. An anecdote about *marching*. You all know that the British infantry marches with a different step from our own: with a different frequency and a different stride. ... For nearly six months, in the streets of Bailleul [*provincial town near Dunkerque*], long after the Battle of the Aisne, I often saw the following sight [literally: spectacle]: the regiment had preserved its English march but had set it to a French rhythm. It even had at the head of its band a little French light infantry regimental sergeant major who could blow the bugle and sound the march even better than his men. The unfortunate regiment of tall Englishmen could not march. Their gait was completely at odds. When they tried to march in step, the music would be out of step. With the result that the Worcester Regiment was forced to give up its French buglers. (ToB 71–72)

This series of five short and detailed stories at the beginning of the lecture (well known to scholars of cultural techniques) can be taken as exemplary for numerous *exemples* and *anecdotes* in Mauss's lecture – and not (only) because they share the entertainment value that is often attributed to them. First, they outline a situation of observation; and second, they show that it is Mauss's recourse to *storytelling* that allows him in the first place (namely in the exposition) to speak of techniques of the body and to attempt to define them, as he himself puts it in the beginning of his lecture.

To do so, Mauss draws primarily on memories of the erudition and training of his own moving (or nonmoving) body, which is explicitly marked as a youthful or young body. With this body, Mauss presents himself as a former passionate (extreme) athlete and thus as the first experimental object in his (discursive) experiment. The examples from ethnology represent reported, foreign knowledge, playing a role in the middle part of the argumentation.[40] By contrast, the biographical parts of the lecture substantiate, on the one hand, a very private and

---

[40] Mauss himself never conducted any field research (see Moebius, *Marcel Mauss*, 9–10), but he did study the idiosyncrasies of various nations (such as Australia, TdC 374; ToB 77), especially during the war. It can be assumed, however, that "Les Techniques du Corps" testifies to a contemporary colonial discourse. Yet this is not easy to verify since Mauss's examples

personal research interest and, on the other, the status of how this (discursive) experiment is set up. Mauss proves himself to be an attentive observer and a witty narrator in determining those techniques of the body that he explicitly describes as a "spectacle" at the end of the series named above.

Mauss's own passion is focused on Alpinism, which he narrates spectacularly; the series of examples he gives is found in the last part of his remarks, where it belongs to the systematic categories of enumeration and judgment, the basis of which he illustrates again using individual cases. Hence their narrative effort is extremely limited, and as a small-scale series they vividly counterbalance the narrative effort from the beginning of the lecture:

> 1. "J'ai dormi debout en montagne. / I have slept standing up [upright] in the mountains." (TdC 379; ToB 81)
>
> 2. "Je peux vous dire que je suis très mauvais grimpeur à l'arbre, – passable en montagne et sur le rocher. / I can tell you that I'm very bad at climbing trees, though reasonable on mountains and rocks." (TdC 381; ToB 83)
>
> 3. "L'histoire des méthodes d'alpinisme est tout à fait remarquable. Elle a fait des progrès fabuleux pendant mon existence. / The history of mountaineering methods is very noteworthy [absolutely remarkable]. It has made fabulous progress in my life-time." (TdC 381; ToB 83; my emphasis)
>
> 4. "La principale utilité que je vois à mon alpinisme d'autrefois fut cette éducation de mon sang-froid qui me permit de dormir debout sur le moindre replat au bord de l'abîme. / The main utility I see in my erstwhile mountaineering was this education of my composure [sang-froid; literally: "cold blood"], which enabled me to sleep upright on the narrowest ledge overlooking the abyss." (TdC 385; ToB 86).

This small-scale series has its own narrative appeal: it revolves around the highly personal limit experiences that Mauss had as an extreme athlete, which – as sleeping on the edge of the abyss – provide the framing and reason for this narrative event, ultimately being hyperbolically exaggerated ("on the narrowest ledge") while being distanced from Mauss's own self ("erstwhile"). Interestingly, the series also raises the question of histories of sport. At the same time, it underpins Mauss's most important conclusion: that techniques of the body consist of an education in "sang-froid" as Mauss says.

Methodologically, "Les Techniques du Corps" is therefore particularly useful in analyzing the relationship between literature and sports, as well as for narratives of the war out of which sports emerged (both historiographically and in

---

in this regard extend to tribes from Australia and New Zealand (Maori) outside the French colonial area.

Mauss's logic).[41] For the storytelling occasions Mauss uses to determine techniques of the body enable us to see the very spectacle of techniques of the body that appear in literary storytelling. I believe that – especially when bodies no longer function as constants, but are hazarded in (literary) experimental arrangements – techniques of the body are not only told but reflected in their possibilities and meanings. If one follows Mauss, such experimental arrangements determine stories of (threatening) damage to the body ("I was *ill* in New York" as a hospital scene [my emphasis]; war situations), on the one hand, and from this perspective, narratives of (extreme) sports, on the other. For instance, Mauss does not initially speak of techniques of the body that he had observed in English and French regiments, but of "elementary [and] sporting techniques" (TdC 367; ToB 72).

César Aira's *Un episodio en la vida del pintor viajero* tells of such an experimental arrangement: the narration is based on the story of the Rugendas family of painters from Augsburg. The great-grandfather – with whom the novel begins as the origin of a dynasty of painters – must re-educate himself from right-handedness to left-handedness following a war injury, which also entails swapping the profession of clockmaker for that of painter. What is described in the rest of the novel is the story of a similarly injured great-grandson, the protagonist of the narrative, who enhances the inherited manual techniques in a highly modern way: transforming them into *écriture automatique*.

4 Techniques of the Body and Maimed Literary Figures: César Airas *Un episodio en la vida del pintor viajero* (2000)

With his cinema anecdote at the beginning of his lecture, Mauss touches, via the apparatus of film, on the question of how techniques of the body are represented. Surprisingly, he does not here establish distance from his story, as he does elsewhere. Rather, he reads the movie as proof of a "trendy" style of walking – the popularity of which ("interesting and famous," in Pethes' terms) he designates with the term "fashion" (*la mode*, TdC 368; ToB 72). Yet in Mauss's Hollywood

---

41 This was shown by an interdisciplinary workshop on "sports and techniques of the body" held at the Universität Erfurt in the summer of 2018; on stories of war and techniques of the body, see my essay "Workaround am eigenen Leib: Das Narrativ der Prothese bei Otto Dix und in der französischen Gegenwartsliteratur," *Ilinx: Berliner Beiträge zur Kulturwissenschaft* 4 (2017): 23–39.

film, the American walking fashion ("la marche américaine") is precisely *not* documented but staged. In this context, a central question concerning techniques of the body and their narrative(s) is the opening and measuring of spaces. It is obvious that, as an ethnologist, Mauss also measures geographical spaces in studying techniques of the body, from the North American city to a provincial city near Dunkerque or the Australian hunting grounds where indigenous people are observed climbing trees (TdC 370; ToB 171).

The novel of the Argentinian writer César Aira, *Un episodio en la vida del pintor viajero* (*An Episode in the Life of a Landscape Painter*), deals with such a measuring of new spaces and fields of knowledge in the sense of a case history. In this text, the main character Moritz Rugendas, a painter and explorer, sketches a *malón* ("Indian raid") in numerous fragments of various scales and dimensions on numerous sheets of paper. At the beginning of the novel, we learn of the "traveling painter" that he invented the new painting technique of the oil sketch that anticipated Impressionists (ELP 157).[42] His art was to be understood in the sense of a "science of landscape":

> Rugendas fue un pintor de género. Su género fue la fisionómica de la Naturaleza, procedimiento inventado por Humboldt …, una suerte de geografía artística, captación estética del mundo, ciencia del paisaje. (EPV 10)

> Rugendas was a genre painter. His genre was the physiognomy of nature, based on a procedure invented by Humboldt …, a kind of artistic geography, an aesthetic understanding [literally: capturing] of the world, a science of landscape. (ELP 147)

To carry out his method of "artistic geography" borrowed from Humboldt, the *pintor viajero* sets out, in the late 1830s, with a cart "of monstrous size" on the arduous crossing into the yet unsurveyed, unmeasured pampas. Along the "straight line" from Mendoza via San Luis to Buenos Aires, he covers a distance of two hundred meters per day – it would take "a lifetime" to reach the goal (ELP 164–165).[43] But he ultimately manages to travel just a quarter of the distance: because a catastrophic riding accident forces him to turn back.

This accident forms the center of the "episode in the life" of Rugendas (EPV 29–35; ELP 172–179), a real historical figure who was a protégé of Alexander von Humboldt. Rugendas's travels to Latin America, such as the trip to Argentina depicted in Aria's text, are documented in numerous letters, drawings, lithographs,

---

[42] In the following I will refer to both the Spanish original text (EPV – Episodio Pintor Viajero) and the English translation (ELP – Episode Landscape Painter) when details are crucial; "una práctica novedosa, la del boceto al óleo" (EPV 18).
[43] "un artefacto de tamaño monstruoso," "línea recta," "vidas enteras" (EPV 23–24).

and paintings.[44] At its outset, the text claims to be a literary narrative but biographically accurate documentation of the "episode" from which the book takes its title, similar to how the painter set out to document the pampas. As the story unfolds, but at the latest after the riding accident that leaves Rugendas with a grave injury on his face, the narrative pace accelerates to finally depict, amid a great din of surrealistic registers, a hallucinating painter among "savages."[45]

I would now like to examine Aira's *Episodio* as inspirationally similar to Mauss' "Les Techniques du Corps," namely as an experimental arrangement for a scientific, medical, artistic, and – on a poetological level – narrative experiment. Ultimately, I want to show to what extent cultural techniques as a method can enrich the literary view of narrative texts, namely, by not circumventing established orders and concepts such as *the grotesque* and *écriture automatique*, but by rendering them describable (in a different way, anew). Once again, the concept of (biographical) case history will help me investigate how the literary text operates, in this context, with techniques of the body, with their stagings and spectacular compressions in narrative form.

Aira's "episode" is a case history when it tells how an injury occurs, how it is treated, and how the injured person begins, under the constant influence of morphine, to ride out again, to draw sketches as a means of research – ultimately with an effort that is no longer geographical, but ethnological – [46] in order to document a *malón*. The accident destabilizes the body and its techniques to such an extent that the painter has no choice but to accept this new normality.

In this sense, the *episodio* tells of the destabilization and stabilization of habitualized techniques of the body, and it therefore does not end with Rugendas's death but with finally achieving the goal of mixing among the "Indians" as one of their own in order to be able to depict them undisturbed. "This last part of the episode" (ELP 226) culminates in an anachronistic – and hence all the more grotesque – cinematic image: "In the depths of that savage night, intoxicated by drawing and opium" (ELP 230), the "monstrous painter" (ELP 227) sits at the campfire amid drunken "savages" to celebrate an earlier military campaign.[47] As

---

[44] Parts of Rugendas's work can be found in the Buenos Aires City Museum, the Augsburger Stadtarchiv, and the Staatliche Graphische Sammlung München.
[45] Throughout the text, and in its historical context, Native Americans are referred to as "savages."
[46] Lévi-Strauss's *Tristes Tropiques* (English title *Tristes Tropiques*, Spanish title *Tristeza del Trópico*, literally *sad tropics*) are mentioned ("tristeza del trópico"; EPV 65; translated as "tropical sadness," ELP 219). First published in France in 1955, the autobiographical text documents Lévi-Strauss's travels to the heart of Brazil between 1935 and 1938.
[47] "Esta parte final del episodio" (EPV 71); „Drogado por el dibujo y el opio, en la medianoche salvaje" (EPV 74); "pintor monstruo" (EPV 72).

the flames once again illuminate Rugendas's injury, the "Indian" faces repeat and multiply the twitching movements of his face (ELP 229)[48]: The painter becomes a blinking machine.

From the outset, the reader is systematically unsettled as to whether the "episode" is, in fact, a historically certain event or whether the *fait divers* (concrete fact) of Rugendas's injury is fictitious.[49] This unsettling is consistently heightened by language that is itself increasingly infected by the madness of the painter and his grotesque body, as the narrative of the injury and how it is dealt with first leads to the narrative of an "Indian" attack, which Rugendas sketches while riding a horse. The text entertains the reader with a wealth of exaggerations, comparisons, and metaphors, producing the impression, from the anachronistic perspective of the late 1990s in which it was written, of a complete reversal of the world through the invasion of the supposed "savage."[50]

The grotesque play with semantics related to techniques of the body (such as walking and riding) and the injuries of the body and its manual techniques (such as drawing and warfare) begins with the story of a maimed painter's hand and the relearning of techniques of the body. The hand of this painter founded the Augsburg dynasty of painters to which Rugendas belongs, and thus also founds Aira's case history. The passage is reminiscent of Miguel Cervantes, a soldier who famously said about the success of his *Don Quixote* that he had lost the ability to move his left hand to the glory of his right.[51]

---

[48] "A la luz bailarina del fuego, sus rasgos dejaban de pertenecerles. Y aunque poco a poco recuperaron cierta naturalidad, y se pusieron a hacer bromas ruidosas, las miradas volvían imantadas a Rugendas, al corazón, a la cara" (EPV 74).

[49] At one point, in any case, the text expressly asserts its factuality: "haya tantos datos, no sólo de los hechos sino de sus repercusiones íntimas, respecto de todo lo que rodeó a este episodio. La documentación era el oficio de Rugendas el pintor, y en alas de la excelencia lograda se le había vuelto una segunda naturaleza a Rugendas el hombre"; "there is so much information directly or indirectly related to this episode, concerning not only the events themselves but also their intimate repercussions. The artist's mastery of documentation had carried over to the rest of his life, becoming second nature to the man" (EPV 41; ELP 187).

[50] On the connection between laughter, fear, and the reversal of the world, see the centrally exemplary work of Mikhail Bakhtin, *Rabelais and his World*, trans. Hélène Iswolsky (Bloomington: Indiana University Press, 1984 [1968]), and, focusing on the Gothic novel, Wolfgang Kayser, *The Grotesque in Art and Literature*, trans. Ulrich Weisstein (New York: Columbia University Press, 1981 [1963]).

[51] Cervantes writes this in the preface to his *Novelas Ejemplares*. He suffered a gunshot wound to his left forearm during the famous naval battle of Lepanto in 1571, which left his hand permanently paralyzed and gave him the nickname "el manco de Lepanto" (the one-handed man of Lepanto). On Cervantes' hand, see Martin von Koppenfels, "Flauberts Hand: Strategien der

Su bisabuelo Georg Philip Rugendas (1666–1742), fue el iniciador de la dinastía de pintores. Lo hizo por haber perdido en su juventud la mano derecha; la mutilación lo incapacitó para el oficio de relojero, que era el tradicional de su familia ... Debió aprender a usar la mano izquierda, y manejar con ella lápiz y pincel. Se especializó en la representación de batallas, y tuvo un formidable éxito derivado de la precisión sobrenatural de su dibujo ... .

(EPV 8; my emphasis)

It was Johan Moritz's great-grandfather, Georg Philip Rugendas (1666–1742) who founded the dynasty of painters. And he did so as a result of losing his right hand as a young man. The mutilation rendered him unfit for the family trade of clockmaking, in which he had been trained since childhood. He had to learn to use his left hand, and to manipulate a pencil and brush [with it]. He specialized in the depiction of battles, with excellent results, due to the preternatural [supernatural] precision of his draftsmanship ... .

(ELP 144–145)

Aira's text is thus marked by the techniques of destabilization and stabilization of a "series of assembled actions" (TdC 372; ToB 76), and the effective use of the hand is here relearned through those processes that Marcel Mauss calls "sang-froid": processes of focalization and stabilization. In this regard, here the injury of the body transforms primary cultural techniques (writing, drawing, painting) back into techniques of the body, and secondary cultural techniques (introduced by the abandoned profession of the clockmaker) become impossible. From the very beginning of the novel, the newly learned technique is connected with the concept of the "sobrenatural" (literally: "supernatural"), which means not only unnatural but superhuman. With the grandson, the text will take this aspect to the extreme.

This is the genealogical prehistory, but the injury of the great-grandson is even worse: "He was not like his ancestor, who had to start over with his left hand [literally: by educating his left hand]. If only he had been so lucky!" (ELP 181).[52] His facial disfigurement burns itself "straight into his nervous system" (ELP 174),[53] just as the landscape, for the "making" of which the painter first set out,[54] directly burns itself onto his retina. The grotesque semantics that the text emphasizes from

---

Selbstimmunisierung," *Poetica* 34 (2002): 171–191, here 183–184; and Volker Klimpel, "Berühmte Amputierte," *Würzburger medizinhistorische Mitteilungen* 23 (2004): 313–327, here 314.

52 "No era como su antepasado, que había tenido que educar la mano izquierda; ¡ojalá lo hubiera sido!" (EPV 37)

53 "lo absorbió directamente con el systema nervioso" (EPV 31).

54 "... se le abría un mundo ya hecho, y también, a la vez, por hacer, más o menos como le sucedío por la misma época al joven Darwin"; „.... the world that opened before him was roughly mapped out yet still unexplored [literally: a world already made, and also, at the same time, yet to be made], much as it was, at around the same time, for the young Charles Darwin." (EPV 9; ELP 146; my emphasis).

the outset are produced by the logic that techniques of the body and cultural techniques become, in such an automatic process of branding, *superfluous*. However, as a medical case history the "episode" is told spatially and made observable in a reversal of the world, thereby using all the means of the grotesque to produce the spectacularity of accident and injury: carnivalesque slapstick and the horror of the Gothic novel: *Frankenstein*.[55]

On his ride of discovery, Rugendas ends up in a rock formation called "El Monigote: the puppet [on a string]." In this "circo máximo" ("vast amphitheater") of interlayered rocks, rider and horse are struck by two Frankenstein-like lightning bolts[56] – "lit up with electricity, Rugendas witnessed the spectacle of his body shining" – and literally fuses to an assemblage: horse and rider together now turn a "voltereta", a somersault (this notion misses in the translation; ELP 176–174).[57] The panicked animal then drags the fallen painter along, his foot in the stirrup: "a classic riding accident" (ELP 177).[58] However, the sudden appearance of a thunderstorm of apocalyptic proportions turned the sky lead-grey, scribbled a "horseshoe" across the sky ("herradura"; this notion misses in the translation; EPV 31; ELP 174) and thus anticipated a reversal of the worlds that now finally becomes manifest in the painter lying motionless with his face to the earth. His travel companions find him with a horribly disfigured face and exposed nerves (EPV 34; ELP 178). The party is forced to return to San Luis, where Rugendas suffers a medical procedure of assemblage: he is "stiched up," literally: "sewn" back together as well as possible ("lo estaban cosiendo"; EPV 35; ELP 179) – the notion of assemblage stems from the French verb *assembler* which can also mean "to sew together." Yet the hospital is not a place of healing but inhabited by incurable, disabled "monsters" (ELP 180).[59] Rugendas thus remains

---

[55] On the carnival, see Bakhtin, *Rabelais and his World*; on the Gothic novel of German Romanticism, in particular, see Kayser, *The Grotesque*; on slapstick see Ilka Becker, "*Agencement* und *Amusement*: Duchamp, Slapstick und retroaktive Geschichten der Moderne," in *Unmenge: Wie verteilt sich Handlungsmacht?*, ed. I.B., Michael Cuntz, and Astrid Kusser (Munich: Fink, 2008), 93–121.

[56] As in *Frankenstein*, in Aira we also find two flashes of lightning in an apocalyptic thunderstorm. Mary Shelley, *Frankenstein* (Oxford: Oxford University Press, 2008 [1831]), 76.

[57] "Volaron unos veinte metros, encendidos ... la caída no fue fatal ... sino que la magnetización del pelaje de la bestia había hecho imán, y Rugendas quedó montado en toda la voltereta ... hombre y bestia se encendieron de electricidad. Rugendas se vio brillar, espectador de sí mismo por un instante de horror" (EPV 31–32).

[58] "accidente que non por repetido (es un clásico de la equitación de todos los tiempos)" (EPV 33).

[59] "habitado por media docena de monstruos, mitad hombre mitad animales, producto de accidentes genéticos acumulados. Ellos no tenían cura" (EPV 35).

in his "estado semiinválido" (EPV 42; ELP 189), and the expedition is marked by spatial reversal: "The only thing that had changed was Rugendas's face. And the direction" (ELP 186).[60]

The text creates this "semi-invalid monster" by staging techniques of the body. Since *Frankenstein* (1831), the monstrous body, patched together as an assemblage of body parts, has been associated with technological inventions and electricity.[61] As a result, Frankenstein's monster, especially his face, is not only repulsively ugly but also extraordinary in a frightening, grotesque way because of its nonhuman techniques of the body, which culminate in the ability to run with superhuman speed.[62] By contrast, Rugendas's techniques of the body – literally: the "sleepwalking steps of a monster" (ELP 197)[63] – appear exceptionally slow and stiff, disconnected from the brain, like the movements of a marionette (as in "El Monigote"), and the painter's destabilized and therefore grotesque techniques of the body transform his living space into a chaotic laboratory:

> Caminaba con pasos rígidos, pero todo el cuerpo parecía afectado de inconexión ... El cuarto se había transformado, por acción de sus movimientos, en el laboratorio de un sabio loco que se proponía lograr alguna clase de transformación del mundo. (EPV 50–51)

> Rugendas walked stiffly [with stiff steps] but all the parts of his body seemed to be working loose [his whole body seemed disconnected] ... His movements had transformed the room into a mad scientist's laboratory where some transformation of the world was being hatched." (ELP 199)

As a personal union of the monster and Frankenstein, Rugendas tries to establish a new order of his body, i.e., *conexíon*, or in other words: the "sang-froid" claimed by Mauss that bypasses pain and drug. This first affects his habits of seeing, which are severely impaired by both. Upon documenting his first *malón* in the blazing sun, accompanied by his friend Krause, also a painter, Rugendas pulls a mantilla over his head, a lace cloth that fragments his gaze. Instead of experiencing a deterioration – and this is the grotesque turn – Rugendas finds that his vision even improves and, with it, the observing eye of the researcher:

---

60 "Lo único que había cambiado era la cara de Rugendas. Y la dirección" (EPV 40).
61 This is how the "spark of being" that animated the monster through the "instruments of life" has always been read. *Frankenstein* (2008), 57; see also the introduction by J.K. Joseph, viii. For an exemplary overview of the vast secondary literature on *Frankenstein*, see the latest study on this subject by Andrew Smith: "Scientific Contexts," in *The Cambridge Companion to Frankenstein*, ed. A.S. (Cambridge: Cambridge University Press, 2016), 69–83.
62 See, for instance, Shelley, *Frankenstein*, 57–58 and 76.
63 "las evoluciones adormecidas del monstruo"; translated as "the monster's somnolent bumbling" (EPV 49; ELP 197).

"He had never seen better in his life" (ELP 206).[64] This grid of threads laid over the retina repeats the "landscape's structuring grid of horizontal and vertical lines … overlaid by man-made traces" (ELP 155).[65]

Furthermore, this allusion to the cartographic, written, two-dimensional measurement of the world spatially clarifies this *malón*. Body and landscape fuse into a practice of measuring on paper through the staging of techniques of the body and cultural techniques:

> They [Krause and Rugendas] put their papers into the saddlebags and spurred the horses on … . It seemed they would have to resign themselves to seing the Indians in miniature, like lead soldiers. Yet the details were all there, violently impressed on their retinas, magnified [literally: amplified] on the paper [1]. …
>
> It was like wandering from room to room at a party, from the living room to the dining room, from the bedroom to the library, from the laundry to the balcony, all full of noisy, happy, more or less drunk guests, looking for a place to cuddle or trying to find the host to ask him for more beer. Except that it was a house without doors or windows or walls, made of air and distance and echoes, of colors and landforms.
>
> This stream could have been the bathroom. The Indians wanted to charge but they were retreating; the white men wanted to retreat, but in order to do so they had to charge (in order to scare the enemy more effectively with their bangs). This ambivalence was driving the horses crazy; they plunged into the water, splashed about, or simply stopped to drink, very calmly, while their riders yelled themselves hoarse in simultaneous flight and pursuit. The skirmish had an infinite (or at least algebraic) plasticity, and since Rugendas was observing it at closer range this time, his flying pencil traced [literally: he threw/catapulted his pencil at/ toward] details of tense and lax muscles, wet hair clinging to supremely expressive shoulders [2] … Everything sketched in this explosive present was material for future compositions, but although it was all provisional, a constraint came into play. It was as if each volume captured in two dimensions on the paper would have to be joined up with the others, in the calm of the studio, edge to edge, like a puzzle, without leaving any gaps. And that was indeed how it would be, for the magic of drawing [3] turns everything into a volume, even air. Except that for Rugendas the "calm of the studio" was a thing of the past; now there was only torment, drugs and hallucinations.
>
> The savages scattered in all directions, and four or five came climbing up the knoll where the painters had stationed themselves. Krause drew his revolver and fired twice into the air; Rugendas was so absorbed that his only reaction was to write BANG BANG on his sheet of paper [4]. (ELP 206–209; my emphasis)[66]

---

[64] "Nunca había visto mejor" (EPV 55).
[65] "Al cuadriculado de verticales y horizontales que componía el pasaje se superponía el factor humano, también reticular" (EPV 16).
[66] (EPV 56–58). [1] "los detalles estaban hahí, hacían una violenta impresión en sus retinas y se amplificaban en el papel" (EPV 56). [2] "lanzaba el lápiz en escorzos de musculatura distendida y contraída, cabelleras mojadas pegándose a hombros sumamente expresivos …" (Spanish *lanzar*, to throw, to catapult; EPV 58). [3] "la magia del debujo" (EPV 58). [4] "Rugendas estaba tan compenetrado que se limitó a escribir en su hoja: BANG BANG" (EPV 58).

The chaos of the large-scale raid carried out by an "Indian" tribe in the barely developed area in order to steal animals from local farmers is, in being compared to a party in a bourgeois living room, linguistically tamed in a grotesque way, made representable in the first place. The "wandering" from room to room (with Certeau: practicing the map)[67] produces, in a grotesque reversal, a wild ride for the purpose of documenting through drawing. Closely related to this are the themes of fragmentation. The "Indian" bodies transferred into the two-dimensionality of paper as fragments by the manual techniques of the artist (the stock of paper is carried in saddlebags and is as mobile as the painters riding on horseback) spring directly from the maimed painter's body. It is the fragmented view caused by the injury that permits an exclusively fragmented perception of the "Indian" body, which also appears as such on paper. The logic here is that these fragments could be put together at home in the studio like a puzzle into a whole. Rugendas's injury, however, has made this impossible; for him, this fragmented chaos has become permanent. And in this context, the representability of this chaos proves equally impossible on both the textual and narrative levels: it is marked by a number of referential short circuits, as – to use Aria's language – *inconexión* between the signifier and the signified. The main character thereby infects the text. For example, the comparison of the *malón* with a stroll through a house with neither walls nor doors nor windows literally disappears into thin air.

The climax of the novel is marked by a scaling and dimensioning staged by the techniques of an intoxicated body that is both riding and drawing, namely, as the painter depicts the raging "Indians" with his pencil, at first from a great distance and then at close range. The riding warriors, who first appear as "lead soldiers," are depicted in their athletic physicality; the brush itself becomes a piece of war equipment and acquires a stubborn power to act in its detachment from the body, when Rugendas literally "throws" his pencil at the "Indian" body to capture the spectacle of its muscles.

Since the impressions now inscribe themselves directly into Rugendas's retina and the techniques of the painter's maimed body are characterized by a grotesque *inconexión* that extends into the drawing pencil, what is seen comes to paper as if by magic, as if run through an amplifier, or through a technical apparatus (Spanish: *amplificar*). The man struck by lightning is now apparently able to leap forward over seventy years of art history: after the realistic, meticulous style of his drawing before the injury, he leaves behind a work that is almost surrealist in its scale and content. This transformation culminates in a Dadaist-like writing

---

[67] "ir recorriendo los ambientes de una casa" (EPV 57). Michel de Certeau, *Arts de faire: The Practice of Everyday Life*, trans. Steven Rendall (Berkeley: University of California Press, 1984).

scene, with a pencil replacing the depiction of the *malón* with all its agents and components by the registration of its sounds, captured in a doubling of four capital letters: BANG BANG.

The impression of a surrealist "picturesque" [68] *écriture automatique* is produced by a staging of techniques of the body that, at the end and as the climax of the case history, no longer appear here at all. The pencil has taken on a life of its own. Rugendas's accident was much worse than that of his ancestor, who had to laboriously train his left hand, and yet Rugendas not only no longer needs to practice any techniques of the body: he no longer needs any hands at all. The spectacle of the cultural technique of writing is based on a technique of the body that is no longer present. This absence culminates in the claimed invention of the avant-garde *per se*, literally *avant la lettre*: Impressionism, Expressionism, Surrealism, and Dadaism together, from the machine to the paper. The "nonhuman" cultural technique of the monstrous painter Rugendas corresponds, in this respect, to the superhuman technique of the body of Shelly's raging monster: "The body is a strange thing, and when it is caught up in an accident involving nonhuman forces, there is no predicting the result" (ELP 179).[69]

The result of my thoughts, starting with Mauss's lecture "Les Techniques du Corps" up to this reading of a literary text, could be summarized as follows: I have read both Mauss's lecture and Aria's novel as case histories to show that techniques of the body, as is presumably true for all cultural techniques, constitutively depend on storytelling. This is because they are obviously generally based on an individual case and an associated case study. Both are constitutively characterized by an anthropological, or more pointedly: by a more or less personal, biographical core, which can be analytically generalized and abstracted to a scientific exploration and documentation as (with Pethes) "measuring new fields of knowledge."

The dilemma of media studies to whether (to quote Harun Maye) "the techniques of the body can be subsumed under cultural techniques or, vice versa, cultural techniques can be derived from the techniques of the body" could be answered from the perspective developed here: it depends. For here we are not dealing with a logic of either/or, but first of all with a question of focalizing: techniques of the body are essentially cultural techniques; cultural techniques

---

[68] "La escena era sumamente pintoresca. El carboncillo empezó a volar sobre el papel." ("The scene was picturesque in extrem [the scene of the *malón* in de Argentinian landscape, as well as the following scene of drawing]. The stick of charcoal began to fly across the paper"; EPV 56; ELP 207).

[69] "El cuerpo es una cosa extraña, y cuando lo afecta un accidente donde actúan fuerzas no humanas, nunca se sabe cuál será el resultado" (EPV 34–35).

are techniques of the body if the focus is not on the question of differentiation but rather if the analytical spotlight is directed toward the body itself.[70] With the discussion of habitualized cultural techniques or techniques of the body, the question of focusing is then followed by the question of concrete processes of stabilization and destabilization of such techniques, which can clearly be analyzed in many different ways. Mauss' examples of the steamship and of "sang-froid," which at first glance seem so completely different, come together in the allusion to the "trained" body as an ostensibly technical, "cold-blooded" object, whose production (i.e., stabilization as such), as shown by Mauss's Alpinism series and also his example of swimming, requires, however, significant processual efforts of habituation or dehabituation.

Harun Maye's dilemma, however, has even greater implications, because this question necessarily entails discussing a fundamental understanding of media:

> Should cultural techniques be understood primarily as a physically habitualized skill, possibly supported by tools and instruments that then appear as an extension of these *techniques of the body*? or are they primarily *media techniques* that, derived from dominant basic media (writing, image, number), constantly generate new media and cultural innovations?[71]

In Aira's *Episodio*, cultural techniques are finally not supported by an instrument and thus become an extension or exteriorization of techniques of the body, as the body is separated from its instrument (of writing, drawing, arithmetic) in a magical act *(inconexión)*. The instrument becomes autonomous, as if by magic. In this respect, the text introduces the question of precisely this connection, as Maye poses it here, taking it to an absurd conclusion. First, as a stubborn resistance of such tools and instruments (in other words: the *agency* of objects). And second, as a conjunction of techniques of the body and cultural techniques, which are discussed as manual techniques or rather: as techniques of handedness. With the concept of the "result," the effectiveness of both and thus their connection to the medium remains unknown. The sheet of paper on which the drawing and writing take place remains a mass of fragments that cannot be assembled into a meaningful whole. Nothing remains of these so-called basic media in their Dadaist sequence (BANG BANG) but symbolic onomatopoeia.

So what is the conclusion to draw from all of this? With its main character, the text comprehends the immortality of art as manual techniques inherent to human

---

[70] Mauss speaks of the body not as an "instrument," but as "man's first and most natural technical object, and at the same time technical means"; "Le corps est ... plus exactement, sans parler d'instrument, le premier et le plus naturel objet technique, et en même temps moyen technique, de l'homme" (TdC 372; ToB 75).
[71] Maye, "Was ist eine Kulturtechnik," 122.

beings. But at its ending, it stages a scene of the Anthropocene, the beginning of which is marked by the zero hour of midnight. The question of techniques of the body and cultural techniques has been settled inasmuch as the body becomes a machine.

In this context, the text constructs an argument via the idea of algebraic lines, thus abstracting Rugendas's initial route to measure the unknown pampas. For the injured painter, as one reads in the text, there would only be a perpetual beginning in the sense of starting from scratch, symbolized namely by the broken line, the *inconexión*, just as Rugendas's accident forced him to turn back on his journey from San Luis to Buenos Aires. His friend, the painter Krause, on the other hand, "by virtue of his health, was moving along an unbroken line, a continuum, without beginning or end" (ELP 225)[72]: Because of his injury – and "this last part of the episode is even more inexplicable then the rest" – Rugendas thus thinks about his approaching death at the end of the novel (ELP 226).[73] In his imagination, all humankind dies out after his fragments, which indeed cannot be put together like a puzzle, and make the long journey to Europe by sea, where they end up in galleries on the wall.

> They [the Indians] might have killed him. A minor detail. In any case, by the time his correspondents saw the resulting pictures, that is by the time his work reached European galleries or museums, he would certainly be dead. The artist, as artist, could always be already dead. There was something absurd about trying to preserve his life. An accident, big or small, could kill a man, or a thousand, or a thousand million men at once. If night were lethal, we would all die shortly after sunset. Rugendas might have thought, as people often do: "I have lived long enough," especially after what had happened to him. Since art is eternal, nothing is lost. (EPV 71–72; ELP 226–227)

Ultimately, the ending of the story does not emphasize the eternity of the medium but the perpetuated manual technique, which has here become posthuman. Rugendas sits drawing by the campfire:

> [E]staba tan concentrado en los dibujos que no se daba cuenta de nada. Drogado por el dibujo y el opio, en la medinoche salvaje, efectuaba la contigüidad como un automatismo más. El procedimiento seguía actuando por él. (EPV 74)

> [H]e was so absorbed in his work that he remained oblivious to the rest. In the depths of that savage night, intoxicated by drawing and opium, he was establishing contact [literally: contiguity] as if it were simply another reflex. The procedure went on operating through him. (ELP 229–230)

---

[72] "Era Krause, no él, quien por efecto de la salud estaba en una línea única, un continuo, sin comienzo ni fin" (EPV 70).
[73] "Esta parte final del episodio fue más inexplicable todavía que el resto" (EPV 71).

The fragmentation (and abstraction) of the body (as Rugendas imagines it in drawing the "Indians") is historically linked to the synchronization of the body with the machine. Although the working hand lost importance, the concept of artistic creativity developed as a reaction.[74] The "automatismo más," presented here as the coronation of the artist at the end of the novel, consists in making the connection (*conexión; contigüidad*) that was impossible for the human body to accomplish with its manual techniques. These no longer play a role. The body becomes the machine of *el arte por el arte (l'art pour l'art)*.[75]

Already for the observer Mauss, and in how he reproduces them (narratively), techniques of the body are spectacular. This spectacularity is not only grounded in the visual. With his anecdote about the soldiers' gait to marching music, Mauss captures the spectacle of techniques of the body as both an observable and an audible phenomenon, just as Aira makes the techniques of the body, as a writing scene, visible and audible, and with them the connection between techniques of the body and cultural techniques. "BANG BANG." The end of Aira's novel once again makes it clear that a technique of the body, especially in its "medial compressions,"[76] not only is a narrative but also depends on narration. Cinema (as in Mauss) as well as literary texts can be optimal experimental arrangements to discuss techniques of the body and their spaces (and ultimately their mediality).

---

[74] This was a process that is widely considered to have been triggered by the industrial age that emerged in the nineteenth century. See Becker, "*Agencement* und *Amusement*: Duchamp, Slapstick und retroaktive Geschichten der Moderne," 121.
[75] See Bernhard Siegert, "Türen: Zur Materialität des Symbolischen," *Zeitschrift für Medien- und Kulturforschung* 1, 1 (2010): *Kulturtechnik*: 151–170.
[76] Schüttpelz, "Körpertechniken," 114.

Collectives

Bettine Menke
Writing Out – Gathered Up at a Venture from All Four Corners of the Earth: Jean Paul's Techniques and Operations (on Excerpts)

The relationship of poetic works and their hermeneutics to cultural techniques is nothing if not unproblematic, since from the middle of the eighteenth century onwards works are conceived as genuinely existing without any preconditions and thus as self-contained, traceable solely to authorial creation. From the perspective of cultural techniques, by contrast, poetic works are ascribed to their preconditions and marginal conditions. They participate in "the expulsion of spirit from the humanities" ("Austreibung des Geistes aus den Geisteswissenschaften"), to cite and follow Friedrich Kittler.[1]

Of course, philology as such is initially determined by cultural techniques: it emerges in the margins of pages; it is formed in glossaries and commentaries on the text that generate the text to begin with. But this will not be my topic for now. Nor will it be the operations that constitute the text and its always "problematic limit between an inside and an outside"[2] by acting at and from its margins, by repeatedly creating a distinction between text and nontext that is thus compelled to reappear in the text. But when philology mutated into a hermeneutics of literary works in being regrounded in the individual that is supposed to be the creator of the work, all memory of its own techniques, as well as those of literature, were erased in this mutation. One might speak of a hermeneutic oblivion of cultural techniques (and not least of all those of philology).

---

[1] Friedrich Kittler (ed.), *Austreibung des Geistes aus den Geisteswissenschaften* (Paderborn: Schöningh, 1980); on the situation following and reacting against Kittler (under the motto of "rephilologization") see Nicolas Pethes, "Actor-Network-Philology? Papierarbeit als Schreibszene und Vorgeschichte quantitativer Methoden bei Jean Paul," in *Medienphilologie: Konturen eines Paradigmas*, ed. Friedrich Balke and Rupert Gaderer (Göttingen: Wallstein, 2017), 199–224, here 199, 201f.; see Gaderer and Balke, "Introduction," in the same volume, 7–22, here 9, 17f.; see also the contribution by Balke, esp. 59–63, 65; on cultural techniques, see the contributions by Christina Lechtermann, Harun Maye, Dietmar Schmidt, and Julia Kursell.
[2] Jacques Derrida, "This is Not an Oral Footnote," in *Annotation and Its Texts*, ed. Stephen A. Barney (Oxford: Oxford University Press, 1991), 192–205, here 196.

Translated by Michael Thomas Taylor

∂ Open Access. © 2020 Bettine Menke, published by De Gruyter. This work is licensed under a Creative Commons Attribution 4.0 International License.
https://doi.org/10.1515/9783110647044-013

Yet literary texts refer to cultural techniques in many ways, and not only in referring to their own historical a priori conditions. Rather, they reflect this relationship to their cultural techniques – or, to invoke another powerful metaphor, they fold it into themselves: reading and writing, reading that performs itself in writing (excerpting), writing that is reading (citing), organizing by writing, referring to book pages, handling books, turning their pages. These texts thematize such operations and the media that they handle; they embed scenes devoted to their presentation; and they relate themselves to them in many different ways.[3]

This is the context in which I will situate the writing of Jean Paul, which intervenes into the discourse network (*Aufschreibesystem*) 1800 characterized by the linkage of genius and expression, work and authorship, inasmuch as it allows techniques and devices – a "technical system"[4] – to come between reading and writing *and* also asserts that technical system in the texts in various ways. And moreover, such that they process the "problematic boundary" between inner and outer, text and nontext.

Johan Paul Friedrich Richter read by excerpting. He began his first volume of excerpts in 1778 when he was still in school, and by 1823, had compiled nearly 110 quarto notebooks comprising around 12,000 pages.[5] Intending to become

---

[3] This has become such a frequent topic in literary studies and its media history and theory (going back even behind the letters to the operational breaks of writing [Derrida], to the paper, the gesture of scribbling, etc.) that an overview will not even be attempted.

[4] Hans-Walter Schmidt-Hannisa, "Lesarten: Autorschaft und Leserschaft bei Jean Paul," *Jahrbuch der Jean-Paul-Gesellschaft* 37 (2002): 35–52, here 41.

[5] See Eduard Berend, "Anmerkungen zu den Exzerpten," in *Jean Pauls Sämtliche Werke, Historisch-kritische Ausgabe*, ed. E.B. et al. (Weimar: Böhlau, 1927ff), here HKA II.1 (1928), 23–31, XVII-XX (bibliographic reference is given as HKA, with division indicated with Roman numerals and the volume with Arabic numerals); Berend, "Jean Pauls handschriftlicher Nachlaß: Seine Eigenart und seine Geschichte," in *In libro humanitas: Festschrift for Wilhelm Hoffmann*, ed. Ewald Lissberger, Theodor Pfizer, and Bernhard Zeller (Stuttgart: Klett, 1962), 336–346. Partial documentation has been offered by Götz Müller, *Jean Pauls Exzerpte* (Würzburg: Königshausen & Neumann, 1988); see also Müller, "Jean Pauls Privatenzyklopädie: Eine Untersuchung der Exzerpte und Register aus Jean Pauls unveröffentlichtem Nachlaß," *International Archive for the Social History of German Literature* 11 (1986), 73–114, here 76. The excerpt notebooks are held by the Manuscript Department of the Staatsbibliothek Berlin (Preußischer Kulturbesitz); see *Der handschriftliche Nachlaß Jean Pauls und die Jean-Paul-Bestände der Staatsbibliothek zu Berlin – Preußischer Kulturbesitz*, part 1 (Fasc. I–XV), ed. Ralf Goebel, part 2 (Fasc. XVI–XXVI), ed. Markus Bernauer and Ralf Goebel (Wiesbaden: Harrassowitz, 2002). Götz Müller has provided a bibliographical overview of the excerpts from the tables of contents, which Jean Paul himself put at the end of many of the volumes; Michael Will notes the "blatant incompleteness" of this list: "Jean Pauls Exzerpthefte elektronisch," http://computerphilologie.uni-muenchen.de/jg02/will.html (visited on January 29, 2020). For the history and situation of the edition, see Andreas B. Kilcher, *mathesis und poiesis: Die Enzyklopädik der Literatur 1600–2000* (Munich: Fink, 2003), 38–86; the philological problems posed by the edition

an author, he began in 1782 to excerpt "unsorted excerpts of texts that one encounters by chance while shifting quickly between parallel readings" of many books at once ("ungeordnete Textauszüge, wie man ihnen zufällig bei schnell wechselnder Lektüre [von mehreren Büchern parallel] begegnet").[6] His practice is reminiscent of early modern forms of *miscellanea*. As in the tradition of *loci communes*, material is apparently drawn from sources of all kinds: religion, philosophy, natural history, medicine – neither empirically gleaned knowledge nor knowledge verified by criticism, but a "convolute of observations and opinions since antiquity."[7] The "technical system" of writing out passages, of administering (*Verwaltung*) and "handling" ("Handhabung")[8] the excerpts, of writing as operations on them,[9] forms a "heterogeneous ensemble" of writing.[10] It requires recursive writing operations: "Excerpts of excerpts" ("Exzerpte aus Exzerpten") that add to the collections and "registers" ("Register") providing access to what has been compiled in writing.[11] The first diary that records the fruits of these readings without any order is doubled in a second, learned accounting (*Buchführung*) that sorts things by assigning subject headings.[12] "Tables of contents"

---

concern in particular the excerpts ('before' the work); their edition as transcripts was produced electronically: http:// www.jp-exzerpte.uni-wuerzburg.de (visited on January 29, 2020).
6 Müller, *Jean Pauls Exzerpte*, 322; see also Will, "Jean Pauls Exzerpthefte elektronisch," 4f.
7 Müller, *Jean Pauls Exzerpte*, 323, 326f.; Müller, "Jean Pauls Privatenzyklopädie," 80f.
8 These act out mobility; see Jean Paul, *Vorschule der Ästhetik*, in *Sämtliche Werke*, ed. Norbert Miller, (Munich: Hanser, 1974ff.), here I.5, 199f. (bibliographic reference is given with Roman numerals for the division and Arabic numerals for the volume).
9 Jean Paul provides technically operative *fictions*, such as in *Vorschule der Ästhetik*, I.5, 202f. (also quoted in the contribution by Nicolas Pethes); *Titan*, I.3, 167, as well as in several novels; see below.
10 Rüdiger Campe, "Die Schreibszene, Schreiben," in *Schreiben als Kulturtechnik*, ed. Sandro Zanetti (Berlin: Suhrkamp, 2012), 269–282, here 271 (originally published in *Paradoxien, Dissonanzen, Zusammenbrüche: Situationen offener Epistemologie*, ed. Hans Ulrich Gumbrecht and Karl Ludwig Pfeiffer [Frankfurt am Main: Suhrkamp, 1991], 759–772, here 760); see also Martin Stingelin, "'Schreiben:' Einleitung," in *"Mir ekelt vor diesem tintenklecksenden Säkulum": Schreibszenen im Zeitalter der Manuskripte*, ed. M.S. (Munich: Fink, 2004), 7–21, here 8f., 13, 17f.; see also Pethes, "Actor-Network-Philology," 208, on Jean Paul 214ff.
11 For example, in Fasc. VIIIb (17).
12 This learned bookkeeping of excerpts (see Vincentius Placcius, *De Arte Excerpendi: Vom gelahrten Buchhalten* [Hamburg: Liebezeit, 1689]) with reference to the double-entry bookkeeping from Italy, is documented from Comenius through the seventeenth century to Lichtenberg; see Helmut Zedelmaier, "Buch, Exzerpt, Zettelschrank, Zettelkasten," in *Archivprozesse: Die Kommunikation der Aufbewahrung*, ed. Hedwig Pompe, and Leander Scholz (Cologne: DuMont, 2002), 38–53, here 44f.; Zedelmaier, "Lesetechniken. Die Praktiken der Lektüre in der Neuzeit," in *Die Praktiken der Gelehrsamkeit in der Frühen Neuzeit*, ed. Martin Mulsow and H.Z. (Tübingen: Niemeyer, 2001), 11–30, here 22; Bernhard Dotzler, *Papiermaschinen: Versuch über Commu-*

and "registers" present points of access (*Zugriffe*) to what has been stored, they make this access[13] possible and write themselves down beside it – in a convolute labeled "register" itself comprising 1244 pages and ordered alphabetically by keyword, without any categorizing systemization of what it conveys.[14] And of course, this requires a "register of registers" ("Register der Register.")[15] The apparatuses duplicate themselves and multiply. Repertoires of self-instructions such as the "register [or list] of what I have to do" ("Register dessen was ich zu thun habe") are evidence not only of the circularity of the recursions, but also of their tendency toward paradox (at least in a temporal sense).[16] The "register" of that "what I have to do" just in the first place lists: "1 *This* register to be made *now*" ("1 *Dieses* Register *ietzt* zu machen")[17] – an instruction that opens up a paradoxical circularity because to follow it "now" defers all other tasks ("what I have to do") infinitely or results in an irresolvable blockage.

---

nication & Control in Literatur und Technik (Berlin: Akademie-Verlag, 1996), 561–566; Wolfgang Schäffner, "Nicht-Wissen um 1800: Buchführung und Statistik," in *Poetologien des Wissens um 1800*, ed. Joseph Vogl (Munich: Fink, 2010), 123–144, here 132.

13 Regarding Jean Paul's "methods of organizing knowledge" and "retrieval": "In addition to concise tables of contents at the conclusion of the notebooks, which have a length of sixty to eighty pages," and a "reference system … within the various preliminary versions and collections of texts," [one has] "the carefully managed index of keywords" (Will, "Jean Pauls Exzerpthefte elektronisch," 3) as well as further excerpts registering excerpts: "Auszüge aus den Exzerpten (Studien I)," Fasc. VI (HKA II.9, in preparation); Fasc. VII: "Actio (1–2): Bausteine und Wörterlisten," 4. "Sammlung von Bausteinen," etc. "Stoff zu satirischen Erfindungen," "Launestoff," 15–18, "Einfälle" (1–4)," Fasc. VI, vol. 5: "Synonyme" (48–50); Fasc. VIIIb (16): "Album": "Metaphern (Einfälle und Arbeitsnotizen)," and others.

14 See the list of the "Registerartikel," Fasc. IIIa and IIIb, in *Der handschriftliche Nachlaß Jean Pauls*, vol. 1, 26–33; in *Jean Pauls Exzerpte*, Götz Müller lists more than 150 such register entries, documenting several in exemplary form (with several misreadings, omissions, and unmarked addenda from Jean Paul's abbreviations). "In the course of time these indexes grew to comprise nearly 2000 manuscript pages, and were themselves made accessible for finer differentiation by additional indexes" (Will, "Jean Pauls Exzerpthefte elektronisch," 3). See, for example, the register entry "machine(s)" (Manuscript Dept. Staatsbibliothek Berlin), call number: NL Jean Paul fasc. IIIb2, sheets 7–13; electronic transcription: http://www. jp-exzerpte.uni-wuerzburg.de/index.php?seite=register/maschine&navi=_navi/reg05 (visited on January 29, 2020); and Fasc. VIIIb: "Wörterbücher und Register (Studien IV)"; Fasc. VIIIb: "Unalphabetisches Register" (15); Fasc. VIIIb: "Register" (17); and Fasc. VIIIb: "Wörterbücher und Register (Studien IV)"; Fasc. VIIIb: "Unalphabetisches Register" (15); Fasc. VIIIb: "Register" (17).

15 Fasc. VIIIb. (18f.); on the ways in which the various registrations of the excerpts possibly functioned, with examples of Jean Paul's working methods, see *Der handschriftliche Nachlaß Jean Pauls*, vol. 1, 72f.

16 See *Studier-Reglement*, HKA II.6, 561f., 563f., 566, 568, 574; *Register dessen was ich zu thun habe*, HKA II.6, 551, 558.

17 Jean Paul, *Register dessen was ich zu tun habe*, HKA II.6, 551f. (emphasis B.M.).

The "register" entries are organized as a list, without any hierarchy, to foster recombination, connections that transpose and transect linear sequentiality, and the potential of contact in transverse (re)reading between heterogeneous entries that are set apart from each other – contact that is capable of bringing forth effects of witty (*witzig*) invention in the combinations of moveably joined discrete elements – which is how Jean Paul actually used his compilations, wandering through them in reading and digressing, in order to write his texts, mainly his novels.[18]

The word "baroque" is used again and again to categorize how Jean Paul processes knowledge.[19] "The excerpting system that Jean Paul develops and that fundamentally shapes his reading – by his own admission, he 'hardly [reads] anything anymore ... except what is to be excerpted' ('er lese fast nichts mehr ... als was zu exzerpieren ist') – clearly functions according to the model of the baroque collectanea,"[20] treasure troves of *topoi*, a source for *inventio*.[21] Jean Paul's texts are thus characterized by the *obsolete* (*veraltete*) form of knowledge of the "polyhistorians" ("Polyhistor'n"), the outmodedness (*Veraltetsein*) of which is made clear by the opposing postulate of the lyric I that Christian Fürchtegott Gellert posits in his poem "Der Polyhistor," (1746): "I studied nothing but myself / Nothing but my

---

[18] The excerpts served to support "the invention of witty similarities" and "of central motifs in the work" (Müller, *Jean Pauls Exzerpte*, 338ff.); see also Hendrik Birus, *Vergleichung: Goethes Einführung in die Schreibweise Jean Pauls* (Stuttgart: Metzler, 1986), 52ff. In *this* sense, the excerpts have been consulted by research for quite some time, for example, by Wilhelm Schmidt-Biggemann, *Maschine und Teufel: Jean Pauls Jugendsatiren nach ihrer Modellgeschichte* (Freiburg: Alber, 1975), 104–111; Peter Sprengel, "Herodoteisches bei Jean Paul: Technik, Voraussetzungen und Entwicklung des 'gelehrten Witzes,'" *Jahrbuch der Jean-Paul-Gesellschaft* 10 (1975): 213–248, references 221f., 234ff., among others.
[19] See Georg Wilhelm Friedrich Hegel, *Vorlesungen über die Ästhetik I*, in G.F.W.H., *Werke*, ed. Eva Moldenhauer and Karl Markus Michel (Frankfurt am Main: Suhrkamp, 1970), vol. 13, 382. "Jean Paul's texts are considered baroque ... in their numerous allusions, digressions, and associations, in a word: in their 'polyhistoric,' encyclopedic dispositive" (Kilcher, *mathesis und poiesis*, 381, see also 389ff.).
[20] Schmidt-Hannisa, "Lesarten," 38 (with a quote from Jean Paul's letter to Christian Otto, HKA III.3, 56). The dictum "He did not read anything that he would not have excerpted" belongs to the excerptable *loci communes* deriving from Pliny the Elder; see Helmut Zedelmaier, "De ratione excerpendi: Daniel Georg Morhof und das Exzerpieren," in *Mapping the World of Learning: The Polyhistor of Daniel Georg Morhof*, ed. Francoise Waquet (Wiesbaden: Harrassowitz, 2000), 75–92, here 84.
[21] "The Baroque author ... was required to compile and record 'facts' (Realien) from the most diverse areas of erudition, which would then offer the *poeta doctus* material for rhetorical-poetic inventio" (Schmidt-Hannisa, "Lesarten," 38; see also Kilcher, *mathesis und poiesis*, 381).

heart."²² Jean Paul cites not only polyhistorians²³ but also the *outdated* (*veraltete*) ordering and processing of knowledge and its forms.²⁴

If Hegel finds, in Jean Paul, only "baroque combinations of things which are laying incoherently asunder and whose relations into which his humour brings them together are almost indecipherable" ("barocke Zusammenstellungen von Gegenständen, welche zusammenhanglos auseinander liegen, und deren Beziehungen, zu welchen der Humor sie kombiniert, sich kaum entziffern lassen"),²⁵ then *baroque* – it goes without saying – does not denote an epoch but the quality of being askew and grotesque.²⁶ The outdatedness, the obsoleteness of the form

---

22 "Ich habe nichts als mich studiert / Nichts als mein Herz" (Christian Fürchtegott Gellert, *Fabeln* [Carlsruhe: Schmieder, 1774], 222f.); the polyhistorian in this text does not even understand that things are getting serious: "he hears it and laughs" ("hörts und lacht"); on the perspective of the eighteenth century toward "'polyhistory' ... as a practice that has been left behind" ("Polyhistorie' ... als eine überlebte Praxis"), see Helmut Zedelmaier, "Von den Wundermännern des Gedächtnisses: Begriffsgeschichtliche Anmerkungen zu 'Polyhistor' und 'Polyhistorie,'" in *Die Enzyklopädie im Wandel vom Hochmittelalter bis zur frühen Neuzeit,* ed. Christel Meier (Munich: Fink, 2002), 421–450, here 422, 421–424, 435–450.

23 See Daniel Georg Morhof, who – in *De excerpiendi ratione* (Lübeck, 1688, 4th ed. 1747) – once again presented instructions for how to excerpt correctly; see Zedelmaier, "Von den Wundermännern des Gedächtnisses," 433f., 441, 446; Zedelmaier, "De ratione excerpendi," 78ff.; references in Jean Paul, *Freiheitsbüchlein*, II.2, 833; *Leben des Quintus Fixlein*, I.4, 88, see 126; see Müller, *Jean Pauls Exzerpte*, 193–210; Schmidt-Hannisa, "Lesarten," 38f.; Kilcher, *mathesis and pioesis*, 381, 384f.; Robert Stockhammer, "Zeichenspeicher: Zur Ordnung der Bücher um 1800," in *Das Laokoon-Paradigma. Zeichenregime im 18. Jahrhundert,* ed. Inge Baxmann, Michael Franz, and Wolfgang Schäffner (Berlin: Akademie-Verl., 2000), 45–63, here 53; Magnus Wieland, "Jean Pauls Sudelbibliothek: Makulatur als poetologische Chiffre," *Jahrbuch der Jean-Paul-Gesellschaft* 46 (2011): 97–119, here 116f. See also the "last polyhistorian," the Viennese librarian Michael Denis, *Einführung in die Bücherkunde* (1778); see also Jean Paul, *Exzerpte*, Fasc. IIa, "Geschichte" 1, vol. 1 (1782), 3f.; see Müller, *Jean Pauls Exzerpte*, 345; Lothar Müller, *Weiße Magie: Die Epoche des Papiers* (Munich: Hanser, 2012), 180.

24 In addition to the operations of collecting and storing, the assemblage of knowledge must also enable organization, recovery and invention as the allocation of general material ... with its particular object of speech, as well as access and retrieval as needed, forms and practices of circulation, see Frank Büttner, Markus Friedrich, and Helmut Zedelmaier, "Zur Einführung," in *Sammeln, Ordnen, Veranschaulichen: Zur Wissenskompilatorik in der Frühen Neuzeit,* ed. F.B., M.F., and H.Z. (Münster: Lit, 2003), 7–12, here 7–10; Kilcher, *mathesis und poiesis*, 381, 384f.; Zedelmaier, "De ratione excerpiendi," 78–90; Stefan Rieger, *Speichern/Merken: Die künstlichen Intelligenzen des Barock* (Munich: Fink, 1997), 42–72, here 68–72.

25 Hegel, *Vorlesungen über die Ästhetik I*, 382; Hegel, *Aesthetics: Lectures on Fine Art*, trans. T.M. Knox, vol. 1 (Oxford: Clarendon Press, 1975), 295, translation modified. Hegel's talk of "humor" (especially in its Romantic, "subjective" form) includes irony and wit (*Vorschule der Ästhetik*, especially §§ 31–35).

26 For example, this is how Camille Mélinand uses the word "baroque": "ce qui fait rire," in/as a transgression of the customs of thinking: "Pourquoi rit-on? Étude sur la cause psychologique du

of knowledge and the disfigured, misshapen, coincide. Kant, too, called the "erudition" that wants to know everything "gigantic ["g i g a n t i s c h e"], which is ... often cyclopean, that is to say, missing one eye: namely the eye of true philosophy, by means of which reason purposive uses this mass of historical knowledge, the load [of books] of a hundred camels."[27] As "cyclopean," this knowledge – which is not grounded in principles of reason – is monstrous measured by the metaphorically invoked anthropomorphism, while conversely its regime (*Regierung*) would be figured in the human-like two-eyed face.[28] In 1798, "baroque" can evidently mean "thrown apart in a tumble" („durcheinander geworfen"), as Bouterwek writes of Jean Paul in 1798: "Querfeldein wird erzählt, phantasiert, philosophiert, sarkastisiert, gerührt und amüsiert" ("All across the country, his work recounts, fantasizes, philosophizes, sarcasticizes, affects, and entertains").[29] At the beginning of *Les mots et les choses*, Michel Foucault cites the

---

rire," *Revue des Deux Mondes*, (February 1, 1895): 612–630, here 613ff.). Friedrich Schlegel speaks of Jean Paul's "groteske[n] Porzellanfiguren seines wie Reichstruppen zusammengetrommelten Bilderwitzes" ("grotesque porcelain figures that his pictorial wit drums together like imperial soldiers") (Friedrich Schlegel, *Athenäum-Fragmente*, in *Kritische Friedrich-Schlegel-Ausgabe* [= KFSA], ed. Ernst Behler et al. [Paderborn: Schöningh, 1958ff.], vol. 2, 246 [no. 421]; *Philosophical Fragments*, trans. Peter Firchow [Minneapolis: University of Minnesota Press, 1991], 85).

27 "die ... oft zyklopisch ist, der nämlich ein Auge fehlt: nämlich das der wahren Philosophie, um diese Menge des historischen Wissens, die [Bücher-]Fracht von hundert Kamelen, durch die Vernunft zweckmäßig zu benutzen"; Immanuel Kant, *Anthropologie in pragmatischer Hinsicht*, in I.K., *Schriften zur Anthropologie, Geschichtsphilosophie, Politik und Pädagogik 2*, Werkausgabe XII, ed. Wilhelm Weischedel (Frankfurt am Main: Suhrkamp, 1977), 546f. [hereafter: WW XII]; Immanuel Kant, *Anthropology from A Pragmatic Point of View*, trans. and ed. Robert B. Louden (Cambridge: Cambridge University Press, 2006), 122. With the metaphor of the "load [of books] of a hundred camels," Kant characterizes the polyhistorians as "prodigies of memory" ("Wundermänner des Gedächtnisses") who "did not possess the power of judgment suitable for choosing among all this knowledge in order to make appropriate use of it" ("weil sie vielleicht die, für das Vermögen der Auswahl aller dieser Kenntnisse zum zweckmäßigen Gebrauch angemessene, Urteilskraft nicht besaßen") (Kant, *Anthropology*, 78; Kant, *Anthropologie*, WW XII, 489).
28 Diderot takes the monster as a "model for his 'combinatoire'" and compares, in the *Encyclopédie* article, "the work of a *homme de genie* who ... produces new discoveries and generates new connections with a monstrous act" (Inge Baxmann, "Monströse Erfindungskunst," in *Das Laokoon-Paradigma*, ed. I.B., Franz, Schäffner, 404–417, here 412 and 414f.).
29 Friedrich Bouterwek, cited from Peter Sprengel (ed.), *Jean Paul im Urteil seiner Kritiker: Dokumente zur Wirkungsgeschichte Jean Pauls in Deutschland* (Munich: Beck, 1980), 25, XXI. However, see Friedrich Schlegel: "Die wichtigsten wissenschaftlichen Entdeckungen sind *bon mots* der Gattung. Das sind sie durch die überraschende Zufälligkeit ihrer Entstehung, durch das Kombinatorische des Gedankens und durch das Barocke des hingeworfenen Ausdrucks" ("The most important scientific discoveries are *bon mots* of the kind – are so by the surprising contingency of their emergence, the combinatoric of thought, and the baroqueness of their casual expression") (Schlegel, *Athenäum-Fragmente*, KFSA, vol. 2, 200 [no. 220]; *Philosophical Fragments*, 47).

"'certain Chinese encyclopedia'" feigned by Jorge Luis Borges less as an example of the old order of knowledge but because, in "the amazement at this taxonomy," "we apprehend in one leap" what "is demonstrated as the exotic charm of another system of thought ... the stark impossibility of thinking *that*."[30] Things look very similar in the scholarly meta-list of (lists or registers of) scholarly types that is drawn up in Jean Paul's novel *Quintus Fixlein*:

> [Daß] *Bernhard* [ein Register] von Gelehrten [gegeben], deren Fata und Lebenslauf im Mutterleibe erheblich waren – daß *Bailet* die Gelehrten zusammengezählt, die etwas hatten schreiben wollen – und *Ancillon* die, die gar nichts geschrieben – und der Lübecksche Superintend *Götze* die, die Schuster waren, die die ersoffen usw. Das ... sollte ... uns zu ähnlichen Matrikeln und Musterrollen von andern Gelehrten ermuntert haben ... – z.B. von Gelehrten, die ungelehrt waren – von ganz boshaften – von solchen, die ihr eignes Haar getragen – von Zopfpredigern, Zopf-Psalmisten, Zopfannalisten etc. – von Gelehrten, die schwarzlederne Hosen, von andern, die Stoßdegen getragen – von Gelehrten, die im eilften Jahre starben – im zwanzigsten – einundzwanzigsten etc. – im hundertundfunfzigsten, wovon er gar keine Beispiele kenne, wenn nicht der Bettler Thomas Parre herangezogen werden solle – ...

> [That] *Bernhard* [provided] an [register or list] of learned men whose fate and life in the womb were considerable, that *Bailet* tallies the learned men who had wanted to write something, and *Ancillon* those who wrote nothing at all, and the Lübeck superintendent *Götze* those who were shoemakers, and those who drowned, etc. That ... ought to have ... encouraged us to create similar registers and catalogues of other learned men ... for example, of learned men who were unlearned – of especially wicked ones – of those who wore their own hair – of pigtail preachers, pigtail psalmists, pigtail annalists, etc. – of learned men who wore black leather pants, and others who carried rapiers – of learned men who died in their eleventh year – in their twentieth year – in their twenty-first, etc. – in their hundred-fiftieth year, of which he claims not to know any examples unless the beggar Thomas Parre should be considered – ...[31]

When Jean Paul's texts cite this form of knowledge that had just become outdated (*veraltet*), they *cite* it *as* the outdated or obsolete (*veraltete*) practice of reading that excerpts text in writing it out, of writing of and through excerpts: in a circle of reading and writing.[32] The erudition bound to "dead words," as is said, is called "pedantry," which – as Montaigne already wrote – circulates

---

[30] Michel Foucault, *The Order of Things: An Archeology of the Human Sciences* (New York: Vintage Books, 1994), xv (translation modified); the reference is to Jorge Luis Borges's "El idioma analítico de John Wilkins."
[31] Jean Paul, *Leben des Quintus Fixlein*, I.4, 82f.
[32] Schmidt-Hannisa, "Lesarten," 39.

words fruitlessly as money of account or game tokens.[33] As pedantry, erudition gains "ethnological reality"[34] in Jean Paul's novels, to use Rüdiger Campe's phrase, in that the "scripturality and historicity of literature" ("Schriftlichkeit und Geschichtlichkeit der Literatur") "emerges" "in the interior of (the literariness of) literature" ("im Inneren des Literarischen") as the "Komik" of the pedants, whereas it is written forth/away in writing (*fortgeschrieben*) in puns, which accentuate writing in its materiality and signs as signifiers contrary to the spirit of meanings.[35]

The form of knowledge organized by the practice of excerpting "began [so one reads] to become obsolete in the eighteenth century,"[36] inasmuch as knowledge now presupposed the critical distinction[37] between what it determines to be "true" science and the "history of doctrines and authorities."[38] What took the place of polyhistoricism, one also reads, were the new *Encyclopedias*.[39] Encyclopedias

---

[33] The "mere circulation" of "the exchanged citations of knowledge" marked the 'old order of knowledge,' according to Michel de Montaigne's "Du pédantisme" (in : M. de M., *Œuvres complètes*, ed. Albert Thibaudet and Maurice Rat [Paris: Gallimard, 1962], vol. 1, *Essais* I, 25, 132–143): it's no good to anything, "qu'à compter et jeter" (136); "to reckon with, or to set up at cards," trans. Charles Cotton, in *Montaigne's Essays* (London: Barker, 1743), vol 1, 144; see also Rüdiger Campe, "Schreibstunden in Jean Pauls Idyllen," in *Fugen: Deutsch-Französisches Jahrbuch für Text-Analytik* 1 (1980): 132–170, here 143f. Montaigne apparently does not believe that they have the force to usure/proliferate that comes from the difference in the citation; see, among others, Kilcher, *mathesis und poiesis*, 381, 389ff.; Wieland, "Jean Pauls Sudelbibliothek," 102, 100–104.
[34] "[E]thnologische Wirklichkeit," for example, in *Leben des Quintus Fixlein* und *Wutz*, see Campe, "Schreibstunden in Jean Pauls Idyllen," 156ff.
[35] Campe, "Schreibstunden in Jean Pauls Idyllen," 157, 160.
[36] Müller, *Jean Pauls Exzerpte*, 333.
[37] See Foucault, *The Order of Things*, 74–75, 80–81.
[38] Müller, *Jean Pauls Exzerpte*, 333; on the tension between the power of judgment and the *memoria* of the polyhistorians (Kant, *Anthropologie in pragmatischer Hinsicht*, WW XII, § 31, 489), see Büttner, Friedrich, and Zedelmaier, "Zur Einführung," 7, 9f.
[39] This assertion is hardly correct. On the one hand, "encyclopedia" has an older history before the modern revival of the Greek term and meant the area of the *artes liberals*; see Joseph von Hammer-Purgstall, "Encyklopädie," in *Allgemeine Encyclopädie der Wissenschaften und Künste in alphabetischer Folge von genannten Schriftstellern bearbeitet*, ed. Johann Samuel Ersch and Johann Gottfried Gruber (Leipzig: Gleditsch, 1818–1889 [discontinued after 167 volumes were completed]); unchanged reprint, Graz 1971), 204–208. A 'boom' of 'encyclopedias' occurred in the early modern period, Johann Heinrich Alsted's encyclopedia article also includes: "representation ... b) of all 'philosophical' disciplines ..., c) everything that can be taught"; quoted from Jürgen Mittelstraß, "Enzyklopädie," in *Enzyklopädie: Philosophie und Wissenschaftstheorie*, ed. J.M., vol. 1 (Stuttgart and Weimar: Metzler, 1995), 557–562, here 558f., see 557; Christel Meier, "Enzyklopädie," in *Reallexikon der deutschen Literaturwissenschaft*, vol. 1 (revised edition of the *Reallexikon der deutschen Literaturgeschichte*), ed. Klaus Weimar (Berlin and

shape the completeness of knowledge in two competing models: totality can be conceived as a sum, an aggregate, or (the tendency of philosophy) as the systematicity of classification.[40] The encyclopedic endeavors of the seventeenth and eighteenth centuries continue to be caught up in the conflict between the complete gathering of knowledge and its (possible) ordering. Jean Paul was an "enthusiastic reader of encyclopedias" (as he was of the polyhistorians), which he also "excerpted all along"[41]: Zedler's *Universal-Lexikon* (1732–1752), Bayle's *Historisches und Critisches Wörterbuch*,[42] the famous *Encyclopédie ou dictionnaire raisonné des sciences, des arts et de métiers* by Diderot and others (1751–1780),[43] Krünitz's *Oeconomische Encyclopädie oder allgemeines System der Land- Haus- und Staatswissenschaft* (1773–1858),[44] Hederich's *Gründliches Mythologisches Lexikon* (1770), the *Enzyklopädische Wörterbuch oder alphabetische Erklärung aller Wörter aus fremden Sprachen, die im Deutschen angenommen sind, wie auch aller in den Wissenschaften, bei den Künsten und Handwerkern üblichen Kunstausdrücke*, begun

---

New York: De Gruyter, 1997), 450–453, here 451f. On the history of the encyclopedia, see Ulrich Ernst, "Standardisiertes Wissen über Schrift und Lektüre, Buch und Druck: Am Beispiel des enzyklopädischen Schrifttums vom Mittelalter zur Frühen Neuzeit," in *Die Enzyklopädie im Wandel vom Hochmittelalter bis zur frühen Neuzeit*, ed. Christel Meier (Munich: Fink, 2002), 451–494, here 451f., 453ff.; Gilbert Hess, "Enzyklopädien und Florilegien im 16. und 17. Jahrhundert: Doctrina, Eruditio und Sapientia in verschiedenen Thesaurierungsformen," in *Wissenssicherung, Wissensordnung und Wissensverarbeitung: Das europäische Modell der Enzyklopädien*, ed. Theo Stammen and Wolfgang E. J. Weber (Berlin: Akademie-Verlag, 2004), 39–57, here 44f., 49f. On the other hand, the transition is inherently contradictory – with singular, incommensurable effects.
40 See Mittelstraß, "Enzyklopädie," 560, and Christoph Meinel, "Enzyklopädie der Welt und Verzettelung des Wissens: Aporien der Empirie bei Joachim Jungius," in *Enzyklopädien der Frühen Neuzeit*, ed. Franz Eybl et al. (Tübingen: Niemeyer, 1995), 163–187, here 177; for the perspective of philosophy, see Ulrich Dierse, *Enzyklopädie: Zur Geschichte eines philosophischen und wissenschaftstheoretischen Begriffs* (Bonn: Bouvier, 1977).
41 See Müller, *Jean Pauls Exzerpte*, 345.
42 Pierre Bayle, *Dictionnaire historique et critique*, originally 1694–1697. Gottsched provided the translation into German: *Herrn Peter Baylens, weyland Professors der Philosophie und Historie zu Rotterdam, Historisches und Critisches Wörterbuch, nach der neuesten Auflage von 1740 ins Deutsche übersetzt; auch mit einer Vorrede und verschiedenen Anmerkungen sonderlich bey anstößigen Stellen versehen, von Johann Christoph Gottscheden* (Leipzig: Breitkopf, 1741; repr. Hildesheim et al.: Olms, 1997).
43 *Encyclopédie ou dictionnaire raisonné des sciences, des arts et de métiers par une sociéte de gens de lettres: mis en ordre & publ. par Diderot & par d'Alembert*, nouvelle impression en facsimilé de la première edition de 1751–1780 (Stuttgart-Bad Cannstatt: Frommann, 1988 and 1995).
44 Jean Paul excerpted the parts edited by Krünitz, 1–73 (1773–1796), according to Kilcher, *mathesis und poiesis*, 128; in "Studier-Reglement," Jean Paul also names the reading of "part 76 to part 200" ("76. Th[eil] bis 200"), HKA II.6, 563; see Apparat 144.

by Gottlob Heinrich Heinse (1793–1805), Ersch's und Gruber's *Allgemeine Encyclopädie der Wissenschaften und Künste* (begun in 1818),[45] and others. All these works are ordered alphabetically by keyword and not systematically by subject.[46] They manifest a "lexical diversity" of collective authorship, a multi-handedness (*Vielhändigkeit*) of writing[47] that conflicts with systematic order.[48] On the one hand, they assertively revert to the contingency of alphabetic order.[49] In Roland Barthes' saying: "L'ordre alphabétique ... refoule toute origine" ("The alphabetical order ... banishes every origin").[50] And on the other hand, via this order they are able to provide selective access (*Zugriff*) to their assemblage, without any hierarchy,

---

45 See Jean Paul, *Blumen-, Frucht- und Dornenstücke oder Ehestand, Tod und Hochzeit des Armenadvokaten F. St. Siebenkäs* (I.2, 283); Müller, *Jean Pauls Exzerpte*, 345; Andreas B. Kilcher, "Enzyklopädische Schreibweisen bei Jean Paul," in *Vom Weltbuch bis zum World Wide Web: Enzyklopädische Literaturen*, ed. Waltraud Wiethölter, Frauke Bernd, and Stephan Kammer (Heidelberg: Winter, 2005), 129–147, here 141f.; Kilcher, *mathesis und poiesis*, 128–131.
46 On the alphabetic as a principle of ordering(ing) that replaces restrictive, obsolete or failing systematic order, see Foucault, *The Order of Things*, 37–38; Kilcher, *mathesis und poiesis*, 179, 287, 386; Stockhammer, "Zeichenspeicher," 54f.
47 Diderot emphasizes this in the article on "encyclopédie," *Encyclopédie*, vol. 5 (1755), 635–648, here 635f.
48 See Mittelstraß, "Enzyklopädie," 559f., who nevertheless characterizes only the nineteenth century in this way. The new development, for which the *Encyclopédie ou Dictionnaire raisonné* edited by Diderot and d'Alembert is exemplary, follows a program that is conflicting with itself, for which an old "understanding of order and system" of knowledge was (still) "effective": "the project of the *Encyclopédie*, with its systematic networking of knowledge by adherance to universal operations of knowledge, simultaneously leans on it and differentiates itself from it" (Baxmann, "Monströse Erfindungskunst," 412).
49 This, Neumeister argues (about Bayle's *Dictionnaire*), is disorder "hidden by the alphabet": Sebastian Neumeister, "Unordnung als Methode: Pierre Bayles Platz in der Geschichte der Enzyklopädie," in, *Enzyklopädien der Frühen Neuzeit*, ed. Fyhl et al., 188–199, here 192ff.
50 *Roland Barthes par Roland Barthes* (Paris: Seuil, 1975), 151; *Roland Barthes, by Roland Barthes*, trans. Richard Howard (Berkeley: University of California Press, 1977), 148. Barthes writes: "Tentation de l'alphabet: adopter la suite des lettres pour enchainer des fragments, c'est s'en remettre à ce quit fait la gloire du langage ...: un ordre immotivé (hors de toute imitation)" (*Roland Barthes par Roland Barthes*, 150); "Temptation of the alphabet: to adopt the succession of letters in order to link fragments is to fall back on what constitutes the glory of language ... : an unmotivated order (an order outside of any imitation)" (*Roland Barthes, by Roland Barthes*, 147). The dissolution of the logic of inherent coherence is the "euphoria" of the alphabet. In Jean Paul's *Leben Fibels*, one can see an "experiment" with the "form ... of knowledge of the alphabet"; see Kilcher, *mathesis und poiesis*, 287, 390; Bettine Menke, "Alphabetisierung: Kombinatorik und Kontingenz, Jean Pauls Leben Fibels, des Verfassers der Bienrodischen Fibel," *Zeitschrift für Medien- und Kulturforschung* 1, 2 (2010): 43–59, here part I.

avoiding a systematicity that had become obscure, and/or would represent knowledge as a completed enclosedness that had become problematic.[51]

Reading by excerpting is *Stellen-Lektüre*: a dissociative reading of isolated *lieux*; it happens in browsing/leafing through the pages.[52] Jean Paul reads encyclopedias as he does all books: he comprehends them as lists, as conglomeration (or gatherings) of equivalent sections, which is what the *encyclopedias* already were, in regards to their alphabetic ordering, in their incomplete diversity of equally accessible entries.[53] He opens up space for the *euphoria of the alphabet*, described by Barthes: "L'alphabet est euphorique: fini l'angoisse du 'plan', l'emphase du 'développement', les logiques tordues."[54] Reading, which browses, leafs and skips (pages), realizes the "recombinatory force of connection" that is set free in what it disassembles, and its "excesses" ("Überschüsse"),[55] the model for which was provided by the *Encyclopédie*,[56] with its cross-references that loose slips and slip boxes will translate into apparatuses and operations for handling them.

Jean Paul gave himself the prescription: "Alle Morgen in einem Gedanken- und Geschichtsbuche nur blättern, nicht lesen" ("Every morning, just leaf through, don't read, in a book of thoughts and of history").[57] The act of turning the pages to which Jean Paul commits himself is a discontinuous, dissociative

---

[51] "All particles are subject to the same conditions: the times and rights for access ensure on equal terms that nothing is not, or everything is, easily accessible and thus remains unforgotten" (Rieger, *Speichern/Merken*, 73).

[52] Jean Paul: "Leaf through your dictionaries daily to scout out your riches" ("Blättere täglich deine Wörterbücher zu[r] Auskundschaftung deines Reichthums durch") (HKA II.6, 558); Jean Paul forces himself to "leaf through excerpts" ("Blättern in Exzerpten") in "Studier-Reglement" (1795) (under 2., 15.) (HKA II.6, 574).

[53] On Bayle's *Dictionnaire*, see, for instance, Neumeister, "Unordnung als Methode," 194f., 198.

[54] "... et pour la suite de ces atomes, rien que l'ordre millénaire et fou des lettres françaises (qui sont elles-mêmes des objets insensés – privés de sens)" (*Roland Barthes par Roland Barthes*, 150); "The alphabet is euphoric: no more anguish of 'schema,' no more rhetoric of 'development,' no more twisted logic ... and as for the succession of these atoms, nothing but the age-old and irrational order of the French letters (which are themselves meaningless objects – deprived of meaning)" (*Roland Barthes, by Roland Barthes*, 147).

[55] Markus Krajewski, *ZettelWirtschaft: Die Geburt der Kartei aus dem Geiste der Bibliothek* (Berlin: Kulturverlag Kadmos, 2002), 69.

[56] The multitude of *cross-references* in the *Encyclopédie* and its double indexes (of *mots* and of *choses*) manifest an cross-linking of entries, oblique to the alphabet, but precisely not to its system of classification; see Diderot, "Encyclopédie," 136–141; Baxmann, "Monströse Erfindungskunst," 415; Winfried Menninghaus, "Vom enzyklopädischen Prinzip romantischer Poesie," in *Vom Weltbuch bis zum World Wide Web*, ed. Wiethölter, Bernd, and Kammer, 149–163, here 159.

[57] Jean Paul, *Register dessen was ich zu thun habe*, HKA II.6, 559.

practice designed to effect (re)combinations.[58] Reading as wandering through (*Durchschweifen*),[59] roaming through (*Durchstreifen*), or rummaging through (*Durchstöbern*) (to allude to metaphors from Michel de Montaigne)[60] comprises "planless" ("planlos") movements that, precisely as such, enable *Einfälle* (incidences of inventions, or sallies),[61] the same way that "prey is roused" ("Wild aufgejagt")[62] from the thicket in the forests (*silvae*) of the materials, or that "oddities like butterflies" ("Sonderbarkeiten wie Schmetterlinge") may be rummaged

---

[58] Jean Paul, HKA II.6, 558, as well as HKA II.6, 574. Jean Paul: "Sometimes, I leaf through one book, sometimes through another, without any order, plan, or context" ("Da blättere ich einmal in diesem Buch, ein andermal in einem andern, ohne Ordnung, ohne Plan und ohne Zusammenhang"); cited from Götz Müller, "Mehrfache Kodierung bei Jean Paul," *Jahrbuch der Jean-Paul-Gesellschaft* 27 (1992): 67–91, here 77. On leafing through as a nonlinear mode of reading, see the contributions to Jürgen Gunia, Iris Hermann (ed.), *Literatur als Blätterwerk: Perspektiven nicht-linearer Lektüre* (St. Ingbert: Röhrig, 2002), in particular, Ecckehard Schumacher's introduction, "Aufschlagesysteme 1800/2000," 23–45. Dietmar Schmidt emphasizes the turning of the page (*umblättern*) as nonreading, "Umblättern statt lesen," in *Medienphilologie*, ed. Balke and Gaderer, 146–155.

[59] Jean Paul, *Levana*, I.5, 843.

[60] Kilcher, *mathesis und poiesis*, 138f.; on Montaigne's *Essais*, see Jean Paul, *Essays de Montaigne: Jean Paul oder meine letzten und unaufhörlichen Werke*, *Merkblätter 1816/17*, no. 131, HKA II.6, 315, 350, 369.

[61] The word 'Einfall', in spite of dictionaries indicating otherwise, cannot be translated at all by inspiration or idea; it is (an witty) incident, and always also has a physical implication of violence and contingency deriving from its literal meaning "to fall in", also to invade. Translator's note: Jakob and Wilhelm Grimm's *Deutsches Wörterbuch* has it ranging from physical acts—("ruina, einsturz"), (violent) interruption, attack, or incursion ("irrupti, incursus"), and others —to a sudden cognitive insight: "subita cogitatio, ein plötzlicher, schneller, kluger, guter, glücklicher einfall": a sudden, quick, smart, happy *einfall* (reprint of the first edition [Leipzig 1862], Munich: dtv, 1984), vol. 3, col. 170–172.

[62] "In wandering planlessly in the planless forays of the fantasy, the game is often roused that well-planned philosophy can use in its well-ordered household" ("Durch das Planlose [sic] Umherstreifen durch die planlosen Streifzüge der Phantasie wird nicht selten das Wild aufgejagt, das die planvolle Philosophie in ihrer wohlgeordneten Haushaltung gebrauchen kann") (Georg Christoph Lichtenberg, "Sudelbücher II: Materialhefte, Tagebücher," in Lichtenberg, *Schriften und Briefe*, vol. 2, ed. Wolfgang Promies [Munich and Vienna: Zweitausendeins, 1994], 286: J 1550). Jean Paul rejects reservations against wit that "hunts for wit": "For is there anything in art that need not be hunted, but glides to one's tongue already caught, plucked, and fried? [Yet] in the moment that the effort becomes visible, it was in vain; and wit that is sought out can no more count as being found than the hunting hound can (count) for the game" ("Gibt es denn etwas in der Kunst, wonach man nicht zu jagen habe, sondern was schon gefangen, gerupft, gebraten auf die Zunge fliegt? Wo [aber] die Anstrengung sichtbar ist, da war sie vergeblich; und gesuchter Witz kann so wenig für gefunden gelten als der Jagdhund für das Wildpret") (I.5, 198); Kant decries the "hunt for witty sayings" ("Jagd auf Witzwörter"; *Anthropologie in pragmatischer Hinsicht*, WW XII, 539f.; *Anthropology*, 116).

("aufgestöbert"),[63] falling to the aimless by chance or accident. Wandering happens in writing in digression and divagation.[64] The act of leafing through (instead of reading) models reading as an operation of separating and joining, of dissociating and linking across distances, which – in leaping across the distances between excerpts,[65] and in inserting gaps into reading – makes excerpts a scene for invention.

Even if Jean Paul cites this outdated (*veraltete*) form of knowledge, it is not left unchanged.[66] Jean Paul's writing is not aimed at a great deal of *knowledge*, even if it makes an appeal to "a certain poly-knowledge" ("gewisses Vielwissen") that art certainly demands ("zumute[t]") and is allowed to demand[67] – because,

---

[63] Jean Paul, *Die Taschenbibliothek*, II.3, 769–773, here 771; they designate the contingency of the "selection" of what becomes part of the learned collection; see Krajewski, *ZettelWirtschaft*, 11, 66f., 74.

[64] See Stefan Matuschek, "Exkurs," in *Historisches Wörterbuch der Rhetorik*, ed. Gert Ueding (Tübingen: Niemeyer, 1996), vol. 3, 126–136, here 131f.

[65] For the witty jumps or hops, see Jean Paul, *Vorschule der Ästhetik*, I.5, 171, 175ff., 187; I.3, 68. The German translation of Montaigne (1753/1754), with which Jean Paul was familiar, spoke of "Schweifen" (devagating) as "jumping and hopping" ("Hüpfen und Springen") (Kilcher, *mathesis und poiesis*, 138f.).

[66] "Because ... modes of procedure (Verfahrensweisen) that are especially highly esteemed in the Baroque come to be discredited in the course of an anthropologization of final results; because fixed inventories of data and an equally fixed *ars inveniendi* are no longer the bases for invention, and hence discovery; and because, to put it succinctly, the human being itself is finally the source feeding all traffic of innovations and data, to the higher glory of its inexhaustibility – the arts of invention assume or are assigned new locations in the system." They "experience ... discursive 'reoccupations' ('Umbesetzungen') ... through which they are declared to be incompetent, overestimated, or obsolete" Stefan Rieger, "'Scientia intuitiva' und Erfindungskunst: Zu einer Theorie des Einfalls und der Entdeckung," in *Homo inveniens: Heuristik und Anthropologie am Modell der Rhetorik*, ed. Stefan Metzger, and Wolfgang Rapp (Tübingen: Narr, 2003), 179–196, here 180. This difference is expressed in various ways: Dotzler writes that Jean Paul's texts "*still* live *entirely* from a florilegia (Blütenkultur) fed ... by excerpts and excerpts of excerpts. But these are *extremes* of a practice habitualized in repetition (eingeschliffene Praxis) that *is summoned in citing*" (Dotzler, *Papiermaschinen*, 561; emphasis B.M.). Or, Stockhammer writes: "In Jean Paul's novels, literature develops a dancerly way of dealing with procedures that had been developed by forgotten scholars" (Stockhammer, "Zeichenspeicher," 53).

[67] Jean Paul, *Vorschule der Ästhetik*, I.5, 205. Concerning the "now" ("jetzo") dominant, "special *Vielwisserei*, indeed a greater omniscience and encyclopedia in Germany" ("ja eine größere Allwissenheit und Enzyklopädie in Deutschland"), he points pejoratively to "our general literature journals and libraries that transform everyone, without their knowing it, into a *Vielwisser*" (206) – specifically to Nicolai's *Allgemeine Deutsche Bibliothek* (see Jean Paul, I.5, 377ff., as well as *Leben Fibels*, I.6, 371, 388f.). Conversely, Nicolai reproached Jean Paul for his apparent "Vielwisserey": "He constantly shows off an erudition that he does not have. ... [O]n nearly

already and in any case, there is no longer any general stock (*Fundus*) once known by the name of topics to be found all over the world, across all the seas, because no common criterion of closeness or distance, accessibility or inaccessibility of knowledge, is given/granted.[68] If wit should "form" ("bilden") or rather "demand" ("fodern") the "center of all" ("Mittelpunct aller"), it would, as the joining "sea" ("Meer"), be the epitome of what is unstructured, and without any stable differences.[69] But what emerges from Jean Paul's aggregations of excerpts and knowledges, those heterogeneous "difficult [or witty] combinations" ("schwere[n] [oder witzigen] Kombinazionen"),[70] is "unstable."[71] This distinguishes Jean Paul's productions from rhetorical *inventio* and the baroque circulation of knowledge among writing and reading, of reading as excerpting and of writing as reading by citing.[72] Citations that do not repay or return what and (to) where they have borrowed (it), free their citations from the fixed classifications of topics; they deregulate, dehierarchize, and delimit topical order.[73] The stock of excerpts is now released from this order and is set free in being joined in witty combinations to become rampant: wildly proliferating.[74]

Hence "every reader" – I cite a review from 1797 – "who lacks broad erudition" finds themselves "compelled" by "the material" Jean Paul "gathers on witty thoughts from all four corners of the earth, from all three kingdoms of nature, physics, and chemistry ... to have a series of philosophical, physical, historical dictionaries at hand," which they must then consult, digressively interrupting their

---

every page, one is supposed to be amazed by his erudite, witty allusions and similes"; Friedrich Nicolai, "Jean Pauls Vorschule der Aesthetik," in *Neue allgemeine deutsche Bibliothek* 96 (1805): 208–227, here 209.
68 Jean Paul, *Vorschule der Ästhetik*, I.5, 205.
69 Jean Paul, *Vorschule der Ästhetik*, I.5, 205; see also 358.
70 Jean Paul, *Merkblätter 1816/17*, No. 100, HKA II.6, 164.
71 Müller, "Mehrfache Kodierung bei Jean Paul," 91.
72 Kilcher, by contrast, (still?) finds this circulation in Jean Paul; *mathesis und poiesis*, 382–86, 391, 396f.
73 This already occurs by means of combinatorics; see Renate Lachmann and Elisabeth von Samsonow, "Magieglaube und Magie-Entlarvung," in *Magie und Religion*, ed. Jan Assmann and Harald Strohm (Munich: Fink, 2010), 93–133, here 97ff.
74 "Wild wuchernd"; see Ekkehard Knörer, *Entfernte Ähnlichkeiten: zur Geschichte von Witz und ingenium* (Munich: Fink, 2007), 198, 203ff.; Ricardo Nicolosi, "Vom Finden und Erfinden: Emanuele Tesauro, Athanasius Kircher und die Ambivalenz rhetorischer inventio im Concettismus des 17. Jahrhunderts," in *Homo inveniens: Heuristik und Anthropologie am Modell der Rhetorik*, ed. Metzger, and Rapp, 219–236.

reading, "in order to understand [Jean Paul's] allusions."[75] Yet all of this will not then be entered "into the data storage [of Jean Paul's readers']" – as Hans-Walter Schmidt-Hannisa puts it.[76] Jean Paul's texts *are no* encyclopedic texts.[77] Rather, the "erudition that [Jean Paul's texts] convey consumes itself in … its witty-poetic effect," to again cite Schmidt-Hannisa,[78] and this is precisely what makes for the *hypertrophy* of their encyclopedic tendency.[79] Wit uses up (*vernutzt*) knowledge by transforming its traditional obsolete practices into ecstasies (*Ekstasen*);[80] it pulverizes/blows (*verpulvert*) the knowledge it cites, causes the images (*Bilder*) it calls forth to crepitate or blow out (*verpuffen*).[81]

The procedures expose to heterogeneity what ought to become the work of an author according to the new prescriptions of poetry – the heterogeneity that

---

75 Anonymous review of *Siebenkäs* and *Biographische Belustigungen* (1797), cited from Sprengel (ed.), *Jean Paul im Urteil seiner Kritiker*, 12.
76 Schmidt-Hannisa, "Lesarten," 22f. Jean Paul's texts "can longer … be excerpted" ("Lesarten," 40). This did not prevent them from *being* excerpted, for example, in the *Pädagogisches Florilegium*: "Pädagogische Goldkörner aus anderen Schriften Jean Pauls" ("pedagogical grains of gold …"); Konrad Fischer, *Jean Paul* (Langensalza: Schulbuchhandlung Greßler, 1894), part 2, 144–251; Carl Wilhelm Reinhold, *Wörterbuch zu Jean Pauls Schriften, oder Erklärung aller in dessen Schriften vorkommenden fremden Wörter und ungewöhnliche Redensarten: nebst kurzen historischen Notizen von den ausgeführten Stellen im Zusammenhange: Ein nothwendiges Hülfsbuch für alle, welche jene Schriften mit Nutzen lesen wollen* (Leipzig: Eurich, 1809); or Berend's planned "Jean-Paul-Lexikon," which however was beaten to the punch by Jean Paul himself: "In his third *Gedanken*-Heft from 1803, he subsumed his lexicographical works and plans of writing under the title *Lexion Jean-Paullinum*" (Kilcher, *mathesis und poiesis*, 289, 385).
77 By contrast, see Kilcher, "Enzyklopädische Schreibweisen bei Jean Paul," 119, 130ff., 139–143; Kilcher, *mathesis und poiesis*, 118–136, 383–395. According to Kilcher, what we are dealing with in Jean Paul is the *old* "circularity of reading and writing," so "the relationship of encyclopedia and literature" is "reversible or replaceable" (*mathesis und poiesis*, 390f.).
78 Schmidt-Hannisa, "Lesarten," 40.
79 Jean Paul's "writing [was regarded] as a highly haphazard [willkürliche] data processing proceeded by an exorbitant amount of reading, so that what is read can then be processed into an equally exorbitant number of thick books" (Kilcher, "Enzyklopädische Schreibweisen bei Jean Paul," 130, see 129–132, 143f.; Kilcher, *mathesis und poiesis*, 136–144). Through the circulation of reading and writing, Jean Paul "did not want to enrich knowledge but to cause 'devastation in the realm of erudition'" (Kilcher, *mathesis und poiesis*, 397). The *Fibel* (spelling book) (that Fibel the protagonist of Jean Paul's novel is supposed to have written) is an "experiment" with the "form of knowledge of the alphabet" (Kilcher, *mathesis und poiesis*, 287) and an encyclopedic work that lets us grasp "how the ancients could find, in the thick volumes of Homer, the encyclopedia of all sciences" (Jean Paul, *Leben Fibels*, I.6, 490f.).
80 The operative practice "only attracts attention at all because it operates with the exploitation (Vernutzung) of what had, until then, been regarded in common practice … as the benefit (Nutzen) of *properly handling books*," according to Dotzler, *Papiermaschinen*, 561.
81 Hegel, *Vorlesungen über die Ästhetik I*, 382.

was encountered in the encyclopedias, in their Babelish multi-handedness, as it were, that was rebuffed by contemporaries.[82] This heterogeneity was conceived as the "obscure and crazy (*folle*) polygraphy," by Barthes, as "the work's anti-structure," for which it supposedly suffices "to consider any work as an encyclopedia": "l'œuvre exténue une liste d'objets hétéroclites" ("the work exhausts a list of heterogeneous objects").[83] This polygraphy invokes the "heterogeneous ensemble" of cultural techniques, of agents and practices that is the "Schreib-Szene" ("writing scene").[84]

The heterogeneous multiplicity of what is cited in Jean Paul's texts is *not* integrated by an author, not integrated into a meaningful whole. Hegel labels this plainly, pejoratively, as the double exteriority of this writing:

> [W]enn es nun darauf ankam, selber ans Erfinden zu gehen, [habe Jean Paul] äußerlich das Heterogenste – brasilianische Pflanzen und das alte Reichskammergericht – zueinandergebracht.
>
> Dergleichen hat selbst der größte Humorist nicht im Gedächtniß präsent, und so sieht man es denn auch den Jean Paul'schen Kombinationen durchaus an, daß sie nicht aus der Kraft des Genie's hervorgegangen, sondern äußerlich zusammengetragen sind.
>
> [W]hen it was a matter of actually inventing himself, he brought together the most heterogeneous [things] – Brazilian plants and the old Supreme Court of the Empire.
>
> Even the greatest humourist has not things of this kind present in his memory and so after all we often see in Jean Paul's combinations that they are not the product of the power of genius but are brought together externally.[85]

---

82 On the Babelish modeling of the multi-handed authorship of the *Lexikon*, see Nicola Kaminski, "Die Musen als Lexikographen: Zedlers Grosses vollständiges Universal-Lexicon im Schnittpunkt von poetischem, wissenschaftlichem, juristischem und ökonomischem Diskurs," *Daphnis* 29, 3–4 (2000): 649–693. This is posted as "decay" in relation to "original works" (Herder) (Menninghaus, "Vom enzyklopädischen Prinzip romantischer Poesie," 143f.; see Waltraud Wiethölter, Frauke Bernd, and Stephan Kammer, "Zum Doppelleben der Enzyklopädik – eine historisch-systematische Skizze," in *Vom Weltbuch bis zum World Wide Web*, ed. W.W., F.B., and S.K., 1–51, here 46f.).
83 *Roland Barthes par Roland Barthes*, 151; *Roland Barthes, by Roland Barthes*, 148 (translation modified).
84 Campe, "Die Schreibszene, Schreiben," 271; see Stingelin, "'Schreiben,'" 13ff., 17f.; for a discussion of writing-scene as a scene of editing, see Uwe Wirth, "Die Schreib-Szene als Editions-Szene: Handschrift und Buchdruck in Jean Pauls Leben Fibels," in *"Mir ekelt vor diesem tintenklecksenden Säkulum,"* ed. Stingelin, 156–174, here 161; Campe, "Schreibstunden in Jean Pauls Idyllen," 132f., 142f., 158; it is a "collective or network of hybrid actors" (Pethes, "Actor–Network–Philology," 207f., [on Jean Paul] 210–217).
85 Hegel, *Vorlesungen über die Ästhetik I*, 382; *Lectures on Fine Art*, 295–296 (translation modified. The order of the passages cited has been altered from the source; translator's note). In contrast to this "schiefe Orginalität" ("awry originality"), the "wahrhafte[s] Kunstwerk" ("true work

The preceding texts (as such) are external (*äußerlich*) to the work, and the emerging connections remain "external" (by this Hegelian model) because they are not integrated into a totality from within (*von innen*). If writing nourishes itself from writings, writing, and the written in such a way that it submits and suspends itself to it (script, writing, and the handling of papers) *as* exteriority, then it resists the concept of aesthetics established since the middle of the eighteenth century, according to which works internally constitute themselves in being based in the presupposed "interiority of a subject which generates coherence."[86] Contrary to this concept (of genius as *natura naturans*) all knowledge prior to the so-called work, all operations and apparatuses, the fact of their having been written down, are themselves – in the aesthetic terms of the late eighteenth century – only external, i.e., contingent.[87] These preconditions are kept secret in the discourse network (*Aufschreibesystem*) 1800.[88] Jean Paul's "work," Jens Baggesen notes in 1797 (in praise!), looks like "an assemblage of all the ruins from Babylon, Persepolis, Rome, and Nuremberg, heaped together among each other in one place, at a venture [or by chance]" ("eine Sammlung aus allen Trümmern Babylons, Persepolis', Roms und Nürnbergs, auf einem Platz auf gut Glück untereinander zusammengehäuft").[89] And Jean Paul stylizes his novels much the same way.[90]

---

of art") "erweist seine echte Originalität nur dadurch, daß es als die eine eigene Schöpfung eines Geistes erscheint, der nichts von außen her aufliest und zusammenflickt, sondern das Ganze im strengen Zusammenhange aus einem Guß, in einem Tone sich durch sich selbst produzieren läßt, wie die Sache sich in sich selbst zusammengeeint hat" ("proves its genuine originality only by appearing as the own creation of one spirit which gathers and compiles nothing from without, but produces the whole by itself, in one piece, just as the thing itself has united them in itself") (Hegel, *Vorlesungen über die Ästhetik* I, 383; *Lectures on Fine Art*, 296; translation modified).
86 "Kohärenz stiftenden Innerlichkeit eines Subjekts," see Schmidt-Hannisa, "Lesarten," 42f.
87 "[Was] bloß von außen her zueinander [findet, sei] nur als zufällig durch ein drittes ... verknüpft" (Hegel, *Vorlesungen über die Ästhetik* I, 383) (That which finds "together ... just from the outside ... is merely linked accidentally by a third," Hegel, *Lectures on Aesthetics*, 296); "Then literary procedures [Verfahren] ... are accidental constructions of an allusive form of writing" (Kilcher, *mathesis und poiesis*, 398).
88 This "secrecy" alone makes that "geniuses in the Goethe period generate texts" (Krajewski, *ZettelWirtschaft*, 75ff.); see Stingelin, "'Schreiben,'" 9–12, and many others.
89 Letter from Jens Baggesen to Johann Benjamin Erhard from May 17, 1797, cited from Sprengel (ed.), *Jean Paul im Urteil seiner Kritiker*, XXXIV. This corresponds to the loads of books for "a hundred camels," that 'polyhistorians' "carry around in their heads"; but here, no "other heads come along to process it with judgment" (Kant, *Anthroplogy*, 78; Kant, *Anthropologie*, WW XII, § 31, 489).
90 Regarding his novel *Hesperus*, it is "not at all possible to say" "what it ... is if you consider it to be a work on coats of arms or on insects – or a dictionary for the illitterati – an old codex – or a Lexicon homericum or a bundle of inaugural disputations – or an ever-ready clerk – or heroic poems and exposé – or for murder sermons ..." ("für ein Wappen- oder ein Insektenwerk ansehen

Reading, which relegates Jean Paul's texts to knowing *as* the heteronomy of the work, results (if ever) only in a flash that consumes (itself), or blows out (*verpufft*) *and* disperses into the *Dictionnaires*, readers in interrupting their reading and digressing from it are forced to take recourse to.[91]

Operating in the space of poetry newly conceived by the "purification" ("Reinigung") of everything external,[92] by the "cut separating knowledge and poetry" ("Schnitt zwischen Wissen und Poesie"),[93] Jean Paul's texts give a new function to the obsolete (*veraltet*) writing techniques they cite,[94] as fictions of their emergence[95]: they mark the materialities of writing and their operations in these techniques and non-worklike forms, and as such they intervene in opposition to the closure of poetry that was established in the concepts of author and work around 1800 (and then solidified during the nineteenth century).

– oder für ein Idiotikon – für einen alten Codex – oder für ein Lexikon homericum oder für ein Bündel Inaugural-Disputationen – oder für einen allezeit fertigen Kontoristen – oder für Heldengedichte und Expose – oder für Mordpredigten ...") (Jean Paul, *Siebenkäs*, I.2, 22).

91 See Jean Paul, *Vorschule der Ästhetik*, I.5, 206.

92 Jean Paul replies to the concept of "literary creation from nothing, namely from itself" of "many recent poets": "in der Tat ist das Leere unerschöpflich" ("in fact the emptiness is inexhaustible"), with the "counsel" ("Rat") that "a young artist of writing and poetry" should "pursue sciences, such as astronomy, botany, geography, etc." (*Kleine Nachschule*, § 1 I.5, 459).

93 Dotzler, *Papiermaschinen*, 637f. Conversely, this corresponds to the purification of knowledge from literariness and fiction. It goes without saying that such an exclusion will be unstable, that is exists in a stripped away, repeatedly retraced relation to the excluded preconditions; the relationship of literature to knowledge (Wissen) and ignorance (Nichtwissen) remains to be negotiated; see Campe, "Ereignis der Wirklichkeit," 269; and Michael Gamper, "Einleitung," in *Literatur und Nicht-Wissen: Historische Konstellationen 1730–1930*, ed. M.G. (Zurich: diaphanes, 2012), 9–21.

94 The shifts in the relationship between poetry and knowledge are often insufficiently considered: authorship distinguishes the poet from constitutional lawyer J.J. Moser (but see Krajewski, *ZettelWirtschaft*, 69–74), as it does literary men such as Jean Paul from Büttner (but see Stockhammer, "Zeichenspeicher," 52ff.). Secondary literature on Jean Paul usually categorizes this "useless erudition" ("nutzlose Gelehrsamkeit") as parody (Birus, *Vergleichung*, 52ff. as many others). According to Dotzler, the outdated attitude was "all the more vehemently quoted as its excessive demonstrations precisely aim at the quality of a poetry that was beyond it" (Dotzler, *Papiermaschinen*, 561). As Campe sees it, the pedant comically becomes the object of "criticism" (in the "idylls"), but the texts continue the criticized remainder (das kritisierte Überständige), writing it into wit, into the materiality of its signifiers (Campe, "Schreibstunden in Jean Pauls Idyllen," 145f., 157, 160).

95 For example: the reading history of Aubin (*Die Taschenbibliothek*), who excerpts and memorizes free of any purpose; the writing history of *Schulmeisterlein Wutz*, who writes his own library; and Fibel, who, by imprinting his name on the title pages, fictitiously makes himself an author of many books, and overwrites the 'instance of the author original.'

In citing "hybrid forms of text" "close to the boundaries of the book" ("hart an der Grenze des Buches"), as Armin Schäfer puts it,[96] that belong to the external organization of knowledge and refer to this limit from its outside, from the outside of the problematic limit between inside and outside the book – i.e., lexicons, compilations of all kinds, lists that comprehend their items as elements separated from each other, displaceable and augmentable,[97] slip boxes,[98] as objects or devices for handling learned knowledge – Jean Paul methodically and willfully "unsettles" ("verunsichert") "the evidence of the concept of the work."[99] The heterogeneity of the materials from the excerpts, coming from elsewhere, has dissociatively inscribed itself into the texts: as insertions, asides, appendices, digressions and divagations,[100] thus inscribing the turnings of its "own" writing out from itself: into potentially *all* other books, to which it refers reading.[101] The overmature or outlived practice (*überständige Praxis*) manifests in a remainder of writing. With recourse to Foucault, the texts demonstrate that the "borders of book are never clear-cut" but extend "beyond the title, the first lines, and the last full stop"; a book is "caught up in a system of references to other books, other

---

[96] Armin Schäfer, "Jean Pauls monströses Schreiben," *Jahrbuch der Jean-Paul-Gesellschaft* 37 (2002): 216–235, here 221; Rieger, *Speichern/Merken*, 88.

[97] For the list see Sabine Mainberger, *Die Kunst des Aufzählens: Elemente zu einer Poetik des Enumerierens* (Berlin and New York: De Gruyter, 2003), 12ff., 19, 30–36.

[98] The program of Jean Paul's *Leben des Quintus Fixlein, aus funfzehn Zettelkästen gezogen* (1796) is its representation "in 14 slip boxes and one last chapter" ("in 14 Zettelkästen und einem letzten Kapitel"); and in the novel Fixlein's "boxes for reminder slips from the twelfth, thirteenth, fourteenth etc., from the twenty-first year and so on" ("Kästen für Erinnerungszettel aus dem zwölften, dreizehnten, vierzehnten etc., aus dem einundzwanzigsten Jahre und so fort") (I.4, 83f.); writing out slip boxes is also metapoetically understood as a procedure of its own (I.4, 84, 165).

[99] To quote Schäfer, Jean Paul's "books [are] experiments with the discursive form of the book" (Schäfer, "Jean Pauls monströses Schreiben," 221), this applies, for example, to *Leben Fibels* in its relation to the *Fibel*, the spelling book, that is glued on – as the book's fictive 'matrix.'

[100] On the noncoherence of what has been combined, disparately put together, the "encyclopedic poetics of digressions" (of digressions), their noninclusion and that of the appendixes, see, among other things, *Clavis Fichtiana*, part of the 'comic appendix' to *Titan*, the "appendix of the appendix" of *Jubelsenior, Register der Extra-Schößlinge*, the (virtually complete) *Flegeljahre* as a 'ample appendix,' etc.

[101] It is argued that the "physical-spatial movement" of the writing – its turning away, referring to the notebooks of excerpts, and interrupting itself – finds "its literary counterpart in the published text in the excursus leading away from the main thought"; Christian Helmreich, "'Einschiebeessen in meinen biographischen petits soupers:' Jean Pauls Exkurse und ihre handschriftlichen Vorformen," in *Schrift- und Schreibspiele: Jean Pauls Arbeit am Text*, ed. Geneviève Espagne and C.H. (Würzburg: Königshausen & Neumann, 2002), 99–122, here 121.

texts, other sentences: it is a node within a network."[102] Jean Paul's texts refer reading to what preceded the writer and the respective emerging book, and refer the future reading of the texts to the world of all the books as the text's margin, abyss, and exterior. More specifically, the texts refer to what, from among all of his amassed notes, he was not able to "get rid of" ("loszuwerden") and "carry away" ("wegzubringen") into the "printed works" (however many there might have been!) despite the immense multiplication of digressions; they refer to their remainders before and beyond themselves; "only a mere tenth" was incorporated into "the published works," as Eduard Berend, the first editor of the historical critical edition, stated.[103] "So läuft der Lotto-Schlagsatz meiner ungedruckten Manuskripte höher auf, je mehr ich dem Leser Auszüge und Gewinste gedruckter daraus gönne" ("Thus the lottery-seignorage of my unprinted manuscripts increases the more I indulge the reader with excerpts and winnings in printed form from it"), Jean Paul writes in a digression to *Titan*.[104] The materials that might be incorporated into a respective printed work then are (respectively) drawn or sorted from the "lottery wheel filled with treasures" ("[dem] mit Schätzen gefüllte[n] Lottorad") of stored excerpts, loosened (to combine freely) as lots (*sortes*) are, in a lottery of winning or losing bets. And conversely, to every actual drawing or lot, to every (more or less) fortunate contingency, to every happenstance "the whole lottery wheel filled with treasures" presents a vague space of what is held in reserve, in the backdrop of the particular occurrence.[105]

Contrary to assumptions about poetic works around 1800, the *exteriority* from which the texts emerge exposes all of their inventions to chance. And conversely, this exteriority is represented and made possible by operations that rely on chance. "Auf gut Glück" – "at a venture," "on the off chance" – all is heaped together among each other, Jens Baggesen commented on Jean Paul's writing – a fortunate phrase.[106] Jean Paul feigns to hazard his findings (*auf gut Glück zu setzen*) in feigning his operations as techniques of *hand*ling discrete elements, that gives space for chance: in shuffling cards, or throwing dice, or drawing lots.

---

[102] Michel Foucault, *The Archaeology of Knowledge*, trans. A.M. Sheridan Smith (New York: Pantheon Books, 1972), 23; translation modified.
[103] See Helmreich, "'Einschiebeessen in meinen biographischen petits soupers,'" 113; on what can be found in the real of the archive, see the catalogue of the *Handschriftlicher Nachlaß Jean Pauls*, vol. 2, Xf.
[104] Jean Paul, *Titan*, I.3, 167; hence the fantasy: "Grace of God be unto the world, should I ever find a vehicle < craft > that could carry everything I would give to < throw on > it" ("Gnade Gott der Welt, wenn ich einmal ein Vehikel <Fahrzeug> finde, das alles trägt, was ich ihr geben <vorschütten> kann") (Jean Paul, *Merkblätter 1816/17*, No. 58, HKA II.6, 175).
[105] Jean Paul, *Titan*, I.3, 167.
[106] Cited from Sprengel, *Jean Paul im Urteil seiner Kritiker*, XXXIV.

The notably displaced footnote, which is appended to the last sentence of § 54 before, or at the foot of, § 55 (on "Scholarly Wit," "gelehrte[r] Witz") is well known:

> Es wäre daher die Frage, ob nicht eine Sammlung von Aufsätzen nützete und gefiele, worin Ideen aus allen Wissenschaften ohne bestimmtes gerades Ziel – weder ein künstlerisches noch ein wissenschaftliches – sich nicht wie Gifte, sondern wie Karten mischten und folglich, ähnlich dem Lessingschen geistigen Würfeln, dem etwas eintrügen, der durch *Spiele* zu *gewinnen* wüßte, was aber die Sammlung anbelangt, so hab' ich sie und vermehre sie täglich, schon bloß deshalb, um den Kopf so frei zu machen, als das Herz sein soll.
>
> Hence the question would be whether a collection of essays might not be useful and appealing in which ideas from all sciences were shuffled together like cards, not mixed together like poisons, without any certain straight aim – be it artistic or scientific – , like Lessing throwing spiritual dice, thus bringing gains to the one who knows how to *win* by *gaming*; but as far as the collection is concerned, I have it and increase it daily, alone for the sake of making my head as clear as my heart should be.[107]

These cards to be shuffled as dice are to be cast to generate a specific combination might make one think here of the cards that serve to register (*verzetteln*) the holdings of libraries and of (excerpts of) knowledge, of loose slips and index cards – conceive a fictitious aleatorics of the operations carried out upon them.[108] As or like *loose* slips, they present the recombinability and displaceability of elements: this is how the dissociated parts of excerpted knowledges are handled. According to this technical fiction, the excerpts and "excerpts of excerpts", registers and "registers of registers", that must be continually reread, browsed through, or leafed through as a vague space for latent possibilities of joinings out of which inventions (*Einfälle*) come (*zukommen*): namely in linking *Einfall* with *chance* (*Zufall*), as a happenstance (*Zu-Fall*) of coincidences that are technically generated by *aleatorics*. If we trace the significant relations, on the one hand, of (Latin) *sors* / (French) *lot* /(German) *Los* to sorting or allocating, by chance, and, on the other hand, of (German) "*los*" to releasing or freeing up (*lösen, freimachen*) but also to (German) *lose* in the pejorative sense of loose (*liederlich*), or *lottern*, the leading of

---

107 Jean Paul, *Vorschule der Ästhetik*, I.5, 202f. Writing is hazarding itself when – this is the editor's fiction of *Leben Fibels* – it is modeled as a materialization of excerpting: pulling out, and reusing paper, the contributions by the hands of others, the contingency of those who contribute and the fact that these contributions happen; see Bettine Menke, "Alphabetisierung: Kombinatorik und Kontingenz," especially part II.

108 Playing cards are part of the history of paper (Müller, *Papier*, 51f). They came to be used in the registration of knowledge and of library holdings by way of their "unprinted reverse sides" (Müller, *Papier*, 178); see Stockhammer, "Zeichenspeicher," 52f. They suggested themselves to this use because of their standardized "format, which was easy to mix and sort" ("misch- und sortierfreudiges Format") (Krajewski, *ZettelWirtschaft*, 43ff., 62f., see 37–41).

a wasting, debauched life, both tracks join ambiguously in *lottery*.¹⁰⁹ Cards that are allotted the way *lots* (*sortes*) are drawn issue connections as coincidences. These cards are loose (*lose*) like pages torn loose (*los-gemacht*) from the binding of a book, detached (*gelöste*) sheets, or loose slips that in their spatial arrangements (as in card files and slip boxes) can be spatially displaced, rearranged and re-sorted, always allowing for its displacements, de- and re-allocations, that in each case change the places of all cards, as well as for the assignment of more cards to come, into the assemblage that in a strict sense is incompletable.¹¹⁰ The looseness (*Losigkeit*), in all senses assembled, sets free the *force* for making new connections, which is modeled as *chance* or *accident* (*Zufall*) – a potential that is given space by fictitious techniques and devices that suspend the regime of conscious knowledge (as the supposed unity of inner coherence). Their handlings also conceive reading in a different way, since reading encounters, in every element, which would be determined only in a context, its possibility of displacement, of its citation elsewhere, of its iterability, its virtual alterity, by its inherent virtual relations.

---

109 Grimm, *Deutsches Wörterbuch*, vol. 12: "los" in the meaning of free, freed (col. 1153–1156); "Los" in the meaning of lot (col. 1156–1168); "lose" in the sense of loose (col. 1181ff.), "lösen" in the sense of to loosen, release (col. 1190–1196); "Lotterie," "sortes, sortitio, ... sortilegium," (col, 1213); and "lottern," related to German: "lose": in the sense of (English) loose, (German) "liederlich" (col. 1214).
110 See Krajewski, *ZettelWirtschaft*, 66–69, 74; Meinel, "Enzyklopädie der Welt und Verzettelung des Wissens," 169–179; Zedelmaier, "De ratione excerpendi," 88, 85ff.

Nicolas Pethes
Collecting Texts: Miscellaneity in Journals, Anthologies, and Novels (Jean Paul)

I

The cultural technique of collecting is relevant for the history of books and literature in a fundamental way: already in antiquity, texts were not only published as coherent works by individual authors, but also as compilations of various authors or genres. In a sense, these collections of texts are second-order books, as it were, that select and rearrange published material, mostly short genres such as poems or essays. In doing so, they may also reinforce the concepts of the singular work of art and individual authorship (as in the case of a *best of*-collection). At the same time, however, collections of texts emphasize the importance of different agents within the literary field such as editors (as well as readers) who, by selecting and recombining texts, contribute to a notion of the book apart from the idea of the completed and isolated work (as in the case of a *Reader's Digest*).[1]

In premodern and early modern times, such collections of texts were either labeled *collectanea* or, in metaphorical reference to the practice of collecting flowers, *anthologies, florilegia,* or *Blütenlese*.[2] In a similar way, the metaphor of *silvae* (or *Wäldchen*) was used, which refers to the different trees in a mixed forest. But the idea of mixture is most explicitly represented by the concept of "miscellany," a term that is also used for the section of short news found in journals and papers, which Roland Barthes characterized as *faits-divers* and which Sara Danius and Hanns Zischler identified as one of the previously overlooked starting points for the composition of the modern novel, namely James Joyce's *Ullyses*.[3]

---

[1] See Barbara Benedict, "Literary Miscellanies: The Cultural Mediation of Fragmented Feeling," *ELH* 57 (1990): 407–430. The role of the reader is emphasized in Benedict's monograph *Making the Modern Reader: Cultural Mediation in Early Modern Literary Anthologies* (Princeton: Princeton University Press, 1996). See also Emily Wilkinson, *The Miscellaneous: Toward a Poetics of the Mode in British Literature: 1668–1759* (PhD diss., Stanford University, 2008).
[2] See the special issue of *German Life and Letters* 70, 1 (2017): *Das Erblühen der Blumenlesen: German Anthologies 1700–1850*, ed. Nora Ramtke and Sean M. Williams.
[3] See Roland Barthes, "Structure du fait divers," in R.B., *Œuvres complètes*, vol. 1 (Paris: Seuil, 1995), 1306–1319; Sara Danius and Hanns Zischler, *Nase für Neuigkeiten: Vermischte Nachrichten von James Joyce* (Vienna: Zsolnay, 2008).

Translated by Michael Thomas Taylor

And "miscellany" is also the term for the various genres and formats of collections of text, ranging from bookish compilations to periodical journals and almanacs.[4]

According to Barbara Benedict, the success of these miscellaneous collections of texts in bookish formats, i.e., miscellanies in the sense of anthologies, was based on the increasing demand for easily accessible reading material and the lack of copyright regulations. As a result, cheap portable paperback books were published that contained a "selection of short, light pieces, often humorous or satirical, into which the educated reader could dip at convenient moments for a literary lift."[5] While this made literature easily accessible in the form of extracts, Benedict also emphasizes the dialectical downside of this democratization process, the canonization of particularly marketable authors:

> Anthologies contribute in several ways to the dialectical movement of canon construction and deconstruction that marks the eighteenth century. They disseminate particular texts by making copies of fashionable works reasonably affordable and by recycling unsold copies with additions sewn into them. At the same time, by grouping unknown publications under the rubric of a famous name, either accurately or not, and by providing expansive and flexible envelopes for pirated pieces, they continually test the limits of an author's popularity and authority.[6]

But as such a heterogeneous and unstable endeavour, anthologies were also subject to criticism in the eighteenth century – mainly due to their lack of uniformity, which was opposed to the classicist understanding of art, as well as because of the libertine tendencies that collections of popular texts seemed to support.[7] Most notably, Shaftesbury and Pope lamented the lack of form and the arbitrary contents of anthologies, labelling them as fragmentary, shapeless, grotesque, disgusting, and dangerous.[8] This criticism, of course, only reflects the genre's success. In this sense, there is a connection between collections of texts and the monographic genre of the novel, which was subject to the same twofold reception: novels were also popular on the growing book market while being received critically with respect to their lack of form, relevant content, and seriousness. In

---

[4] See Paul G. Klussmann and York-Gotthard Mix (eds.), *Literarische Leitmedien: Almanach und Taschenbuch im kulturwissenschaftlichen Kontext* (Wiesbaden: Harrassowitz, 1998). For the practical dimension of collecting within these formats, see Carlos Spoerhase: *Das Format der Literatur: Praktiken materieller Textualität zwischen 1740 und 1830* (Göttingen: Wallstein, 2018), 40–46.
[5] Benedict, "Literary Miscellanies," 407.
[6] Benedict, *Making the Modern Reader*, 17.
[7] See Benedict, "Literary Miscellanies," 424: "By incapsulating in the literary miscellany the infamous triad oft he feminine, the sentimental, and the modern, the author implies that the very form of the miscellany reflects the sins of decadent culture."
[8] Wilkinson, *The Miscellaneous*, 23.

addition, there is a structural relation between novels and anthologies: novels often inlcude digressions as well as changes in genre, style, and narrative perspective, so that – as in the cases of Swift, Defoe, and Sterne – eighteenth-century books appear miscellaneous and polyphonic in content and structure in spite of their monographic format.[9] The epistolary novel, for instance, appeared as a collection of letters by various writers. By integrating further subgenres, novels could be read as collections of miscellaneous texts that sometimes tended to overrule the linear narrative of the plot.

II

In what follows, I will pursue this interrelation between the anthology and the novel through an exemplary analysis of a satirical short novel by the popular German writer Johann Paul Friedrich Richter, better known as Jean Paul, which was first published 1809: *D. Katzenbergers Badereise; nebst einer Auswahl verbesserter Werkchen* ("Dr. Katzenberger's journey to the spa; with a selection of revised minor works"). As I will argue, Jean Paul published this novel *as* an anthology, i.e., he not only added a selection of articles he had previously published in various periodicals to the novel's 'monographic' narrative but designed the entire publication as a heterogeneous arrangement of miscellaneous texts, including the chapters (called "Summula") of the actual story. This miscellaneous structure of a novelistic publication implies a number of challenges and difficulties both with respect to the analysis of the network of references between the different texts as well as with respect to conventional methods of reading and hermeneutically reducing such heterogeneities and differences. Moreover, this structure differs from Jean Paul's earlier incorporations of fictional newspaper and magazine articles in his novels such as the "Extrablättchen" ("Special supplement") in his first novel *Die unsichtbare Loge* from 1893[10] or the *Pestitzer Realblatt* in the appendix to *Titan* in 1800.[11] At first glance, *D. Katzenbergers Badereise* seems much more conventional compared to these experiments: it tells the humorous story of a medical anatomist and vivisectionst who travels to the spa of Maulbronn, where

---

9 Leah Price, *The Anthology and the Rise of the Novel: From Richardson to George Eliot* (Cambridge: Cambridge University Press, 2000).
10 See Magnus Wieland, *Vexierzüge: Jean Pauls Digressionspoetik* (Hanover: Wehrhahn, 2013).
11 Shortly before his death in 1825, Jean Paul even planned to write and publish a novel with the title *Der Apotheker* entirely as a "Wochenschrift" ("Weekly"); see Dennis Senzel, "Werkchen, die zum Werk werden: Zu Jean Pauls 'Wochenschrift,'" *Colloquia Germanica* 49 (2018): 119–136.

he intends to beat up the author of a critical review of his book on the vivisection of animals. He is accompanied by his daughter Theoda and an incognito playwright whom Theoda passionately admires (but does not recognize), resulting in series of mix-ups and entanglements that are to be expected in a satirical novel.

However, these confusions not only result from the novel's content. They are intensified by the paratextual presentation of the first edition in 1809: the subtitle announces a selection of "revised minor works" as an addition to the novel's core narrative; the table of contents presents these eleven "minor works" on the same hierarchical level as the two sections of the "spa story" ("Badgeschichte"), inasmuch as they are listed by the same Roman numerals as the two parts of the novel – although a glance at the page numbers on the right shows that the two installments of the *Katzenberger* narrative comprise almost 300 pages, while the eleven "minor works" cover only roughly half of this amount.

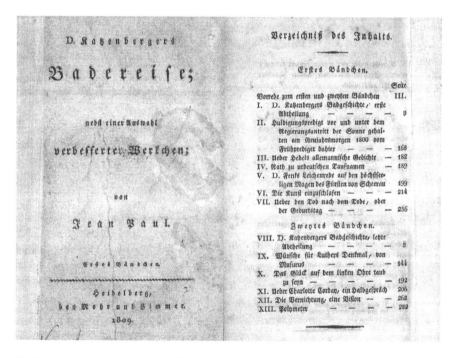

Fig. 1: *D. Katzenberger's Badereise*, title page and table of contents.

This paratextual framing controls the reader's expectation to the effect that the *Katzenberger* narrative seems to be merely one element among others – so that it is not considered as a major "work," but one of the "minor" works next to which it appears in the table of contents. Thus, the design of the table of contents suggests that Katzenberger's journey is only one topic within a heterogeneous collection of texts: it is published, among others, alongside a sermon that honours the sun as a symbol for political government on the first day of the new century, which Jean Paul had originally written for Friedrich Heinrich Jacobi's *Überflüssiges Taschenbuch auf das Jahr 1800* ("Superfluous almanac for the year 1800"); a review of Johann Peter Hebel's *Allemannische Gedichte* ("Alemannic Poems") from the *Zeitung für die elegante Welt* ("Newspaper for the elegant world") of 1803; and an essay on German christening names ("Urdeutsche Taufnamen") also from the *Zeitung für die elegante Welt* in 1803. And finally, the "minor works" consist of a number of articles Jean Paul had contributed to annual almanacs such as the eulogy on the stomach of a dead prince from 1801, metaphysical reflections on "Death after Death" from 1802, a short biography of Charlotte Corday, a visionary fantasy on "Destruction," and a series of aphorisms under the headline "Polymeter."

Research on Jean Paul's *Katzenberger* has nevertheless mostly treated the "minor works" as an unnecessary appendix and focused solely on the spa story – probably due to the fact that Jean Paul chose a different presentation for the second edition of his novel in 1823. In the second edition, he not only distributed the text into three volumes and rearranged the minor works accordingly but also clearly separated the previous periodical articles from the spa story in the table of contents by listing all of its forty-five chapters individually. Thus, the second edition clearly highlights Katzenberger's journey as the actual novelistic content of the book opposed to the symmetrical arrangement of "spa story" and "minor works" in the anthological format of 1809. Through this arrangement, the first edition presents itself not as a novel but as a miscellaneous collection much in the sense of a book project that Jean Paul had already briefly mused about in a footnote to his *Preschool of Aesthetics* in 1804:

> Hence the question would be whether a collection of essays might not be useful and appealing in which ideas from all sciences were shuffled together like cards, not mixed together like poisons, without any certain straight aim, be it artistic or scientific, like Lessing throwing spiritual dice, thus bringing gains to the one who knows how to *win* by *gaming*; but as far as the collection is concerned, I have it and increase it daily, alone for the sake of making my head as clear as my heart should be.[12]

---

12 Jean Paul, *Vorschule der Ästhetik*, in Paul, *Sämtliche Werke*, vol. IX, ed. Norbert Miller (Munich: Hanser, 1989), 202: "Es wäre daher die Frage, ob nicht eine Sammlung von Aufsätzen nützete und gefiele, worin Ideen aus allen Wissenschaften ohne bestimmtes gerades Ziel – weder ein künstlerisches noch wissenschaftliches – sich nicht wie Gifte, sondern wie Karten mischten und folglich, ähnlich den Lessingschen geistigen Würfeln, dem etwas einträgen, der durch *Spiele zu gewinnen wüßte*; was aber die Sammlung angeht, so hab' ich sie und vermehre sie täglich,

In addition, there is immediate evidence for my reading of the *Katzenberger*'s first edition as a miscellaneous "collection of essays" in Jean Paul's correspondence with various publishers in the spring of 1808. Already on April 6, 1807, Jean Paul announced to his publisher Cotta a book by the title *Vermischte Schriften* ("Miscellaneous writings") for the next book fair, and on November 13 of the same year he announced to the publisher Geßner in Zurich:

> They are miscellaneous minor writings (I don't know the title yet), of which 2/3 are completely new (e.g., a journey to the bath by Doctor Katzenberger, a comic novel of about 12 printed sheets) and the last third has been revised and improved.[13]

It is remarkable that the hierarchy between the Katzenberger-story and the miscellaneous articles seemed to be in favour of the latter at this early stage, when Jean Paul continues: "The book begins with serious essays, e.g., on immortality and about Corday, the satirical ones as well as the journey to the spa follow in between, distributed on the two volumes."[14]

Receiving no answer from Zurich, Jean Paul keeps looking for a publisher and finally succeeds in convincing Mohr & Zimmer in Heidelberg,[15] as well as his usual publisher Vieweg, with whom he also negotiates the publication of *Vermischte Schriften*. After he receives an offer from Heidelberg on March 15, 1808,[16] he withdraws his offer to Vieweg. But while the latter would have accepted the title *Vermischte Schriften*, Mohr & Zimmer ask Jean Paul to change the title for strategic reasons,[17] so that the book comes out with the main title *D. Katzenbergers Badereise* on the book fair on Easter 1809.

---

schon bloß deshalb, um den Kopf so frei zu machen, als das Herz sein soll." For an analysis of Jean Paul's practice of compliling his excerpts, see Bettine Menke, "Ein-Fälle aus Exzerpten: Die inventio des Jean Paul," in *Rhetorik als kulturelle Praxis*, ed. Renate Lachmann, Ricardo Nicolosi, and Susanne Strätling (Munich: Wilhelm Fink, 2008), 291–307, as well as Menke's article in the present volume.

13 Jean Paul, *Sämtliche Werke: Historisch-Kritische Ausgabe*, part 3, vol. 5, *Briefe 1804–1808*, ed. Eduard Berend (Berlin: Akademie Verlag, 1961), 178: "Es sind vermischte Schriftchen (den Titel weiß ich noch nicht), wovon 2/3 ganz neu (z.B. eine Badereise des Dokor Katzenberger, ein komischer Roman von etwa 12 Druckbogen) und das letzte Drittel erneuert und verbessert ist."
14 Jean Paul, *Briefe 1804–1808*, 178: "Mit ernsthaften Aufsätzen z.B. über die Unsterblichkeit und über Corday fängt das Buch an, dazwischen treten die scherzhaften und die in beide Bändchen verteilte Badereise."
15 Jean Paul, *Briefe 1804–1808*, 198 (letter to Mohr & Zimmer from February 28, 1808).
16 Jean Paul, *Briefe 1804–1808*, 204f.
17 Jean Paul, *Briefe 1804–1808*, 243. See Uwe Schweikert, "Jean Paul und Georg Zimmer: Mit einem teilweise unbekannten Brief Jean Pauls zur Druckgeschichte von Dr. Katzenberbers Badereise," *Jahrbuch des Freien deutschen Hochstifts* (1973): 347–353.

III

The letters to various publishers prove that Jean Paul did not simply add a series of random articles to an otherwise self-contained work of art. Instead, the book was planned as an anthology from the start, so that *D. Katzenbergers Badereise* emerges from the collection of miscellaneous articles rather than the other way around – in the same sense of Leah Price's argument that anthological formats were the foundation for eighteenth-century novels.

In his preface to *D. Katzenbergers Badereise*, Jean Paul establishes a similar connection:

> With the pocket almanacs and journals, minor miscellaneous works have to increase in number – because authors have to support them with their best contributions – so that in the end hardly anybody writes a major one anymore. Even the author of this work (though also the author of several major ones) has established himself in eight journals and five almanacs by way of minor branch offices and other properties.[18]

According to this confession, the changes on the book market bring about a shift toward the production of miscellaneous articles, which influences the status of the literary artwork as an autonomous monography. And the table of contents of the first edition, in its ahierarchical presentation of the two parts of the novel and the eleven articles, visually represents this shift toward an equality of literature and journalism. The notion of autonomous literature is challenged by the economic requirements of filling periodicals with ongoing content. It is, in other words, the economic success of serial and miscellaneous formats that forces literary authors to orient themselves toward multiple "minor works" instead of singular major ones.

The opening sentence of the preface thus uncovers the hybrid conditions of producing texts at the turn of the nineteenth century, which literary history usually hides within a 'black box' in the sense of Bruno Latour's reconstruction of the hybrid production of scientific facts through multiple layers of inscriptions, paperwork, and mobilizations[19] – so as to maintain the idea of the creative

---

18 Jean Paul, *D. Katzenbergers Badereise; nebst einer Auswahl verbesserter Werkchen* (Heidelberg, 1809), I: "Mit den Taschenkalendern und Zeitschriften müssen die kleinen vermischten Werkchen so zunehmen – weil die Schriftsteller jene mit den besten Beiträgen zu unterstützen haben –, daß man am Ende kaum ein großes mehr schreibt. Selber der Verfasser dieses Werks (obwohl noch manches großen) ist in acht Zeitschriften und fünf Kalendern ansässig mit kleinen Niederlassungen und liegenden Gründen."
19 See Bruno Latour, "Visualization and Cognition: Drawing Things Together: Die Macht der unveränderlich mobilen Elemente," *Knowledge and Society: Studies in the Sociology of Cultures*

sovereignty of the individual author in relation to his publishers and editors. It may well be that the contemporary criticism of miscellanies I referred to above was also based on an attempt to maintain this idea, including the idea of the self-contained work of art. But if we adopt Latour's method of reverse blackboxing, i.e., if we unfold the actual process of producing and publishing a book, it turns out that this process is not a linear creation of consistent form, but prompted and influenced by heterogeneous precursors and preliminary stages as can be found in miscellaneous formats. Writing a text, then, is not so different from collecting texts, insofar it is based on compiling and transforming existing linguistic material rather then discovering and representing an entirely new and homogeneous idea.

In this sense, Jean Paul's preface opens the black box from which his concept of publishing a novel as a collection of texts emerged. In the same way that Barbara Benedict refers to the "pirated pieces" contained in anthologies because of lacking copyright regulations,[20] Jean Paul reacts to the reprint of his periodical articles in an unauthorized anthology in the *Katzenberger* preface:

> This inspired the Voigtian book publishers in Jena in 1804 to reprint "small writings by Jean Paul Friedrich Richter," without asking me and their conscience. In return, this inspires me to bring to the light here their small writings of J.P. likewise without asking. Calmly, I let the publisher complain about reprinting the reprint, reediting the reedition.[21]

Hence, by designing his *Katzenberger*-novel as an anthology of "miscellaneous writings," Jean Paul claims to steal back his intellectual property, as it were. But the "minor works" from Voigt's reprint make up only "one sixth of this book," as Jean Paul continues:

Past and Present 6 (1980): 1–40, as well as my article "Actor Network Philology? Papierarbeit als Schreibszene und Vorwegnahme quantitativer Methoden bei Jean Paul," in *Medienphilologie: Konturen eines Paradigmas*, ed. Friedrich Balke and Rupert Gaderer (Göttingen: Wallstein Verlag, 2017), 199–214.

20 Benedict, *Making the Modern Reader*, 17.

21 Jean Paul, *D. Katzenbergers Badereise* (1809), I: "Dieß frischte im Jahre 1804 in Jena die Voigtische Buchhandlung an, 'kleine Schriften von Jean Paul Friedrich Richter,' ohne mich und ihr Gewissen zu fragen, in den zweyten Druck zu geben. Sie frischt wieder mich an, ihre kleinen Schriften von J.P. gleichfalls ohne zu fragen, hier ans Licht zu stellen. Gelassen lass' ich hier die Handlung über Nachdruck des Nachdrucks, über Nachverlag des Nachverlags schreien." The unauthorized collection of altogether 28 articles by Jean Paul was published in two editions in 1804 und 1808 by J.G. Voigt. For his *Katzenberger,* Jean Paul used only four of these articles. See Nicola Kaminski, "'Nachdruck des Nachdrucks' als Werk(chen)organisation oder Wie D. Katzenberger die *Kleinen Schriften von Jean Paul Friedrich Richter* anatomiert," *Jahrbuch der Jean Paul Gesellschaft* 52 (2017): 29–70.

The second sixth I collected from magazines, from which he has not yet collected anything from me. The second and third thirds of the book are completely new, namely Dr. Katzenberger's journey to the spa and history, as well as the final polymeters.[22]

D. Katzenbergers Badereise is thus, at the same time, a new novel, a reprint of already published material by Jean Paul, a reprint of a previous reprint of this material, and a compilation of all these elements within one book, which makes for a highly heterogeneous collection of texts both in terms of genre, content, origin, and textual history.

The miscellaneous status of the *Katzenberger* is further supported by a work that Jean Paul published in 1810 (before adding two further volumes in 1815 and 1820) and which picks up the subtitle of the *Katzenberger* novel: *Herbst-Blumine, oder gesammelte Werkchen aus Zeitschriften* ("Autumn flowers, or collected minor works from journals"). Besides their subtitles, the two collections of texts – *Katzenberger* and *Herbst-Blumine* – share the same layout of the table of contents, which lists miscellaneous articles by Roman numerals.

In addition, the first volume of *Herbst-Blumine* contains a text that reflects the genre and structure of miscellaneity, "Meine Miszellen" ("My miscellanies"), which Jean Paul had first published in an almanac in 1807, and which Voigt had also included in his pirated edition, although the collection of aphorisms is clearly marked by a posesive pronoun. The three sections of these miscellaneous aphorisms repeatedly reflect the scattered thoughts that miscellaneous writings present to their equally absent-minded readers, for instance:

> How insatiable man is, especially when he is reading. Even scattered thoughts are read in a scattered manner, and instead of starting from the beginning, he browses and eyes through aphorisms here and there, as everybody will recall from these miscellanies.[23]

The concept of miscellaneity refers here to a reading practice on a self-referential level, insofar Jean Paul's miscellanies not only reflect on but also provoke the scattered reception they talk about. In a similar manner, Jean Paul reflects the

---

[22] Jean Paul, *D. Katzenbergers Badereise* (1809), II: "Das zweite Sechstel sammelte ich aus Zeitschriften, woraus er noch nichts von mir gesammelt. Das zweite und das dritte Drittel des Buchs sind ganz neu, nämlich *Dr. Katzenbergers Badereise* und Geschichte, so wie die Schluß-Polymeter."
[23] Jean Paul, "Meine Miszellen," *Herbst-Blumine, oder gesammelte Werkchen aus Zeitschriften*, vol. 1, (Tübingen: Cotta, 1810), 25–28, here 12: "Wie unersättlich ist der Mensch, besonders der lesende! Sogar zerstreuete Gedanken lieset er wieder zerstreuet, und blättert und schauet in Sentenzen, anstatt sie von vorn anzufangen, zuerst ein wenig herum, wie jeder noch von diesen Miszellen sich erinnern wird." See Bryan Klausmeyer, "Fragmenting Fragments: Jean Paul's Poetics of the Small in 'Meine Miszellen,'" *Monatshefte* 108 (2016): 485–509.

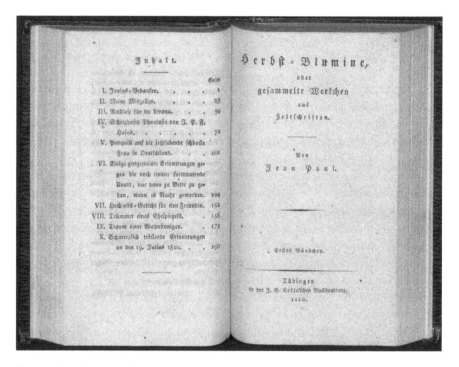

Fig. 2: *Herbst-Blumine,* vol. 1, title page and table of contents.

implications of miscellaneity in his opening article to the new *Morgenblatt für gebildete Stände* ("Morning journal for the educated ranks") on January 1, 1807. Ironically labeled as a retrospective farewell address from the future to this journal, Jean Paul recollects the eight categories the *Morgenblatt* consists of, emphasizing the key role of the category of miscellanies for this classification:

> Article seven, *Miscellanies,* is actually the universal-monarchic directorial article not only of the other seven categories, but of all journals in general, if not of the entire *quodlibet* century itself. Every decent periodical … works toward establishing a considerable number of miscellanies, and presents anything from anybody to anyone.[24]

Here, Jean Paul not only declares the genre of miscellanies to be the textual form that best represents the general structure of miscellaneity within periodicals but

---

24 Jean Paul, "Abschiedsrede bey dem künftigen Schlusse des Morgenblatts," *Morgenblatt für gebildete Stände* 1, 1 (1807): 1–4, here 3: "Der 7te Artikel *Miszellen* ist eigentlich der universalmonarchische Direktorial-Artikel, nicht nur der 7 andern, sondern aller Zeitschriften überhaupt, ja des Quodlibets-Säkulums selber. Jedes Zeitblatt […] arbeitet sich mit der Zeit zu einer schätzbaren Niederlage von Miszellen aus, und reicht allen allerley vom All […]."

also figuratively transfers this structural feature of periodical media onto a basic attribute of the entire nineteenth century, which he labels as the "Quodlibet-Säkulum." These attributes of heterogeneity and arbitrariness are productive features insofar as they are the foundation for the success of miscellaneous articles – as well as for the attempt to design a novel as a collection of texts.

IV

The connection between Jean Paul's *Katzenberger* and his theory of miscellaneous articles is established by the many references to periodicals in the spa story: Katzenberger's journey is initiated by a critical review of his book on animal vivisection in a medical journal; he learns about the whereabouts of the reviewer from the same journal; and this journal also reports that his daughter Theoda's favorite playwright, who goes by the name of Theudobach, is expected at the spa in Maulbronn as well. Finally, in order to find a fellow traveller with whom he can split the coach fare, Katzenberger places an advertisement in a journal to which a certain Mr. Niess (or "Mr. Sneeze") responds, who happens to be no other but Theoda's favorite playwright, although as an author he uses the euphonic name of "Theudobach" as a pseudonym (instead of the actual "Sneeze").

The proximity of the *Katzenberger* novel to periodical formats is further emphasized on a formal level, most of all by the serial structure of its chapters – for example, in the case of the coach's departure as related through a "Fortsetzung" ("Continuation") and a subsequent "Fortgesetzte Fortsetzung der Abreise" ("Continued Continuation of the Departure").[25] Another series of "Summula" is dedicated to Katzenberger's much feared "Mittags-Tischreden" ("dinner conversations") at Maulbronn that treat topics of medical anatomy and, as such, result in chapters that are actually essays very similar to some of the "minor works," such as on the prince's stomach or the art of sleeping, that Jean Paul reprints in *Katzenberger*.[26]

But one might even go so far as to call the entire setting – the coach ride in the first part as well as the conversations among the guests at the spa in the second part – "journalistic," insofar as coaches and baths are heterotopias of nineteenth-century culture that provide polyphonic semiospheres of news, gossip, rumor, and popular culture. This structural relation that connects periodicals with nineteenth-century spa culture is also illustrated by the fact that many journals

---

25 Jean Paul, *D. Katzenbergers Badereise* (1809), 29.
26 Jean Paul, *D. Katzenbergers Badereise* (1809), 141 as well as Summulas 33 and 38.

include a section on spas, e.g., the category "Nachrichten von Bädern" ("News from spas") in the *Zeitung für die elegante Welt*, in which several of Jean Paul's articles were published – including, on March 4, 1809, a preprint of one of the *Katzenberger* chapters.[27] Hence factual and fictional "news from the spas" actually came to stand alongside one another.

Most notably, however, the practice of collecting miscellaneous articles is reflected on a thematic level in *D. Katzenbergers Badereise*: Katzenberger repeatedly stops on his journey to add new specimens to the various collections of animals and monstrosities he passionately maintains: he "hunts" for spiders at an inn where the coach party rests and buys an eight-legged "double hare"[28] from a local pharmacist; he collects human bones from a cemetery, while Theoda and Niess/Theudobach have their first rendevouz at this ghostly location.

Thus, Katzenberger is a medical doctor whose scientific theories are based on collections of highly heterogeneous material, which is, in addition, considered grotesque and disgusting by all nonprofessional protagonists. And since Katzenberger reports these theories during his dinner conversation, the spa story in itself is composed of a series of grotesque digressions, i.e., as a collection of miscellaneous "minor works." From this perspective, the articles that Jean Paul reprints at the end of the novel's two volumes seem less alien to its structure. Rather, the presentation of the "minor works" reflects the miscellaneity of the spa story itself.

More than four decades ago, Peter Horst Neumann already argued that these "minor works" should not be read as a negligible supplement to the actual text of the "spa story."[29] However, this approach has remained without response for a long time. It was not until Armin Schäfer connected Katzenberger's collection of anatomical monstrosities with the compilation of miscellaneous genres that the "minor works" were acknowledged as an essential part of Jean Paul's book, which Schäfer characterized as "monstrous writing" due to this generic hybridity: the monsters from Katzenberger's anatomical collection are metaphors for the miscellaneous format of the book, which escapes conventional genre classifications.[30]

---

27 Jean Paul, "Triumph der Theaterdichter," *Zeitschrift für die elegante Welt* 9, 45 (1809): 353–357. This article becomes Summula 28 "Darum" ("That's Why") in the first edition of *D. Katzenberger*.
28 Jean Paul, *D. Katzenbergers Badereise* (1809), 84.
29 Peter Horst Neumann, "Das Werkchen als Werk: Zur Form und Wirkungsgeschichte des Katzenberger-Korpus von Jean Paul," *Jahrbuch der Jean Paul-Gesellschaft* 10 (1975): 151–186.
30 Armin Schäfer, "Jean Pauls monströses Schreiben," *Jahrbuch der Jean Paul-Gesellschaft* 37 (2002): 216–234.

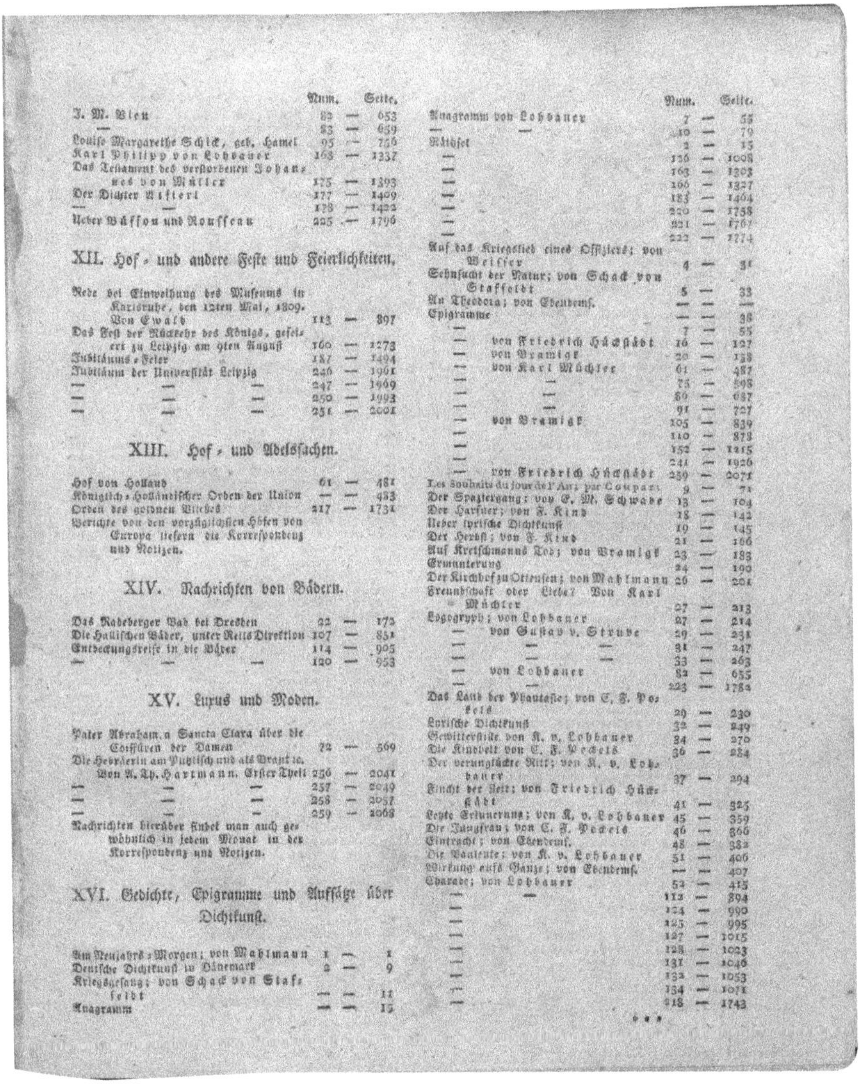

	Num.		Seite.
J. M. Bien	82	—	653
—	83	—	659
Louise Margarethe Schick, geb. Hamel	95	—	756
Karl Philipp von Lohbauer	163	—	1337
Das Testament des verstorbenen Johannes von Müller	175	—	1393
Der Dichter Alfieri	177	—	1409
— —	178	—	1422
Ueber Büffon und Rousseau	225	—	1796

XII. Hof- und andere Feste und Feierlichkeiten.

Rede bei Einweihung des Museums in Karlsruhe, den 12ten Mai, 1809. Von Ewald	113	—	897
Das Fest der Rückkehr des Königs, gefeiert zu Leipzig am 9ten August	160	—	1273
Jubiläums-Feier	187	—	1494
Jubiläum der Universität Leipzig	246	—	1961
— — —	247	—	1969
— — —	250	—	1993
— — —	251	—	2001

XIII. Hof- und Adelssachen.

Hof von Holland	61	—	481
Königlich-Holländischer Orden der Union	—	—	483
Orden des goldnen Vliesses	217	—	1731
Berichte von den vorzüglichsten Höfen von Europa liefern die Korrespondenz und Notizen.			

XIV. Nachrichten von Bädern.

Das Radeberger Bad bei Dresden	22	—	172
Die Hallischen Bäder, unter Reils Direktion	107	—	851
Entdeckungsreise in die Bäder	114	—	905
— — —	120	—	953

XV. Luxus und Moden.

Pater Abraham a Sancta Clara über die Coiffüren der Damen	72	—	569
Die Hebräerin am Putztisch und als Braut ic. Von A. Th. Hartmann. Erster Theil	256	—	2041
— — —	257	—	2049
— — —	258	—	2057
— — —	259	—	2068
Nachrichten hierüber findet man auch gewöhnlich in jedem Monat in der Korrespondenz und Notizen.			

XVI. Gedichte, Epigramme und Aufsätze über Dichtkunst.

	Num.		Seite.
Am Neujahrs-Morgen; von Mahlmann	1	—	1
Deutsche Dichtkunst in Dänemark	2	—	9
Kriegsgesang; von Schack von Staffeldt	—	—	11
Anagramm	—	—	15
Anagramm von Lohbauer	7	—	55
— —	10	—	79
Räthsel	2	—	15
—	126	—	1008
—	163	—	1303
—	166	—	1327
—	183	—	1464
—	220	—	1758
—	221	—	1767
—	222	—	1774
Auf das Kriegslied eines Offiziers; von Weisser	4	—	31
Sehnsucht der Natur; von Schack von Staffeldt	5	—	33
An Theodora; von Ebendems.	—	—	—
Epigramme	—	—	38
—	7	—	55
— von Friedrich Hückstädt	16	—	127
— von Bramigk	20	—	158
— von Karl Müchler	61	—	487
— —	75	—	598
— —	86	—	687
— —	91	—	727
— von Bramigk	105	—	839
—	110	—	878
—	152	—	1215
—	241	—	1926
— von Friedrich Hückstädt	259	—	2071
Les Souhaits du jour de l'An; par Coupart	9	—	71
Der Spaziergang; von G. M. Schwabe	13	—	104
Der Harfner; von F. Kind	18	—	142
Ueber lyrische Dichtkunst	19	—	145
Der Herbst; von F. Kind	21	—	166
Auf Kretschmanns Tod; von Bramigk	23	—	183
Ermunterung	24	—	190
Der Kirchhof zu Ottensen; von Mahlmann	26	—	201
Freundschaft oder Liebe? Von Karl Müchler	27	—	213
Logogryph; von Lohbauer	27	—	214
— von Gustav v. Struve	29	—	231
— — —	31	—	247
— — —	33	—	263
— von Lohbauer	82	—	655
— —	223	—	1782
Das Land der Phantasie; von C. F. Pockels	29	—	230
Lyrische Dichtkunst	32	—	249
Gewitterstille von K. v. Lohbauer	34	—	270
Die Kindheit von C. F. Pockels	36	—	284
Der verunglückte Ritt; von K. v. Lohbauer	37	—	294
Flucht der Zeit; von Friedrich Hückstädt	41	—	325
Letzte Erinnerung; von K. v. Lohbauer	45	—	359
Die Jungfrau; von C. F. Pockels	46	—	366
Eintracht; von Ebendems.	48	—	382
Die Bauleute; von K. v. Lohbauer	51	—	406
Wirkung aufs Ganze; von Ebendems.	—	—	407
Charade; von Lohbauer	52	—	415
— —	112	—	894
—	124	—	990
—	125	—	995
—	127	—	1015
—	128	—	1023
—	131	—	1046
—	132	—	1053
—	134	—	1071
—	218	—	1743

Fig. 3a: News from the Bath in the periodical press.

Fig. 3b: News from the Bath in the periodical press.

This reevaluation of monstrosity precisely matches Foucault's description of the transition from the previous notion, according to which monsters disturb the classification schemes of natural history with the modern theory that monsters are not mistakes of nature but genuinely natural phenomena.[31] And indeed, one of the initiators of this paradigm shift, the anatomist Johann Friedrich Meckel, who was a notorious collector of specimens, recognized himself in Jean Paul's Katzenberger and dedicated his treatise *De duplicitate monstrosa commentarius* to the "viro clarissimo Frederico Richter alias Johanni Paulo," who in return commented on this dedication in the second edition of the novel.[32] And yet, it seems questionable whether the experiment on the format of the book that Jean Paul conducts in *Dr. Katzenberger* should in fact be interpreted as part of Jean Paul's poetic encyclopedia of human knowledge, as Schäfer suggests.[33] Rather, it seems that the miscellaneous format of the book undermines all attempts of a systematic presentation of knowledge. In addition, the fields of knowledge Katzenberger refers to in the novel are all related to what the remaining part of the preface refers to as the specific cynicism of the novel: by chosing a medical doctor as a protagonist, the novel can address phenomena of the human body that would otherwise be excluded from aesthetic discourse and considered inappropriate. Thus, Jean Paul bases his novel on "comic disgust,"[34] and adds a series of references to the various forms, metaphors, and cultural contexts of human excretions such as the "album graecum" (i.e. a remedy extracted from dog excrement) or the "caca de Dauphin."[35] In his preface to the second edition of 1823, this connection between comical effects and bodily effluents is complemented by the reference to vomiting, which "disgusting" texts such as *Katzenberger* may cause in readers.[36]

These references to excretions are of interest not only because they coincide with Mikhail Bakhtin's observations of grotesque scenarios in early modern novels. In the present context, they reflect the way miscellaneous collections of texts were criticized in eighteenth-century literary criticsm: as grotesque,

---

31 See Urte Helduser, *Imaginationen des Monströsen: Wissen, Literatur und Poetik der 'Missgeburt' 1600–1835* (Göttingen: Wallstein, 2016).
32 See Johann Fridericus Meckel, *De duplicitate monstrosa commentarius* (Halle and Berlin, 1815), III, and Jean Paul's comment in the second edition of *Dr. Katzenbergers Badereise* (Breslau: Max und Comp., 1823), 121f. On this exchange of dedications, see Jan Niklas Howe, *Monströsität: Abweichungen in Literatur und Wissenschaften des 19. Jahrhunderts* (Berlin and Boston: De Gruyter, 2016), 56–64.
33 Schäfer, "Jean Pauls monströses Schreiben," 230.
34 Jean Paul, *D. Katzenbergers Badereise* (1809), VI: "das Komisch-Ekle."
35 Jean Paul, *D. Katzenbergers Badereise* (1809), IV.
36 Jean Paul, *Dr. Katzenbergers Badereise* (1823), XIXf.

shapeless, and worthless.[37] In this sense, Jean Paul reflects the notion of an anthology's grotesque lack of form in his multiple digressions on grotesque contents in the novel. Within the aesthetic debate of his time, Jean Paul positions himself on the side of those authors and theorists, above all from England, who do not rely on the classicist understanding of organic work of arts.[38] And as in the case of the paradigm shift within the discourse on monstrosity in natural history, one can say: grotesque monstrosities are metaphors for the format of collections of texts, and yet this format is no longer considered a mistake within the system of classification but recognized as an independent generic form.

V

But if the nineteenth century is in fact a *"quodlibet* century" and if the textual formats it produces are in fact genuinely miscellaneous, how should heterogeneous collections of texts be analyzed? How do we read a novel that is presented as an anthology? Does not reading always imply a homogenizing approach? Are the chapters of *Dr. Katzenberger* entirely miscellaneous or are there cross-references between the spa story and the "minor works"? And then again: would focusing on these connections not undermine the miscellaneity at which Jean Paul is aiming? In the light of these questions, *D. Katzenbergers Badereise* reveals the basic paradox of any collection of objects or texts: as a cultural technique or medial format, collecting strives for unity; but with respect to the individual elements it combines, it is based on difference.

In recent research on the periodical as a genre, this interplay between unity and difference has been metaphorized through the relation of centripetal and centrifugal elements within a journal: insofar as periodicals are serial publications, there is a redundancy of format that allows readers to recognize each individual number as part of a continuous sequence. This centripetal force enables journals to present highly heterogeneous content, which, as such, contributes to the centrifugal miscellaneity of each individual number.[39] But how do we deal with the centrifugal forces of miscellaneity if they are no longer compensated

---

[37] See Wilkinson, *The Miscellaneous*, 23: "Pope identifies the miscellaneous productions of modern authors as literary grotesques whose commitments to excess and variety destroy the uniformity that true art required."

[38] See Magnus Wieland, "Gestörter Organismus: Jean Pauls Ästhetik der Abweichung in der Erzählung 'Dr. Katzenbergers Badereise,'" *New German Review* 24 (2011): 7–25.

[39] See James Musell, "Elemental Forms: The Newspaper as Popular Genre in the Nineteenth Century," *Media History* 20 (2014): 4–20, here 7.

by the centripetal forces of seriality, that is to say: how do we read and analyze anthologies as singular works of art?

I can only hint at the way *D. Katzenbergers Badereise* reflects and mitigates this tension between unity and difference by structuring a seemingly monographic novel as an anthology of periodical articles. Most notably, the very metaphors of centripetality and centrifugality are used by Jean Paul in his satire on absolutistic regimes, the "Sermon of Obeisance to the Sun's Accession to Power," which first appeared in the "superfluous" almanac for 1800 and opens up the sequence of "minor works" in volume one of *Dr. Katzenberger*. In the original version of this article that compares the position of the sun in the solar system to the position of the King in absolutistic regimes, Jean Paul wrote: "According to Newton, *centripetal force* behaves to *centrifugal force* or attraction to omission in the same way as in all cameralistic courts, namely 47000 to 1."[40] The fact that Jean Paul uses the same terminology as current debates on the homogenizing and heterogenizing dynamics within miscellaneous publications is no mere coincidence, since collections of texts also raise the question of a unifying center. But can this question be answered at all? Upon comparing the original version of the "Sermon of Obeisance" ("Huldigungspredigt") with its reprint in the first edition of *D. Katzenbergers Badereise,* one comes across a peculiar omission. The sentence here reads: "According to *Newton*, the *centripetal force* or the attraction to omission behaves as in all cameralistic courts, namely 47000 to 1."[41] The reference to "centrifugal forces" is missing here, so that the word "omission" remains without internal reference. This may well be a mere printing error. But what if Jean Paul deliberately played with the semantics of omitting here – and actually omitted the reference to the very force that causes omission by dissolving unity? In that case, the centrifugal forces of a collection of texts would affect the text on a performative level insofar the replacement of the sun as a "centripetal" ruler results in gaps in the text.

In a similar way, one can easily relate the topic of a dismembered stomach from "D. Fenk's Eulogy on the Stomach of the Duke of Scheerau" both to Katzenberger's dinner conversations on anatomy and the grotesque miscellaneity of

---

[40] Jean Paul, "Huldigungspredigt vor und unter dem Regierungsantritt der Sonne gehalten Neujahrsmorgen 1800 vom Frühprediger dahier," *Überflüssiges Taschenbuch für das Jahr 1800,* ed. Johann Georg Jacobi, 43–54, here 49: "Nach Newton verhält sich bey ihr die *Zentripetalkraft* zur *Zentrifugalkraft* oder das Anziehen zum Weglassen wie bey allen kameralistischen Höfen, nämlich 47000 zu I."

[41] Jean Paul, *D. Katzenbergers Badereise* (1809), 175: "Nach *Newton* verhält sich bey ihr die *Zentripetalkraft* oder das Anziehen zum Weglassen wie bey allen kameralistischen Höfen, nämlich 47000 zu I."

an anthological novel. But it is precisely the temptation to look for connections between the miscellaneous parts of the book that highlights the fundamental methodological problem when dealing with miscellaneous collections of text: whenever such connections are found and identified, the centrifugal structure of the collection is reduced and recentered.

This effect can be demonstrated by examining the original publication context of the "minor works" in *D. Katzenbergers Badereise*: "D. Fenk's Eulogy," for instance, was first published in Leopold von Seckendorffs classicist *Neujahrs-Taschenbuch von Weimar auf das Jahr 1801* ("New Year's almanac from Weimar for 1801"), which contained, among others, contributions by Goethe and was dedicated to the Duchess of Weimar, Anna Amalia. By chosing the topic of the burial of the dismembered stomach of a Duke, Jean Paul's contribution to the *Taschenbuch* seems entirely inappropriate. Through this inappropriateness, however, the grotesque and morbid topic of the essay is transposed to the level of form precisely by not fitting in with its context. Thus, "D. Fenk's Eulogy" is miscellaneous in the twofold sense of the word when it is published in Seckendorff's *Taschenbuch*: it presents a disgusting topic and it interferes with the formal unity of the book.

When Jean Paul reprints "D. Fenk's Eulogy" in *D. Katzenbergers Badereise*, however, the result is exactly the opposite: here, the text blends in among the dinner conversations and other satirical contributions. This blending results in the paradoxical consequence that the grotesque content of the eulogy, which contributed to the article's miscellaneity when it was first published, now reduces this very miscellaneity because it is now contextualized by similar contents. This result demonstrates the interplay of centripetal and centrifugal forces even in the case of monographic collections of texts. And it suggests that this interplay cannot be stopped but must rather be recognized as an ongoing tension between unifying and differentiating forces within cultural techniques of collecting.

This hypothesis is further supported by the reverse relation, i.e., a "minor work" that does not fit into the context of the *Katzenberger* (and, by not fitting, emphasizes its miscellaneity) but was perfectly embedded within its original publication context (which accordingly appears less miscellaneous). Jean Paul's brief biography of the radical revolutionist Charlotte Corday, for instance, was first published in Jean Paul's own almanac in 1801, which consisted almost exclusively of articles on revolutionary uprisings in France, England, and Germany.[42] Transferred to *D. Katzenberger's Badereise*, the sentimental account of Corday's

---

42 Friedrich Gentz, Jean Paul, and Johann Heinrich Voß (eds.): *Taschenbuch für 1801* (Braunschweig: Vieweg, 1800).

childhood, her murder of Jean Paul Marat, and her execution hardly relates to any of the book's satirical contents (unless one is willing to see revolutionary potential in Katzenberger's outrage against his reviewer). Hence in this case, it is the reprint that produces the article's miscellaneity, whereas the original version is dominated by the centripetal force of political ideology.

An exemplary analysis of Jean Paul's novel-anthology thus hints at general aspects of interest when dealing with collections of texts: reading miscellanies means accounting for both the agreement and disagreement of the individual contributions with respect to the respective context from which they were taken and into which they were placed, as well as the differences between the levels of format and content. On the level of format, collections of texts bring together heterogeneous contents and thus highlight the centripetal force of the book; on the level of content, collections of texts highlight the centrifugal forces between the different texts so as not to be mistaken as a homogenous monograph. Thus, collecting is revealed as a cultural technique that simultaneously aims at unifying differences and differentiating unity.

This simultaneity is also the very dynamic of Jean Paul's *D. Katzenbergers Badereise*, as the work emphasizes both the homogeneity and the heterogeneity of the textual elements it combines and thus highlights the hermeneutical paradox of any miscellaneous collections of texts: while a miscellaneous reading of the novel would have to account for the differences between the "minor works" and the "Summula" and in doing so would dissolve the unity of the book in the light of its heterogeneity, the structure of miscellaneity itself can be identified as the unifying principle of Jean Paul's selection of articles as well as his compilation of these articles within the chapters of the spa stories. The literary work of art and the miscellaneous anthology therefore cannot be separated as "purely" as modern literary criticism would have it: by designing a novel as a collection of texts, Jean Paul exposes the basic hybridity of both formats and presents the cultural technique of collecting as a basic operation of literary communication.

Kristina Kuhn
Reading by Grouping: Collecting Discipline(s) in Brockhaus's *Bilder-Atlas*

As a perspective on its material that is documented (and documents itself) in case studies, research into cultural techniques has a decisive advantage over an analysis of multimedia formations and contexts that is grounded purely in media studies: it makes it possible to switch between the perspectives of construction/production and reconstruction/reception while also considering technical, social, and mental constellations. Some of the contributions in this volume are particularly concerned with interrelations and interactions between cultural techniques and literature as a *collection* – though not only of letters (*litterae*) but also of quotations, excerpts, and other media. The scene of writing binds literature to its material, paper, and writing implement (or virtual surfaces), as well as to the movement(s) of the hand (writing, typing, or browsing, which are in a certain sense techniques of the body).

My research material, which draws from encyclopedic corpora, i.e., from a cluster of textual and visual bodies, as well as from bound volumes and their supplements, does not form literature in a classic sense. Even though encyclopedias carry out excessive practices of collecting, collectivizing, or making available to a collective, they find relatively few readers. Neither their articles and indexes nor their legends and lists allow for a linear reading guided by a consistent sense or coherent narrative. Indeed, this is not even their intention. Rather, other procedures of cultural techniques structure how they are perceived. And it is one of these procedures – which we could call a cursory indexing, providing an overview or only a rough scan of contents – that will be the focus of this essay.[1]

---

[1] In this sense, the "distant reading" devised by Franco Moretti could be understood as a non-reading, since it adopts variable macroperspectives and filters texts according to multiple unconnected aspects. Distant reading could therefore be understood as *the* process of encyclopedia reception, which, of course, includes the possibility of going into detail (again) at any time and, in doing so, overcoming (roughly gauging, in skipping over) vast textual distances. With a view to large, intermedial bodies of text, the question of the procedures of distant reading thus certainly makes sense, even if it would have to be treated differently than in the manner proposed by Moretti, for example, by historicizing procedures of distant reading itself. See the contribution by Toni Bernhart, "Quantitative Literaturwissenschaft: Ein Fach mit langer Tradition?" in *Quantitative Ansätze in den Literatur- und Geisteswissenschaften: Systematische und historische*

Translated by Michael Thomas Taylor

Open Access. © 2020 Kristina Kuhn, published by De Gruyter. This work is licensed under a Creative Commons Attribution 4.0 International License.
https://doi.org/10.1515/9783110647044-015

Where does one get stuck in skipping through the pages? where does one pause? and above all, what are the structural characteristics and perceptual effects that make this happen? Which medial combinations imprint themselves or come to attention? How do they become exemplary? My contribution attempts to clarify these questions by examining the interrelationship of textuality, imagery,[2] and collectivity (collectivization). It will investigate how this relationship establishes exemplarity and makes it readable, because an initial observation of this constellation reveals that the collective can only ever be represented in an exemplary way – that it requires a mode of representation-as-substitution (*Stellvertretung*) that makes a whole visible in an exemplar. The practices of taking an overview are forced to disregard the collective that they are supposed to make visible by providing for spatial structures of substitution.

The collective takes up the ambiguity of "collection" as a gathering together and assembling of a (social) collective – an ambiguity that arises a number of times in this volume. In the case of my material, both aspects quite evidently overlap: here, for instance, the social/ethnic collective, as well as the (specific, disciplinary) knowledge collective, are represented by a collection of images that assembles and arranges – that *re*-presents, in one sense, *again and again, anew*.

The material for my case study comes from the *Bilder-Atlas zum Conversations-Lexikon* (Image atlas to the conversation dictionary) published by Brockhaus between 1844 and 1875, described by its subtitle as an *Ikonographische Encyklopädie der Wissenschaften und Künste* (Iconographic encyclopedia of the sciences and arts). This was a kind of intermedial supplement to Brockhaus's *Conversations-Lexikon* (later also called the *Real-Encyklopädie*) – a "supplemen-

---

Perspektiven, ed. T.B., Marcus Willand, Sandra Richter, and Andrea Albrecht (Berlin: De Gruyter, 2018), 207–219. Even if this research is more focused on analytical techniques and tools than on techniques of reading, our task would be asking, from a historical perspective on encyclopedias, about an *analogue* technique of machine reading that emphasizes processes of perception and thus investigates a plurality of cultural techniques of reading. For an introduction, see Helmut Zedelmaier, "Lesetechniken: Die Praktiken der Lektüre in der Neuzeit," in *Die Praktiken der Gelehrsamkeit in der frühen Neuzeit*, ed. H.Z. and Martin Mulsow (Tübingen: Niemeyer, 2001), 11–30. Zedelmaier, for example, describes "Alsted's encyclopedia as a reading grid" (20), notes the recommendation of "'cursory' reading" in the seventeenth century, and interprets reading (aloud) as a kind of technique of the body that involves the entire body and can thus function to animate as well as imprint (22; with reference to Harsdörffer).

2 Sybille Krämer, Eva Cancik-Kirschbaum, and Rainer Totzke (eds.), *Schriftbildlichkeit: Wahrnehmbarkeit, Materialität und Operativität von Notationen* (Berlin: Akademie-Verlag, 2012). Birgit Mersmann, *Schriftikonik: Bildphänomene der Schrift in kultur- und medienkomparativer Perspektive* (Paderborn: Fink, 2015). This contribution is intended as an attempt to build on research into the visuality of text (image-text) by more strongly emphasizing the added value of an approach oriented towards cultural techniques.

tary work," as it is called in the 1875 edition. I hope, however, that some of the tendencies from my observations and theses regarding this image-work can be transferred to other arrangements and series of images, and specifically to convolutes of text and image (such as image databases like Instagram that exhibit rhizomatic encyclopedic structures, as has been discussed many times in relation to the world wide web).

1 Assembling an Atlas

First of all, and before I go into more detail about the compositional characteristics of the *Bilder-Atlas*, the question arises as to its peculiar genre (if we can use that word): an *atlas of images* that, according to a self-description of the book contained in the 1844 edition (*Systematischer Bilder-Atlas zum Conversations-Lexikon* [systematic image atlas to the conversation dictionary]) ought to be systematically arranged. Robert Stockhammer has examined the atlas vogue of the nineteenth century by looking at the relationship between pictorial and cartographic representation that is eminently important for research into cultural techniques.[3] The complex relationship – or, rather, reciprocal crossing – of pictorial representations and nonpictorial records provokes the question of why the atlas's form of assembling its cartographic contents (which does not actually depict anything, although it surely records things) holds such great appeal for other media. In the case of the *Bilder-Atlas*, would not the concept of the panorama, the album, or the catalogue – all of which are forms of organization employed by the atlas on its individual pages, in its overall design, and ultimately as a bound corpus – be more appropriate?

And how is it possible to understand as systematic a catalogue of images that, as at least Gottfried Boehm's theoretical considerations[4] of the images suggest, appears to completely drop out from the semantics and syntax of linguistic (even symbolic) order? The material the atlas shows on its contiguous surface, which is (usually) made up of a double-page spread, presents one more-or-less completed section. By integrating a succession of demarcated surfaces, the atlas thematizes relationships of part and whole; in its entanglement of graphic components

---

[3] Robert Stockhammer, "Bilder im Atlas: Zum Verhältnis von pikturaler und kartographischer Darstellung," in *Der Bilderatlas im Wechsel der Künste und Medien*, ed. Sabine Flach, Inge Münz-Koenen, and Marianne Streisand (Munich: Wilhelm Fink, 2005), 341–361.
[4] Gottfried Boehm (ed.), *Was ist ein Bild?* (Munich: Fink, 1994); G.B., *Wie Bilder Sinn erzeugen: Die Macht des Zeigens* (Berlin: Berlin University Press, 2007); G.B., "Die ikonische Figuration," in *Figur und Figuration*: *Studien zu Wahrnehmung und Wissen*, ed. G.B., Gabriele Brandstetter, and Achatz von Müller (Munich: Fink, 2007), 33–52.

(written characters, schemes, legends, illustrations), it aims to record something complete. In the geographical atlas, this something complete orbits the world or the earth (or in medieval T and O maps, even the entire cosmos of a Christian world order). Hence what claim of relating to the world does the atlas (metaphorically) represent? Does that relation to the world depend on the type of representation, on the qualities of the medium of representation? or does it depend, rather, on the subject matter of what is represented? on what is encompassed by the *Atlas*, representing the world demarcated by natural, geographic as well as political boundaries?

2 Reading Images

Beginning at least with the journal *Bildwelten des Wissens* (Image worlds of knowledge),[5] published by the Helmholtz Centre for Cultural Techniques, the image has become a preeminent object of research into cultural techniques as visibly differing from phenomena of language or writing.[6] In relation to the *Bilder-Atlas*, a question nevertheless arises out of the issues just discussed. Does an atlas comprised of images, or rather with individual map sheets printed with various series of images, not itself follow, in modified form, several rules of language? Or more specifically, does it not follow the rules of linguistic syntax within the representation on the material book page that has been physically opened? In the most fundamental and preliminary sense, this is the direction of reading and writing Latin script from left to right and from top to bottom – in other words: the procedures of a similar kind of cultural technique, namely, reading.

Overall, images and descriptive elements in the *Bilder-Atlas* are formed in many different ways: as rows of images, *tableaux* (some of which are panoramic representations), charts, maps, city maps, etc. The series in the later editions (1875) are often reminiscent of a mixture of photo albums and an ethnological museum

---

[5] Horst Bredekamp et al., "Bildwelten des Wissens," in *Bildwelten des Wissens: Kunsthistorisches Jahrbuch für Bildkritik*, vol. 1, 1: *Bilder in Prozessen*, ed. Claudia Blümle, H.B., and Matthias Bruhn (Berlin: Akademie-Verlag, 2003), 9–20.

[6] Of course, the problematization of this relationship is older, going as far back as Horace's *ut pictura poiesis* and the discussion of this notion in the eighteenth century. Accordingly, the *ut pictura poiesis* and the *Laokoon* debate are to be understood as questions that are eminently related to cultural techniques inasmuch as they examine a shift in media that, at the same time, carried out a shift of dimensions: from the one-dimensionality of the oral telling of myths to the two- or three-dimensionality of the visual (plastic) arts.

in the style of cabinets of curiosity (Fig. 1)[7]: artefacts, expressions of art and cult, furniture, architecture, natural objects, and human artefacts – in short, situative, animated scenes of everyday or cultic life – appear to intermingle seemingly at random.

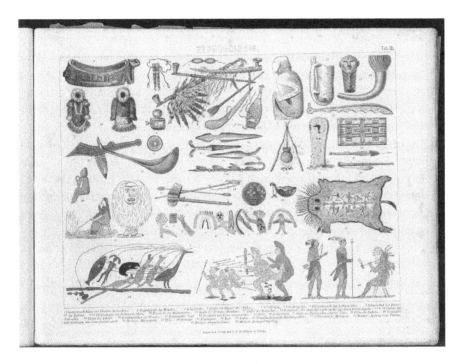

Fig. 1: *Bilder-Atlas*, vol. 7, *Ethnographie* (Ethnography), 1875, plate 13.

This is especially true for the tableau-like representations, which are found in at least two modes of imagery. On the one hand, there are images or busts surrounded by frames that identify them as images and thus immediately categorize

---

7 *Bilder-Atlas: Ikonographische Encyklopädie der Wissenschaften und Künste: Ein Ergänzungswerk zu jedem Conversations-Lexikon*, vol. 1–8, *completely revised second edition*, ed. Karl Gustav von Berneck, Ferdinand Bischoff, Karl Bruhns, Moriz Carriere, Bernhard von Cotta, August Essenwein, August von Eye, Wilhelm Fränkel, Georg Gerland, Wilhelm Hamm, Eduard Hartig, Rudolf Heyn, Henry Lange, Johann Müller, Bernhard Hermann Obst., Otto Prölß, Friedrich Schoedler, Julius Schott, Reinhard Schwamkrug, Alfred Stelzner, Otto Ule, Karl Vogt, Heinrich Adolf Weiske, Theodor Weiß, Reinhold Werner, and Moritz Willkomm, *Fünfhundert Tafeln in Stahlstich, Holzschnitt und Lithographie* (Leipzig: F. A. Brockhaus, 1868–1875), vol. 7, *Ethnographie* (1875), plate 13, origin of digital copy: Klassik Stiftung Weimar, Herzogin Anna Amalia Bibliothek, shelfmark: Th Q 2: 11 (g).

them as illustrations of illustrations. And on the other hand, in the case of the artefacts, there are illustrations that find their place or fit in "directly" on the white background of the book page; they fill in gaps and thus present themselves as if on a table or in a showcase – but they do so, and this is telling, *without framing*. In contrast to the tableaux, other series of images are reminiscent of friezes. They essentially present individual images while often leaving open the possibility of reading the images as the continuation of a series that might even continue across multiple lines: that is, they allow the series to be read as a story told in pictures. Of course, it would be too hasty to read the *Bilder-Atlas* as a graphic novel. Yet, as in the *graphic novel*, the complex of the mutual arrangement and referencing of images and writing, as well as the difference between a strict or more free arrangement of panels, is highly relevant.

The question of image forms and designs, of their arrangements, of directions for (implicit) viewing and reading – and I am making an analogy here to the conventions of the atlas, to its legends that provide the images with paratext via tables of contents[8] – opens up a far-reaching history within European-Western culture and its iconographic memory. The iconographic holdings that the *Bilder-Atlas* retrieves and quotes – even just those that represent "the human" – are already so diverse that they would require an art-historical analysis.[9] And this is not to

---

[8] On the concept of paratext in this sense, albeit in relation to things that have not yet been written, see Kristin Knebel, Cornelia Ortlieb, and Gudrun Püschel (eds.), *Parerga und Paratexte: Steine rahmen, Tiere taxieren, Dinge inszenieren; Sammlung und Beiwerk* (Dresden: Sandstein Kommunikation, 2018).

[9] An associative search for earlier models would include panels since the Middle Ages depicting the Passion of Jesus; the juxtaposition of images and running text on a single manuscript page, split down the middle, of the *Sachsenspiegel* (between 1220 and 1235; see https://digi.ub.uni-heidelberg.de/diglit/ cpg164; visited on August 6, 2019); anatomical illustrations from Vesalius's *De humane corporis fabrica* (1543; numerous scans can be viewed at Wikimedia: https://commons.wikimedia.org/wiki/Category:De_humani_corporis_fabrica_1543 [visited on September 9, 2019]), which founded modern anatomy; and, presumably based in part on Vesalius, convolutes of copper engravings accompanying travel literature, such as Theodor de Bry's *Americae* (ca. 1600; collection available from Heidelberg University Library; https://digi.ub.uni-heidelberg.de/diglit/bry1593ga [visited on September 9, 2019]). It would also include sketches and paintings that were produced as part of James Cook's account of his voyages at the end of the eighteenth century (by Reinhold and Georg Forster, among others). See Rüdiger Joppien and Bernard Smith, *The Art of Captain Cook's Voyages*, vols. 1–3 (New Haven: Yale University Press, 1985–1988). I was in fact able to identify several motifs either from or following Sydney Parkinson (Tahiti, New Zealand) in the *Bilder-Atlas* that exist – presumably after multiple instances of copying – unmarked between stocks of images of other provenances. See also Joppen and Smith, *The Art of Captain Cook's Voyages*, vol. 1 (several pictures) with Brockhaus's *Bilder-Atlas* (1875), vol. 7: "Ethnographie," plate 6 (see note 7).

mention the volumes of the *Bilder-Atlas* dedicated to art history and the history of technology as such, which would certainly be unthinkable without the volumes of plates in the French *Encyclopédie* edited by Diderot and d'Alembert, though this is a comparison I cannot pursue here.

In viewing the images of the *Bilder-Atlas* from the perspective of cultural techniques, my concern here is less with possible iconographic traditions, stocks of motifs, and their citation or modification through replication. The constraints and conditions of a commercial image archive and its purposes of conservation and reproduction, which would render plausible the recycling of stocks of images in different contexts, must also be left aside here – even though these pragmatic factors of visual economics or politics are insightful from the perspective of cultural techniques.

In taking such a perspective, I will focus instead on the highly complex relationships and arrangements of text and image in the *Bilder-Atlas*, which change over the course of the work's three editions from 1844 to 1875. This begins with a brief listing of the plates in tables of contents, complemented in part by volumes commenting in text on each individual volume of plates (1860). Thus degraded to text supplements of the visual supplement to the lexicographical main work, these texts then develop a life of their own. In the next edition (1868–1875), textual elements become more or less extensive image captions making it possible to immediately decipher the images without leafing back and forth in the book or even different volumes. Obviously, it is difficult to follow the text-image references when they are printed relatively far apart, for example, if a reader must connect the table of contents listing different sections (fields of knowledge) and its numbered designation of plates to the subsequent image catalogue that is meant to be browsed (since it has no page numbers); or even in tracing the references between the table of contents, the image catalogue, and the textual commentary that is also structured according to the rationale of sections of knowledge. As noted above, these are not so much commentaries as independently readable volumes of text that essentially function without the illustrations in the atlas they are actually meant to explain. Perhaps because this made it more difficult to read and scan the images, the 1875 edition of the *Bilder-Atlas* increasingly makes use of direct captions in the lower area of each individual page. But even in this case, what is depicted is by no means self-evident.

On the contrary, the images – in combination with their captions and the additional section titles – develop emblematic structures suggesting a message alongside what is shown that is itself, at least, not meant to be read in one, irreversible linear direction or as unambiguously descriptive. Rather than standing in linear relationship structures to each other, the various elements of text and image superimpose different moments of order that overlap, comment upon, and contradict each other.

In this essay, I will thus focus on exemplarily variable directions of reading in the images in the atlas. I would describe these as emblematic and narrative modes to read, i.e. to perceive a book page in its visual appearance, with the assumption that "reading" indicates more than a mere metaphor of interpretation, to then relate them to the *setting* of the collective. I will argue that the reading paradigm (and syntagm) is decisive for a notion of the images. It determines (or makes possible) different directions for seeing the images; it in part creates their visibilities in the first place. A mode of assembly or compression,[10] in the manner characteristic for museums of the nineteenth century, inserts itself here between emblematic and narrative interpretation, although it seems to me that this mode already undergoes a transformation in the short phase between 1860 and 1875.

3 Seeing the Exemplary: "Die Völker Europas" (The peoples of Europe), plate 110

How are the (assembled) collections of various collectives structured in the atlas? One page that can be taken as emblematic, as it were, for the collection process in the *Bilder-Atlas* comes from the section on *Völkerkunde der Gegenwart* (Völkerkunde of the present) (Section 4) from 1860 (Fig. 2).[11] It shows, in three rows stacked horizontally, different types of peoples or ethnic groups in smaller groups, mostly in arrangements of two or three facing each other, that form a common space. (Brockhaus, however, would only publish a volume on the discipline of "Ethnographie" – as opposed to the early discipline of "Völkerkunde" – with the seventh volume of the *Bilder-Atlas* from 1875.[12])

---

10 Annette Graczyk, *Das literarische Tableau zwischen Kunst und Wissenschaft* (Munich: Fink, 2004) has reconstructed the transitions in the tradition of the tableau from the *theatrum* to the museum; see 14, 29, 35.

11 *Bilder-Atlas zum Conversations-Lexikon: Ikonographische Encyklopädie der Wissenschaften und Künste*, vol. 1–10, ed. Johann Georg Heck, *500 Tafeln nebst Text und Universal-Register, fifth edition* (Leipzig: F. A. Brockhaus, 1860), vol. 4, *Völkerkunde der Gegenwart, 42 Tafeln nebst Text*, plate 110 (1), origin of digital copy: Klassik Stiftung Weimar, Herzogin Anna Amalia Bibliothek, shelfmark: N 15667 (f).

12 The discipline of "Völkerkunde" emerged in the nineteenth century as a study of "peoples" that only later developed into the twentieth-century disciplines of ethnography and ethnology. Its methods were closer to what we would today call cultural anthropology. This differentiation is crucial for my argument in this essay.

Reading by Grouping: Collecting Discipline(s) in Brockhaus's *Bilder-Atlas* — **271**

Fig. 2: *Bilder-Atlas*, vol. 4, *Völkerkunde der Gegenwart* (Völkerkunde of the present), 1860, plate 110 (1).

The convention of the book and its rules of reading, from left to right and from top to bottom, seems to require that we dissolve the rectangular space of the page into three rows, but there is nothing preventing a reader from dissolving the three horizontal rows into a continuous single row – or perhaps there is? The individual figures in the course of the series are numbered consecutively so that their denominations can be assigned within thematic subgroups using the table of contents. The numbering does not enumerate the figures in ascending order but seems to follow the inner requirements of the representation that gives the groupings their own regularity. The appearance of grouping results from natural elements, such as shrubs, trees, and rocks, that provide structure in dividing the continuous natural space. The space shown is a land-space through and through, which is striking in that the grouping/dividing element of the world's continents seems rather to slip away as something fluid.

Two principles of order overlap within this linear arrangement of groups of figures, legible as both a spatial continuum and as interior spaces. On the one hand, if one reads the rooms of the world represented by the groups of figures depicted in rows *as a map of the earth,* there is a (roughly traced) geographically contrary movement from west to east, from east to southeast, from southeast to northwest, and back again to southeast. Or one can follow the directional movement from west to east to southeast beyond the map border to north and then again in southeasterly direction *as a diagonal from top left to bottom right, following the line break or change of direction starting again at top left, and from there again to bottom right as the (Eurocentric) direction of reading,* which would also trace a (diagonal) *circumnavigation of the globe,* provided that the change of direction goes beyond the lower right border of the world map. In both cases, the direction begins with a Eurocentric starting point, the old world of antiquity, to then pass through the old world of Asia and Africa to the New World (Australia and America).

And on the other hand, there is a division of "tribes" (see the "Verzeichniß der Erklärung der Abbildungen" [Index of figure captions], in the table of contents after the title page) in a system organized according to "five main races," which are embedded in the two parallel west-east/east-west/west-east or north-south/south-north/north-south orientations, or the diagonal movement over the globe from left to right and from top to bottom, which makes it unnecessary to indicate the cardinal direction. The representation of the globe thus overlaps with the cartographic representation of the world map within the system of the reading directions of the Latin alphabet. Therefore, for *this* page of the *Bilder-Atlas,* the template or metaphor of the atlas, or even of the globe, proves to be quite justified, since the series of images are in fact structured according to a cartographic order. The sequence of continents is based on the established cartographic regime of nineteenth-century European atlases: Europe, Asia, Africa, Australia (New Zealand), and America.

Ironically enough, in the Eurocentric arrangement the "Uebersichtsblatt der fünf Hauptracen" (Overview page of the five main races) forms a subitem under the heading "Die Völker Europas" (The peoples of Europe), thus failing to conform to the originally announced order – probably because this volume of the *Bilder-Atlas* largely offers an ethnography of Europeans themselves. In addition, a marker of civilization has been registered upon the "Uebersichtsblatt" (overview page), insofar as the first upper row, in addition to the extremely barren natural structural elements, is the only one with an architectural background (a keep). The order of "five main races" entered in the index is also superimposed or overdetermined by another designation system, namely by the inclusion of "national costumes," which registers a greater variety of distinguishing features, and not only within the "races" that cannot be integrated into this racial "system."

What is collectively assembled by the series of images and plate legends are not primarily the various "races" and ethnicities of the world – which are abridged or incomplete in any case in relation to each of the recorded characteristics, whether it be "national costumes" or "races." Rather, what is assembled are parallel possibilities of order, oriented toward disciplines of knowledge, and partially determined in terms of media: for geography, this is the globe or map; for anthropology, theories of race or ethnicity; for politics, national or local; and for history, cultural and social history or art history. *The (disciplinary, thematic) knowledge collective shows not only what it shows but above all what it produces and how it is produced.*

The sequence allows for various options of pursuing the collective as a linear continuum, of dividing it into geographical areas along the breaks of the line transitions or other structural elements, or of unfolding the movement globally across the regions of the world.

Because the first row lacks structuring natural elements, it evinces a higher degree of continuity, which simultaneously entails a stronger individualization of the figures. As a continuous space, the first row outlines the geographical world of antiquity in dimensions that are still the basis of most maps of Europe today. The figures in the second and third rows tend to be grouped, while more or less sharply distinguishing one group from another. Interaction almost seems possible between grouped figures, tending to offer narrative (or at least imaginary) potential. The separating elements of nature, which create a (partially) common background to begin with, i.e., a background separated from the other elements, create a connection within groups (which they thus constitute in the first place) and feign a proximity within or between them that is neither spatial, temporal, nor social (the rows produce dislocations and anachronisms). The figures standing in as representatives of their "tribe," of nations or regions, of their "estates" or their social "[c]lass," form internal collectives, while also themselves exemplifying a collective. The exemplarity of the individual figures on the page, by contrast, is highly dependent on the supplementary explanation of the plate, since the attributes given to identify the figures are only readable in combination with their designation. Viewed in isolation, attributes such as clothing, headdress, tools (weapons), or physique seem less specific; and the less dressed the figures are, the less they are characteristic.

Exemplarity consists less in or results less from a catalogue of characteristics that could refer to *one single* generality, since the *respective generality* is split up into different generalities by the various systems of systematization, discourse networks, representation, and reading. Rather, what is meant to be general is indicated *by a single designation* in the table of contents or image captions. The characteristics of visible attribution, for their part, again contribute to larger internal collectives, which form, through mutual reference, a larger collective (within the system of tribes). They do not, however, stand as specific within this larger collective; rather, they function within the collective to introduce minimal differences. Hence it is not only collectives of different sizes, or more or less comprehensive collectives, that are shown – starting from the exemplary type representing and substituting for parallel collectives (a region, nation, "class," or the wearers of a certain ethnic costume); or then as an individual who is in turn integrated into the larger collective of "races" (or even "main races") or of a "tribe"; in order to finally enter the collective of world citizens by overcoming all borders.

Rather, the naive-looking illustrations of the series accord with horizons of diverse knowledge and world systems that by no means intermesh in any contained way or prove to be self-contained. The *Bilder-Atlas* not only hints at various collectives but also presents *various modes of collection*. Who or what is exemplary is permanently reorganized within different assemblage-collections, which is what gives it the status of an exemplar to begin with. The individual exemplar (or exemplum) establishes less a relation to the general than it is taken out of a larger collective to be inserted into another one. With reference to the largest collective visible on the page, the assembly of world citizens, which stretches between the co-presence of linearity and flatness, the question naturally arises of whether there can be an equal arrangement of difference. The parallel diversity created in the spatially movable arrangement acts out conflicting world orders on the layout of a page.

It may be plausible to speak of emblematic structures on the plate analyzed here that enable parallel interpretations of overlapping orders of *inscriptio*, *subscriptio* (in image headings or captions and in indexes) and *pictura* (or perhaps: motivic allusions similar to those in emblem books), and that do not come together to form an overall interpretation but rather open up a disciplinary field from geography to cultural history. At the same time, this can be linked to two further types of image collections (or assemblages) that follow other spatial orders and thus paths of reading: narratively structuring series of pictures, such as the depiction of Egyptian life from the 1875 volume on *Culturgeschichte* (Cultural history) (Fig. 3), as well as an example from the volume on *Ethnographie* (Ethnography) (also from 1875) (see Fig. 1), which looks like a museum assemblage on paper.[13] In both cases we are apparently dealing with quite different systems of order, but also with image effects.

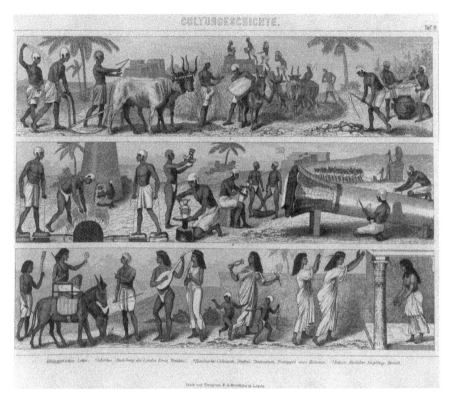

Fig. 3: *Bilder-Atlas*, vol. 6, *Culturgeschichte* (Cultural history), 1875, plate 8, "Altägyptisches Leben" (Ancient Egyptian life).

The page of *Völker der Welt* (Peoples of the world) (Fig. 2) I am analyzing here comes from the volume of the *Völkerkunde der Gegenwart* (Völkerkunde of the present) published in 1860 (Section 4); Section 3 of this issue contains a volume on *Geschichte und Völkerkunde* (History and Völkerkunde). Here, too, the duplication of the modes of order seems to shift time and space against each other and mix them together: (namely) a contemporary and a historical ethnography. Both sections

---

[13] *Bilder-Atlas*, vol. 6, *Culturgeschichte* (1875), plate 8, "Altägyptisches Leben"; *Bilder-Atlas* (1875), vol. 7, *Ethnographie*, plate 13 (museum assemblage).

(volumes) are no longer found in the 1875 edition of the *Bilder-Atlas*. They are replaced by volumes 6 and 7 of the newer edition, published just fifteen years later, which divide the body of ethnographic knowledge ("Völkerkunde") into a scientifically timeless subject, "ethnography," and the (newly emerging) discipline of "cultural history." It seems obvious that this form of division and scientification of the collections (which do not appear so dissimilar when one compares the museum-like illustrations) responds to the problem of the superimposition of the axes of time, space, and knowledge, which itself takes recourse to an earlier and outdated understanding of knowledge that did not meet (or no longer met) the scientific expectations of the late nineteenth century.

The coordinates given by the epistemic order of the atlas thus describe something outdated, a thing of the past. The *Bilder-Atlas* generates (especially in its individual thematic volumes) internal collectives of knowledge; however, it thereby replaces the reference to something complete with an (apparent) system of scientific disciplines that renders this recourse obsolete. In this respect, it is only logical that the *images of the Atlas* are integrated into the *Real-Encyklopädie* as figures (beginning in 1882), where they take on a largely illustrative character without developing their own image-worlds. The reciprocal superimposition of knowledge arrangements able to take shape in the specific spatial arrangement of the work's collection of text and image is no longer desired in the *Reallexikon*; these images now function as a supplement, while the primacy of text serves to discipline the stock of knowledge. As a leading medium, text marginalizes the refuge of the imaginary that the image-supplements open up in the first place.

Stephan Gregory
Patience and Precipitation: Two Figures of Historical Change

1 Enlightenment as Infection

The following deliberations on the temporality of historical change begin with a short passage from Georg Wilhelm Friedrich Hegel's *Phenomenology of Spirit* titled "The Struggle of the Enlightenment with Superstition" ("Der Kampf der Aufklärung mit dem Aberglauben"). In considering this struggle primarily in formal or technical terms, they might at first seem to contradict the intentions of Hegelian philosophy, which is explicitly concerned with the "guises of spirit" ("Gestalten des Geistes"), not those of techniques or technology [*Technik*].[1] But even if Hegel's overall topic is "spirit," when it comes to the details he is an extraordinarily precise and knowledgeable observer of the physical, material, and technical processes in and through which the "movement of the concept" ("Bewegung des Begriffs") is accomplished.[2] In his depiction of the struggle between Enlightenment and superstition, his interest in cultural techniques is even at the forefront: what fascinates Hegel about the Enlightenment is not at all its ideology; it is its artifices, the procedures and dynamics of its implementation. For Hegel, the famous question "What is Enlightenment?" is replaced by the question of how Enlightenment functions: how could the Enlightenment assert itself – so thoroughly and seamlessly that, around 1800, it became possible to speak generally of a "triumph of philosophy" or a "victory of Enlightenment"?

To clarify this question, Hegel offers two models that initially seem to contradict each other, but which ultimately – and this is the dialectic – are shown to complement each other and play into each other's hands. These models are based on two different figures of temporality. Hegel himself does not name them; in order to differentiate them more easily, here they will be called "patience" and

---

1 Georg Wilhelm Friedrich Hegel, "Hegels Selbstanzeige (Intelligenzblatt der Jenaer Allgemeinen Literatur-Zeitung, 28. Oktober 1807)," in G.W.F.H., *Phänomenologie des Geistes*: *Werke*, vol. 3, ed. Eva Moldenhauer and Karl Markus Michel (Frankfurt am Main: Suhrkamp, 1986), 593; translated by Terry Pinkard as "Hegel's Advertisement and Hegel's Note to Himself," in Hegel, *The Phenomenology of Spirit* (Cambridge: Cambridge University Press, 2018), 468 (translation modified).
2 See Hegel, *Phänomenologie*, 38; *Phenomenology*, 22.

Translated by Michael Thomas Taylor

Open Access. © 2020 Stephan Gregory, published by De Gruyter. This work is licensed under a Creative Commons Attribution 4.0 International License.
https://doi.org/10.1515/9783110647044-016

"precipitation." The efficacy of these two figures is not limited to the historical context of the Enlightenment; rather, they represent two fundamental forms of the effect of time in history. As such, "patience" and "precipitation" not only stand for different interpretations of historical change. They can also be regarded as temporal logics that operate in historical events themselves. First, the figure of "patience." Hegel describes it thus:

> Die Mitteilung der reinen Einsicht ist deswegen einer ruhigen Ausdehnung oder dem *Verbreiten* wie eines Duftes in der widerstandslosen Atmosphäre zu vergleichen. Sie ist eine durchdringende Ansteckung, welche sich nicht vorher gegen das gleichgültige Element, in das sie sich insinuiert, als Entgegengesetztes bemerkbar macht und daher nicht abgewehrt werden kann. Erst wenn die Ansteckung sich verbreitet hat, ist sie *für das Bewußtsein*, das sich ihr unbesorgt überließ. ... Sowie daher die reine Einsicht für das Bewußtsein ist, hat sie sich schon verbreitet; der Kampf gegen sie verrät die geschehene Ansteckung; er ist zu spät, und jedes Mittel verschlimmert nur die Krankheit, denn sie hat das Mark des geistigen Lebens ergriffen ... [N]un ein unsichtbarer und unbemerkter Geist, durchschleicht sie die edlen Teile durch und durch und hat sich bald aller Eingeweide und Glieder des bewußtlosen Götzen gründlich bemächtigt, und '*an einem schönen Morgen* gibt sie mit dem Ellbogen dem Kameraden einen Schub, und Bautz! Baradautz! der Götze liegt am Boden' ...[3]

> For that reason, the communication of pure insight is comparable to a peaceful *diffusion* of something like a scent in a compliant atmosphere. It is a pervading infection and is not noticeable beforehand as being opposed to the indifferent element into which it insinuates itself; it thus cannot be warded off. It is only when the infection has become widespread that it is *for consciousness*. ... As soon as pure insight thus is for consciousness, this insight has already made itself widespread, and the struggle against it betrays the fact that the infection has already taken hold. The struggle is too late, and all the remedies taken only make the disease worse, for the disease has seized the very marrow of spiritual life ... [N]ow that it is an invisible and undetected spirit, it winds its way all through the nobler parts, and it has soon taken complete hold over all the fibers and members of the unaware idol. At that point, "*some fine morning* it gives its comrade a shove with the elbow, and, thump! kadump! the idol is lying on the floor" ...

What is Hegel talking about? First of all, we can name the two opponents facing each other in this contest: on the one hand, the form of consciousness of faith, i.e., the traditional Christian world view; and on the other, "pure insight," i.e., the figure of a "pure," self-grounding reason that recognizes no authority but its own[4] – in other words, what we could for the sake of simplification call "modern," "Cartesian"

---

[3] Hegel, *Phänomenologie*, 402f.; *Phenomenology*, 316f.
[4] See Hegel, *Phänomenologie*, 398: "Diese reine Einsicht ist also der Geist, der *allem* Bewußtsein zuruft: *seid für euch selbst*, was ihr alle *an euch selbst seid, – vernünftig*." *Phenomenology*, 313: "This pure insight is thus the spirit that calls out to every consciousness: *Be for yourselves* what you all are *in yourselves – rational*."

rationality.[5] The one party (faith) is overcome by the other (pure insight), and this happens in a way that Hegel finds remarkable, or at least more remarkable than what pure insight itself has to say. Hegel is primarily interested here in the process by which pure insight displaces the system of faith; and he is obviously pleased by the metaphor he has found for it: if pure insight is a disease afflicting faith, the Enlightenment represents nothing but the process of infection through which it spreads.[6] In other words, Enlightenment does not represent a substantial corpus of knowledge, nor does it stand for a particular figure of consciousness (this is rather what "pure insight" does). It primarily denotes a form of dissemination – a specific, highly effective form of "impartion" (*Mitteilung*); or, to put it bluntly, a "communication strategy." Despite all the reluctance that Hegel is compelled to express against such low-content packaging and marketing expertise, this is still what interests him in the venture of Enlightenment: what is decisive is not what the Enlightenment wants (for Hegel, this seems completely banal and not the least bit removed from old prejudices); rather, it is that the Enlightenment seems to have found a new formula for history: a procedure for the production of historical changes, as noiseless as it is effective, which Hegel understands and describes as "infection."

The choice of words itself is not particularly original. Since the idea of infection was first conceived, there has also been a lively metaphorical intercourse between forms of disease transmission and socially operative processes of transmission. The plague never designates just a physical illness; it also always stands for the disintegration of social ties. René Girard, who examined the plague metaphor in the context of his theory of mimetic desire, quotes a famous sixteenth-century doctor, the surgeon Ambroise Paré:

> As soon as the plague becomes evident, it is not unlikely that even the highest authorities will flee ... General anarchy and confusion take hold, and there can be nothing worse for the community, because this is the moment when disorder brings about a plague of a different kind, a plague that is much worse.[7]

The idea of labeling the opposing position as infectious – and thus as particularly insidious – is not new, either. Particularly in the debates about Enlightenment,

---

5 See Michael Pfister, "Aufklärung als Ansteckung: Zu einer Passage in Hegels 'Phänomenologie des Geistes,'" in *Ansteckung: Zur Körperlichkeit eines ästhetischen Prinzips*, ed. Mirjam Schaub, Nicola Suthor, and Erika Fischer-Lichte (Munich: Fink, 2005), 263–274, here 264.
6 Hegel confirms this differentiation; he speaks "of pure insight and of its diffusion, the *Enlightenment*" (*Phenomenology*, 314); "von der *reinen Einsicht*, und ihrer Verbreitung, der *Aufklärung*" (*Phänomenologie*, 400).
7 René Girard, "Die Pest in Literatur und Mythos," in R.G., *Die verkannte Stimme des Realen: Eine Theorie archaischer und moderner Mythen* (Munich: Hanser, 2005), 153–154.

a constant ping-pong of such accusations and counteraccusations plays out. For instance, the proponent of French materialism Paul Thiry d'Holbach speaks in 1768 of a "Contagion sacrée,"[8] a sacred contagion through which Christian superstition obstinately survives; conversely, the propagandists of the counter-enlightenment never tire of hurling epidemiological metaphors at their enemies: these critics speak of insidious poisons that undermine the health of the community, of invisible worms that attack the body of the state and eventually cause its demise.

What distinguishes Hegel's portrayal of the Enlightenment as infection from the polemics of counter-enlightenment is not only the moment of fascination it contains. It is also the analytical precision with which he operates. Hegel does not simply use the metaphor of infection; he analyzes it. What is so captivating about the image of infection? What makes it interesting for use as a political metaphor? Hegel's depiction is focused on two obviously interrelated motifs: first, that of the invisibility of the opponent; and second, that of its particular chronological sequence.

As Hegel shows, religion's confrontation with the Enlightenment must lead to religion's downfall precisely because this conflict cannot be recognized as a confrontation. There is no front, no clear boundary, separating one party from the other, for it is part of the unspoken strategy of Enlightenment to not meet the old system with maximum contradiction, but with a multitude of small advances that do not by themselves provoke resistance. Faith is imperceptibly infected, as it were, by pure insight. Without any compulsion to do so, faith begins to defend its convictions with precisely the same rational means that Enlightenment used against faith. In this way, being drawn onto the field of its opponent, "religion accepts in advance the logic of its enemy."[9] An example of this maneuver is the way faith responds to the arguments of rational Biblical criticism. True faith needs no confirmation; source-critical studies of the historical significance of Holy Scriptures are completely alien to faith.

> In der Tat aber fällt es dem Glauben nicht ein, an solche Zeugnisse und Zufälligkeiten seine Gewißheit zu knüpfen; er ist in seiner Gewißheit unbefangenes Verhältnis zu seinem absoluten Gegenstande, ein reines Wissen desselben, welches nicht Buchstaben, Papier und Abschreiber in sein Bewußtsein des absoluten Wesens einmischt und nicht durch solcherlei Dinge sich damit vermittelt.[10]

---

[8] See Pfister, "Aufklärung als Ansteckung," 265.
[9] Slavoj Žižek, *For They Know Not What They Do: Enjoyment as a Political Factor* (New York: Verso, 2008 [1991]), 65.
[10] Hegel, *Phänomenologie*, 410–411; *Phenomenology*, 322–323.

But in fact it never even occurs to faith to link its certainty to that kind of testimony and those kinds of contingencies. In its certainty, faith stands in an unencumbered relation to its absolute object. It is a pure knowing of that object, and it never lets letters, paper, or copyists interfere with its consciousness of the absolute essence; it does not mediate itself with the absolute essence by those kinds of things.

But it is difficult for faith to evade the burden of proof. Faith becomes increasingly involved in the debates of rational theology, itself becoming interested in the contingencies of written tradition, in the role that "letters, paper, and copyists" play in constituting the truth of faith. Faith thus already finds itself on the field of its opponent, has subjected itself to its opponent's method of argumentation, has already lost:

> Wenn der Glaube sich aus dem Geschichtlichen auch jene Weise von Begründung ... geben will und ernsthaft meint und tut, als ob es darauf ankäme, so hat er sich schon von der Aufklärung verführen lassen; und seine Bemühungen, sich auf solche Weise zu begründen oder zu befestigen, sind nur Zeugnisse, die er von seiner Ansteckung gibt.[11]

> However much faith wishes to substantiate itself ... and if it wishes to do this by drawing on what is historical, and if furthermore it seriously thinks and acts as if something really depended on its doing so, then so has it already let itself be seduced by the Enlightenment. Its efforts to ground itself or to bolster itself in this way only amounts to testimony that attests to its infection by the Enlightenment.

The quotation Hegel has woven into the text also speaks of the unsurpassable effectiveness of an antagonism that does not reveal itself as such. The passage, namely, is taken from Denis Diderot's dialogue *Rameau's Nephew*, which was published for the first time in 1805 in German, in Goethe's translation. Here the strategy of imperceptible land seizure – a common topos of Enlightenment – is traced back to Jesuit missionary practices:

> Dieser fremde Gott setzt sich bescheiden auf den Altar, an die Seite des Landesgötzen. Nach und nach gewinnt er Platz, und an einem hübschen Morgen giebt er mit dem Ellbogen seinem Kameraden einen Schub und Bautz! Baradautz! der Götze liegt am Boden. So sollen die Jesuiten das Christenthum in China und Indien gepflanzt haben, und eure Jansenisten mögen sagen, was sie wollen, diese politische Methode, die zum Zweck führt, ohne Lerm, ohne Blutvergießen, ohne Märtyrer, ohne einen ausgerauften Schopf, dünkt mich die beste.[12]

---

11 Hegel, *Phänomenologie*, 411; *Phenomenology*, 323.
12 Denis Diderot, *Rameaus Neffe: Ein Dialog*, trans. Johann Wolfgang von Goethe (Leipzig: Göschen, 1805), 282–283; *Rameau's Nephew and D'Alembert's Dream*, trans. Leonard Tancock (New York: Penguin, 1966), 101.

> The foreign god takes his place unobtrusively beside the idol of the country, but little by little he strengthens his position, and one find day he gives his comrade a shove with his elbow and wallop! down goes the idol. That, they say, is how the Jesuits planted Christianity in China and the Indies. And the Jansenists can say what they like, this kind of politics which moves noiselessly, bloodlessly towards its goal, with no martyrs and not a single tuft of hair pulled out, seems the best to me.

The intertwining of the temporal figures, which Hegel subsequently develops in more detail, distinctively emerges here: first, the gradual erosion of the opposing position, a strategy of waiting, which can be described with the term "patience"; and then the little shove that – wallop! – brings the position crashing down: a minimal gesture that does not require much effort but nevertheless stands out from among the series of inconspicuous, modest acts that seize space step by step. Insofar as this second movement requires a decision that is not contained in the logic of waiting, we can try to grasp it with the concept of "precipitation." Following Slavoj Žižek, the "double scansion" characterizing the dialectic process as a whole can be found in this joint between two forms of time:

> First, we have the "silent weaving of the Spirit," the unconscious transformation of the entire symbolic network, the entire field of meaning. Then, when the work is already done and when "in itself" all is already decided, it is time for a purely formal act by means of which the previous shape of Spirit breaks up also "for itself." ... The strategy of the New, of the spiritual "illness," must therefore be to avoid direct confrontation for as long as possible; a patient "silent weaving," like the underground tunneling of a mole, waiting for the moment when a light push with the finger will be enough for the mighty edifice to fall to pieces.[13]

As Diderot's example shows, Hegel could have developed the connection between the two forms of time even without an epidemiological reference (Diderot speaks of Christianity being "planted," but this does not necessarily imply the idea of spreading like the plague). When Hegel invokes the metaphor of infection, which is unusual in the context of a philosophy of consciousness, he obviously does so because the natural and social events he describes seem themselves to have a certain logic, that is, in a certain sense, they have the shape of Spirit.

2 Plague and Time

In order to understand how this metaphor of infection functions, it is first necessary to clarify what "infection" was capable of meaning for Hegel, that is, in the

---

[13] Žižek, *For They Know Not What They Do*, 65.

period around 1800. In particular, it seems, we must beware of the temptation of retrodiagnosis, i.e., of a reading that projects today's medical knowledge onto historical descriptions of diseases. For example, Hegel obviously alludes in the passage quoted above to the plague's mode of infection, but as the word "miasma" (*Pesthauch*) that Hegel uses shows, the plague that Hegel has in mind cannot be the same disease we speak of today. It is not even necessary here to investigate the pedantic question of whether the pathogen that causes the plague, *Yersinia pestis*, existed before it was discovered by Alexandre Yersin in 1894[14]; here, it is sufficient to note that "plague" is a term that has been applied to all kinds of epidemics throughout history and that, even in the period around 1800, its meaning was by no means clearly fixed to a particular clinical picture.

Contagio has been an established term in medical discourse since Alessandro Benedetti's *De observatione in pestilentia* of 1493. In his 1546 treatise *De contagione et contagiosis morbis et eorum curatione*, Girolamo Fracastoro distinguishes between three different modes of transmission: an infection based on direct contact ("contactu"), one transmitted by objects such as household effects and clothing ("fomite"), and one that also works at a distance ("ad distans").[15] In 1658, Athanasius Kircher contributed the hypothesis that the plague was caused by tiny worms ("vermiculi") "not perceptible to the unaided eye,"[16] which accumulated in the blood of the sick and spread through contact ("per contactum") or through the vapors discharged by the corpses ("effluvia"). In addition, in the hospitalization and quarantine practices of late-medieval cities, a practical knowledge of infection was developed early on to which "scientific medicine ... [reacted] in part only after centuries of 'delay.'"[17]

Unfazed by the knowledge of transmission established in this way, however, the old miasma theory of the plague connected with ancient theories of humoralism persisted into the nineteenth century, that is, the idea that a concentration of "bad air" or "evil vapors" was responsible for the outbreak of the disease.

In a book published in 1968, the French philosophical historian Jacques D'Hondt set out to find the "hidden sources of Hegelian thought." D'Hondt

---

14 Bruno Latour, "Haben auch Objekte eine Geschichte? Ein Zusammentreffen von Pasteur und Whitehead in einem Milchsäurebad," in B.L., *Der Berliner Schlüssel: Erkundungen eines Liebhabers der Wissenschaften* (Berlin: Akademie-Verlag, 1996), 87–112.
15 Girolamo Fracastoro, *De Sympathia Et Antipathia Rerum Liber Unus: De Contagione Et Contagiosis Morbis Et Curatione*, Libri III (Venice: apud heredes Lucaeantonij Iuntae Florentini, 1546), 29r, 30r, 30v.
16 Athanasius Kircher, *Scrutinium physico-medicum contagiosæ Luis, quæ pestis dicitur ...* (Rome: Typis Mascardi, 1658), 40.
17 Bernhard Siegert, *Passagiere und Papiere: Schreibakte auf der Schwelle zwischen Spanien und Amerika* (Munich: Fink, 2006), 119f.

noticed just how much Hegel's attempt to comprehend his time in thought owes to popular publications, especially the journals, of his day.[18] The notion of "infection," too, can be traced back to common knowledge circulating at the time in the press. D'Hondt refers to a passage from Hegel's writings on the "The Spirit of Christianity" ("Geist des Christentums") written between 1796 and 1800, which refers – completely without any further explanation, as if this were a well-known story – to the "notorious robbers during the plague at Marseille" ("die berüchtigten Diebe während der Pest zu Marseille").[19] D'Hondt's reconstruction makes it plausible that Hegel's knowledge of these thieves probably goes back to a book that had just been translated into German (in 1794) by Baron Knigge, namely *Herrn von Antrechaus merkwürdige Nachrichten von der Pest in Toulon, welche im Jahre 1721 daselbst gewüthet hat* ("Herr von Antrechau's remarkable news of the plague in Toulon, which raged there in 1721").

In the book, Jean d'Antrechaus, the town councilor of Toulon at the time of the disaster, gives a detailed report of the epidemic and the measures taken by the authorities to contain it. With utmost meticulousness, he tries to reconstruct the disease's paths of transmission. What deserves particular attention is the narrative he gives of the disease's origin. It is the story of a theft in Marseille:

> The plague had manifested itself in Marseille in the hospital when the first bales of merchant goods that had been loaded by Captain Chateau in Syria were opened. They were so poisoned that … it was considered advisable to drop off part of this cargo on one of the islands near Marseille … where they were subsequently burned by order of the [royal] court. …
>
> Some inhabitants of Bandol, a small sea port three miles from Toulon … landed that night on the island and took a bale of silk with them, which – unopened and not yet spread out – still contained all of the poison. On their return to Bandol, they shared the spoils among themselves. Since they had committed the crime together, each of them now had to bear the punishment, and these miserable ones … infected their families and the whole village so suddenly that it would not have been possible for any of them to spread the plague any further, almost all of them being infected simultaneously, if a certain barque captain from Toulon named Cancelin, who was in Bandol on that very day, had not had the unfortunate idea of leaving his barque there and returning overland to Toulon. … In this way, and concealing his absence and how he had traveled, he appeared late in the evening on the fifth of October 1720 before the gate of Toulon, where he was admitted without difficulty under the protection of his proper passport.

---

18 See Jacques D'Hondt, *Verborgene Quellen des Hegelschen Denkens* (Berlin: Akademie-Verlag, 1972 [1968]), 157.

19 Georg Wilhelm Friedrich Hegel, "Der Geist des Christentums und sein Schicksal," in G.W.F.H., *Frühe Schriften: Werke*, vol. 1 (Frankfurt am Main: Suhrkamp, 1986), 274–418, here 282; "The Spirit of Christianity," in *Friedrich Hegel on Christianity: Early Theological Writings*, trans. T. M. Knox (New York: Harper & Bros., 1961), 190.

> Our unlucky hour had come without the inhabitants of Toulon having the slightest inkling of it. ... It was not until two days after Cancelin's return that it became clear he was sick. He died on the eleventh of October and was buried in the usual way for one whose death did not appear strange to anyone. There was not the slightest suspicion of the nature of his illness and of his death until the seventeenth, the day his daughter died.[20]

The report reveals a lively interest in the specific temporality of the epidemic. In periods of plague, time also seems to be seized by a plague; at any rate, Antrechaus cannot avoid the depressing realization that evil is always one decisive step ahead, while knowledge principally comes too late. This not only applies to the interval between infection and the visible outbreak of the disease; this natural delay is compounded by all the temporal shifts resulting from the speed of commerce and the inertia of human behavior, and ultimately also from the dysfunctionalities that must be understood as secondary effects of the plague's sway. Even the city authorities responsible for combating the disease react with a culpable delay; since their fear of isolation motivates them to postpone declaring a state of emergency:

> This explanation was sufficient to make the city suspicious; but since no order obligated us to make this declaration ... we let ourselves be moved – by the observation that being abandoned by everyone would be the only fruit we could expect from our frankness ... – to wait and see what the consequences would be of the first moment of the plague, and we still hoped to have hampered its course. But what a delusion! and whom did it not mislead!"[21]

The German translation of Antrechaus's report is accompanied by a detailed foreword written by the Hamburg physician Johann Albert Heinrich Reimarus, which illuminates the medical aspects of the infection of the plague, and which (we can probably say) reflects the most advanced state of epidemiological knowledge at that time. Arguing against the miasma theory, Reimarus takes a position in favor of the new doctrine, according to which "the plague ... does not spread through the air, but only infects by touching contaminated people or things."[22] It is thus mainly "coming close ... that must be avoided, and this must be done with utmost care because even the slightest contact with polluted things or people, even if they are still walking about on the streets, can transmit the disease."[23]

---

20 Jean d'Antrechaux, *Merkwürdige Nachrichten von der Pest in Toulon, welche im Jahr 1721 daselbst gewüthet hat* (Hamburg: Bachmann und Gundermann, 1794 [1756]), 32–34.
21 Antrechaux, *Merkwürdige Nachrichten*, 37.
22 Johann Albert Heinrich Reimarus, "Vorrede: Ueber die allgemeinen Eigenschaften ansteckender Seuchen," in Antrechaux, *Merkwürdige Nachrichten*, III–LXXVI, XI.
23 Reimarus, "Vorrede," XII.

Replacing the atmospheric mythology of the miasma theory, an explanation is now developed here that bases everything on the idea of direct, physical contact, postulating – in search of a medium of transmission – a living vector: because the "pestilent" or "virulent substance" has the special characteristic "that it *multiplies in the body of the living animal* and passes from one body to the next,"[24] it should "be assumed that it must be something living,"[25] smaller perhaps even than the "infusion animalcules" ("Infusionsthierchen")[26] visible under the microscope.

No trace of these subtleties of infection theory can be found in Hegel. On the contrary: he appears to hold fast to the miasma theory, or to mix it with the contagion theory, for example in describing the infection as "*diffus[ing]* ... like a scent in a compliant atmosphere" ("*Verbreiten* wie eines Duftes in der widerstandslosen Atmosphäre"). While the medium of transmission is obviously not so important to him, he is all the more interested in the "dynamic scheme"[27] of its progression. He was able to find information about the progression of the disease in Antrechaus, as well as in Reimarus, whose "Preface" also deals with the specific temporality of the epidemic:

> The plague lasts a long time everywhere, and the suffering and devastation that it causes during this period is utterly horrible. It is also terrible that its early discovery and infection is hindered by various circumstances, since at first it sneaks around clandestinely, to then spread unstoppably afterwards.[28]

Hence Hegel's infection theory of Enlightenment uses the enlightened epidemiological knowledge of his time, albeit selectively. It is not the paths and substances of transmission that interest him but only infection's temporal logic: the combination of a gradual change taking place unseen with a moment of a delayed comprehension that is then left with no other choice but to subsequently attest the completed change. Hegel was thus able to find a model for his philosophy of consciousness in the plague's mode of transmission, because the belatedness of consciousness is not limited to the particular case of the plague so clearly illuminating its temporal structure. Rather, consciousness (or philosophy, the "owl of Minerva") fundamentally and necessarily comes too late; it always takes place after the fact, as the result of an unconscious process. This is why the metaphor

---

24 Reimarus, "Vorrede," XXXIV.
25 Reimarus, "Vorrede," XXXVIII.
26 Reimarus, "Vorrede," XL.
27 D'Hondt, *Verborgene Quellen des Hegelschen Denkens*, 170.
28 Reimarus, "Vorrede," XVI.

of infection is so well suited for Hegel's purposes: it perfectly demonstrates what Slavoj Žižek has called the core of Hegel's dialectic: "This is the signifier's performative 'temporal' reach [maille temporelle], which *retroactively* makes the thing in question ... *what it already was.*"[29]

3 The Patience of Techniques

Let us return to Hegel's initial question, namely how the Enlightenment could so easily triumph over traditional faith. With his epidemiological explanation, his description of Enlightenment as infection, Hegel also provides a general hypothesis about the nature of historical change. The essential point is not so much the apparent fractures in which consciousness becomes aware of a change, but rather the long, continuous, invisible movement that precedes them. "Great revolutions which strike the eye at a glance," he explains in the 1790s, "must have been preceded by a still and secret revolution in the spirit of the age, a revolution not visible to every eye, especially imperceptible to contemporaries, and as hard to discern as to describe in words":

> Den großen, in die Augen fallenden Revolutionen muß vorher eine stille, geheime Revolution in dem Geiste des Zeitalters vorausgegangen sein, die nicht jedem Auge sichtbar, am wenigsten für die Zeitgenossen beobachtbar und ebenso schwer mit Worten darzustellen als aufzufassen ist.[30]

This notion of a subterranean, hidden efficacy corresponds perfectly to the Enlightenment's self-conception, and not only in the secret societies that make a program of imperceptibly conquering the state apparatus. The Enlightenment appreciates artifices and dodges – being, in this regard, merely a faithful imitator of arcane absolutist politics. Paradoxically, even the "public sphere," or the instance opposing the cabals of princely power politics, is conceived of as a countervailing power operating beneath the surface. For example, the philosopher Christian Garve speaks of the widespread "habit of viewing public opinion as an

---

[29] Slavoj Žižek, *The Most Sublime Hysteric: Hegel with Lacan* (Cambridge: Polity Press, 2014), 24; the original French reads: "Voilà la 'maille temporelle' de la performativité du sifnifiant qui fait de la chose en question ... *rétroactivement ce qu'elle était déjà*"; Slavoj Žižek, *Le plus sublime des hystériques: Hegel avec Lacan* (Paris: Presses Universitaires de France, 2011), 49.
[30] Georg Wilhelm Friedrich Hegel, "Die Positivität der christlichen Religion (1795/1796), Zusätze," in Hegel, *Frühe Schriften: Werke*, vol. 1 (Frankfurt am Main: Suhrkamp, 1986), 190–217, here 203; "The Positivity of Christian Religion," in *Hegel on Christianity*, 67–181, here 152.

invisible being of great effectiveness and of counting it among the hidden powers that rule the world."[31]

The space in which Enlightenment spreads is, first of all, a space of discourse: at the beginning of the century, it is a space of learned debate, and later it also becomes a space of popular demands. Accordingly, the infectious or epidemiological character of the Enlightenment comes to attention almost exclusively in the media disseminating this discourse: contemporaries complained about the "flood" of new publications, the rampant "nonsense" of journals, the "plague" of reading societies, the "tempest" of political pamphlets, etc. When Hegel reports on the struggle between Enlightenment and superstition, he too is referring to this world of discourse, the world of public opinion: the undermining of traditional faith and the victory of the Enlightenment takes place upon the field of ideology; it is a "revolution in the spirit of the age."

Of course, the forms of dissemination of the textual universe are particularly suitable for epidemiological interpretation: characters can be repeated and combined at will and are also highly transportable. However, focusing attention on the activity of signs loses sight of another aspect of Enlightenment. In addition to the various forms of discursive dissemination, no less diverse forms of technical modernization can be found. By no means does the infection process of the Enlightenment occur only in the field of discursive debate. It is at least as effective in the field of technical innovation. Precisely if one looks at the pervasiveness of the Enlightenment "in the countryside," one quickly becomes convinced that its actual effectiveness consists not so much in a change of convictions but rather in the slow replacement, repeated again and again in each village, of old technologies by new ones: lightning rods in place of consecrated bells, fire insurance in place of intercessions, smallpox vaccination in place of votive pictures. Like the Christian idol next to the old national idols, technology takes its place as a kind of "new magic" next to traditional magical practices. These continue to exist alongside the new technology for quite some time until, "one fine morning," they are finally knocked off their pedestal.

Hence one can say that the Hegelian model of "patience," or the subliminal change that can be grasped only after the fact, i.e., "too late," corresponds almost exactly to the way in which technical modernization spreads. Technology is not only immediately captivating, its use almost inevitable. Above all, it holds – and this is what made it so interesting for the proponents of Enlightenment – the

---

[31] Christian Garve, "Über die öffentliche Meinung," in Garve, *Versuche über verschiedene Gegenstände aus der Moral, der Litteratur und dem gesellschaftlichen Leben*, vol. 5 (Breslau: Korn, 1802), 291–334, here 294.

promise of irreversible progress. Once the technologies have become established, once they have become a natural way of acting and interacting, the development becomes impossible to reverse. For example, Diderot sees, "in books, a guarantee for the irreversibility of the Enlightenment – 'les lumières conservées par l'Imprimerie'"[32] – and the historian Edward Gibbon is able to dispel the public's fear of a new attack by barbarians. He can do so not only because it is possible to repel invading barbarians with superior war technology, but also because the barbarians would first have to acquire the necessary knowledge in order to resist the technology – meaning they would no longer be barbarians: "before they can conquer, they must cease to be barbarous."[33]

In the lee of ideological battles, the technicians get to work executing the details of the Enlightenment. Beyond the area of discursive dissemination that Hegel examines, the epidemiological process of the Enlightenment can be understood as a chain of technical changes – as a process whose effectiveness is only recognized when it is "too late," when the way back is blocked. Hegel's epidemiological model thus contains the possibility, especially when applied to inconspicuous technical processes, of a new understanding of historical change. It is not the prominent events connected to the banner of revolution that matter; rather, what is truly interesting is the cascade of transmission processes that patiently prepares the way for change.

These processes form a kind of "epidemiological strand" of historical observation, the theoretical application of which consists in a systematic devaluation of large fractures and a corresponding, positive revaluation of small, minor actions and inventions. The infamous "great changes" – or what have been bombastically described as an "epistemological caesura" or "historical break," an "epochal threshold" or "revolution" – are (apart from the drama that people make of them) ultimately merely the effects of a quantitative accumulation of anonymous inventions, of small and minute events that relate to previous, minor events and that, in tirelessly repeating themselves, endow these new structures with the status of inescapable evidence or of a "social fact."

One can find this method of observation in Friedrich Nietzsche, in his suspicion of a "barbarous and shameful confusion"[34] that produces the formations

---

32 Hans Blumenberg, *Die Lesbarkeit der Welt* (Frankfurt am Main: Suhrkamp, 1989 [1981]), 168.
33 Edward Gibbon, *The History of the Decline and Fall of the Roman Empire*, vol. 6 (London, Glasgow, and Dublin: printed for Thomas M'Lean [et alii], 1827), 407.
34 Michel Foucault, "Nietzsche, die Genealogie, die Historie," in M.F., *Von der Subversion des Wissens* (Frankfurt am Main, Berlin, and Vienna: Ullstein, 1982), 83–109, 99; "Nietzsche, Genealogy, History," in M.F., *Language, Counter-Memory, Practice: Selected Essays and Interviews*, ed. Donald F. Bouchard (Ithaca: Cornell University Press, 1980), 139–164, here 155.

of culture in the first place; but one finds it above all in the French sociologist Gabriel Tarde, who thinks of the spread of what is new in terms of an epidemiological model, as a process characterized not least of all by a capricious, unpredictable temporality:

> A cyclone whirls from neighbourhood to neighbourhood ... An epidemic, on the other hand, rages in a zig-zag line; it may spare one house or village among many, and it strikes down almost simultaneously those which are far apart. An insurrection will spread still more freely from workshop to workshop, or from capital to capital. It may start from a telegraphic announcement, or, at times, the contagion may even come from the past, out of a dead and buried epoch.[35]

Gabriel Tarde's esteem for "small inventions" and their unstoppable spread is echoed in Michel Foucault's micropolitical analyses, in his call to base analysis not on large units but on the "infinitesimal mechanisms, which each have their own history, their own trajectory, their own techniques and tactics"[36] and also, of course, in the philosophy of Gilles Deleuze, who developed his own model of causality, which we might call "epidemiological": "It is never filiations which are important, but alliances, alloys; these are not successions, lines of descent, but contagions, epidemics, the wind."[37]

4 Din and Battle Cries

The idea that what is new spreads subliminally, through an accumulation of hardly noticeable changes, as a gradual infection that can only be perceived after the fact, appears irresistible. *Rameau's Nephew* emphasizes the advantage of a "politics which moves noiselessly" – or, as Goethe translates, "without din" ("ohne Lerm") – "bloodlessly towards its goal, with no martyrs and not a single tuft of hair pulled out."[38] But the Enlightenment is not content with the mode

---

[35] Gabriel Tarde, *The Laws of Imitation* (New York: Henry Holt and Company, 1903), 35. The German translation has a slightly different emphasis: "Ein Orkan breitet sich nach und nach aus …. Eine Epidemie wütet anders. Sie schlägt nach links und rechts, verschont dabei dieses oder jenes Haus, die eine oder andere Stadt, erfaßt fast annähernd gleichzeitig viele, weit verstreut liegende Häuser und Städte. Noch freier breitet sich der Aufstand von Stadt zu Stadt, von Fabrik zu Fabrik durch eine Telegraphennachricht aus." Gabriel Tarde, *Die Gesetze der Nachahmung* (Frankfurt am Main: Suhrkamp, 2003 [1890]), 59.
[36] Michel Foucault, "Two Lectures," in M.F., *Power/Knowledge: Selected Interviews and Other Writings 1972–1977*, ed. Colin Gordon (New York: Vintage Books, 1980), 78–108, here 99.
[37] Gilles Deleuze and Claire Parnet, *Dialogues II* (New York: Continuum, 2006 [1977]), 52.
[38] Diderot, *Rameaus Neffe*, 283; *Rameau's Nephew*, 101.

of "extension without opposition" ("gegensatzlose Ausdehnung"), of leisurely spreading.

"[T]his mute weaving of spirit" is, Hegel emphasizes, "only *one* side of the realization of pure insight" ("Dieses stumme Fortweben des Geistes … [ist] nur *eine* Seite der Realisierung der reinen Einsicht"). Its other side consists in reintroducing a moment of conflict and agitation into the process. The Enlightenment becomes spectacular or, in Hegel's words, it makes an entrance as "a noisy ruckus and a violent struggle with its opposite" ("als ein lauter Lärm und gewaltsamer Kampf mit Entgegengesetztem").[39] Hegel is referring not only to the events of the French Revolution but also to the earlier skirmishes in the debate about Enlightenment itself. A Jesuit priest no less, yet one who hardly hesitates to throw himself into the battle, complains in 1787 of the escalation of this war, carried out mostly on paper:

> Boys in primary schools who have not yet studied philosophy but have read the writings of the Enlighteners … already strike out like young horses: they taste the war from afar: they prick up their ears, they tremble with desire to fight.[40]

Thus the "mute weaving of the spirit,"[41] the creeping, imperceptible change grounded in the patience of officials, teachers, and technocrats, cannot be separated from the noise with which literary and political writers simultaneously stylize the "struggle of the Enlightenment with superstition" as a decisive ideological battle. The din of discourses drowns out the silent proliferation of techniques and technologies; the volume of the public debate contrasts with the secrecy under which the "modernization" of modes of cognition and instruments of political control takes place.

But why is open opposition necessary at all? According to Hegel, it is the fact that the two positions are much too close to each other that is to blame. As stubborn beliefs, faith and pure insight are "both … essentially the same" ("beide wesentlich dasselbe"), which is why "their giving and receiving is an undisturbed flow of the one into the other" ("ihr Geben und Empfangen ein ungestörtes Ineinanderfließen"). However, it is precisely this proximity that forces an even more intense movement of distancing: in order to allow the difference from the criticized position to become clear, Enlightenment escalates the quarrel, taking pleasure in "thinking of itself as doing battle with something other" ("etwas anderes

---

39 Hegel, *Phänomenologie*, 404; *Phenomenology*, 317.
40 Joseph Anton Weißenbach, anonymously, "Und das heißt nun aufklären: Ein Versuch von einem unbekannten Verfasser," (n.p.: "printed at Ibi – ubi" ["gedruckt zu Ibi – ubi"], 1787), 35.
41 Hegel, *Phänomenologie*, 404; *Phenomenology*, 317.

zu bekämpfen meint").⁴² Even though this struggle is chimerical because the opponent is no Other, it contributes to the progress of the historical process. It introduces a moment of precipitation or "logical haste"⁴³ driving the movement of history forward – not only by utopically anticipating what does not yet exist, but also because the din it lets loose returns as a form of reentry into historical process, where it generates evidence of an epochal upheaval, or a so-called revolution in the way of thinking.

The "excess of words," the pathos of the revolution, is thus not only decorative embellishment to a process that takes place anyway in silence, subterraneously. Quite to the contrary, in historical events it takes on the not insignificant function of self-surprise: only the anticipatory theatrical staging of the rupture makes it possible for actors to in fact carry it out. Hence the escalating effect of the din, the inherent dynamics of the political-ideological debate, and the reciprocal goading up of forms of radicalism must also be considered, in addition to the "mute weaving of spirit."⁴⁴

Compared to the gradual, subterranean movement of the epidemic spread, this belligerence appears to be a rash movement that gets ahead of itself. The action here seems to come "too early" in a strange way, before the conditions are "ripe" for it. Psychoanalytically speaking, this attitude of rashness or precipitation can be summed up by the concept of hysteria, in contrast to an attitude of cautious waiting, which can be associated with the structure of obsessive-compulsive neurosis. Slavoj Žižek, who follows Jacques Lacan's version of this contradiction, quite obviously sympathizes with the attitude of hysteria, which Žižek attributes equally to Rosa Luxemburg and Hegel, the "most sublime hysteric"⁴⁵:

> The compulsive neurotic hesitates, postpones the act, waits for the right moment and of course never experiences it; but the hysterical woman throws herself "prematurely" into the act in a hurry and thus creates, in [its] failing, the conditions for the right moment of action.⁴⁶

And indeed, if Hegel had been forced to choose between these expressions, he would probably have claimed hysterical precipitation for himself rather than the

---

42 Hegel, *Phänomenologie*, 404; *Phenomenology*, 317.
43 See Jacques Lacan, "Logical Time and the Assertion of Anticipated Certainty: A New Sophism," trans. Bruce Fink and Marc Silver, *Published Newsletter of the Freudian Field* 2, 2 (1988): 4–22, here 19: "what we are trying to demonstrate: the function of haste in logic."
44 On the dynamic of political radicalization in the French Revolution, see Francois Furet, *1789: Vom Ereignis zum Gegenstand der Geschichtswissenschaft* (Frankfurt am Main: Ullstein, 1980 [1978]), 60–61.
45 See Žižek, *The Most Sublime Hysteric*; Žižek, *Le plus sublime des hystériques*.
46 Slavoj Žižek, *Liebe Dein Symptom wie Dich selbst! Jacques Lacans Psychoanalyse und die Medien* (Berlin: Merve, 1991), 13.

compulsive hesitation. In his writings there are numerous scenes in which the position of cautious waiting, stoic toleration, asceticism hostile to the world, and the passivity of a beautiful soul appear as a disgraceful alternative to militant indignation – from the Jews, who escape Egyptian bondage only because the Egyptians are defeated by the "ten plagues,"[47] to the Christians, who believe that "their guilt will be forgiven … because of the merit of another,"[48] to the servant who shuns the fight for life and death, and whose existence will therefore be exhausted in the service of his lord, as "desire *held in check*" and "vanishing *staved off*."[49] Accordingly, Hegel is also upset about the dullness of the Germans, who left the Revolution to the French:

> Wir Deutschen sind passiv erstens gegen das Bestehende, haben es ertragen; zweitens ist es umgeworfen worden, so sind wir ebenso passiv: durch andere ist es umgeworfen worden, wir haben es uns nehmen lassen, haben es geschehen lassen.[50]
>
> We Germans were passive at first with regard to the existing state of affairs, we endured it; in the second place, when that state of affairs was overthrown, we were just as passive: it was overthrown by the efforts of others, we let it be taken away from us, we suffered it all to happen.

But it would be too easy to see in such statements a fundamental rejection of strategies of waiting, abstention, and evasion. What Hegel calls "science" in the *Phenomenology of Spirit* has learned from the tricks and dodges of the Jesuits and Enlighteners: it appears not as a "dogmatism of self-assurance" ("versichernder Dogmatismus") but rather as a cunning observing that waits calmly for all discrete, positive determinations to mutually disassemble each other:

> [S]o ist sie die List, die, der Tätigkeit sich zu enthalten scheinend, zusieht, wie die Bestimmtheit und ihr konkretes Leben darin eben, daß es seine Selbsterhaltung und besonderes Interesse

---

47 See Hegel, "Der Geist des Christentums und sein Schicksal," 282: "Die Juden siegen, aber sie haben nicht gekämpft; die Ägypter unterliegen, aber nicht durch ihre Feinde, sie unterliegen, wie Vergiftete oder im Schlaf Ermordete, einem unsichtbaren Angriff"; Hegel, "The Spirit of Christianity," 190: "The Jews vanquish, but they have not battled. The Egyptians are conquered, but not by their enemies; they are conquered (like men murdered in their sleep, or poisoned) by an invisible attack."
48 Georg Wilhelm Friedrich Hegel, "Fragmente über Volksreligion und Christentum," in G.W.F.H., *Frühe Schriften: Werke*, vol. 1 (Frankfurt am Main: Suhrkamp, 1986), 9–103, here 99; see D'Hondt, *Verborgene Quellen des Hegelschen Denkens*, 165.
49 Hegel, *Phänomenologie*, 153; *Phenomenology*, 115.
50 Georg Wilhelm Friedrich Hegel, *Vorlesungen über die Geschichte der Philosophie III: Werke*, vol. 20 (Frankfurt am Main: Suhrkamp, 2015), 297; *Lectures on the History of Philosophy*, vol. 3, trans. E. S. Haldane and Frances H. Simson (London: Routledge, 1955), 391.

zu treiben vermeint, das Verkehrte, sich selbst auflösendes und zum Momente des Ganzen machendes Tun ist.[51]

In this way, that activity is a kind of cunning which, while seeming to abstain from activity, is looking on to see just how determinateness and its concrete life takes itself to be engaged in its own self-preservation and its own particular interest and how it is actually doing the very opposite.

Even though Hegel repeatedly expresses sympathy for precipitation, his overall philosophical strategy is more in keeping with the attitude of an almost superhuman patience. In the *Enzyklopädie der Wissenschaften (Encyclopedia of Sciences)*, Hegel himself pointed out how closely the idea of a "cunning of reason" ("List der Vernunft") that he formulated is related to the theological idea of divine providence.[52] But this central figure of Hegelian thought obviously has other sources, which are to be sought less in theological tradition than in the contemporaneous debates about the prevalence of the Enlightenment. The "cunning of reason," one could say, can also be traced back to what might be called the cunning of its techniques, to all of the strategies proven in the "struggle of the Enlightenment with superstition" for realizing the indirect, epidemiological, medial, and technical dissemination of "pure insight."

---

51 Hegel, *Phänomenologie*, 53–54; *Phenomenology*, 34.
52 See Georg Wilhelm Friedrich Hegel, *Enzyklopädie der philosophischen Wissenschaften im Grundrisse I: Werke*, vol. 8 (Frankfurt am Main: Suhrkamp, 1986), 365, § 209: "Man kann in diesem Sinne sagen, daß die göttliche Vorsehung, der Welt und ihrem Prozeß gegenüber, sich als die absolute List verhält"; *Hegel's Logic, Being Part One of the Encyclopaedia of Philosophical Sciences*, trans. William Wallace (Oxford: Clarendon Press, 1975), 247: "Divine providence may be said to stand to the world and its process in the capacity of absolute cunning."

Katrin Trüstedt
The Fruit Fly, the Vermin, and the Prokurist: Operations of Appearing in Kafka's *Metamorphosis*

1 Cultural Techniques and the Procedure of Literature

Rethinking objects of the humanities in terms of "cultural techniques"[1] involves a threefold shift: First, it continues a tradition of questioning the privileged position of the subject and relocating it within an institutional, material, and technical framework.[2] Second, it involves a "thinking in verbs"[3] that does not simply focus on the material infrastructures, but more specifically on the *operations* and *techniques* that interconnect actors and objects, media and phenomena, texts and meanings. Such a focus on processes, operations, and practices assumes that practices and operations take priority over both the specific order of phenomena they produce and the concepts that emerge from them, and it challenges the common assumption that cultural practices are intentionally determined by human agency. Thomas Macho suggests that *cultural* techniques are, third, precisely distinguished from other practices and techniques by their self-reflective character: "cultural techniques are *second-order techniques*."[4]

Given these general terms, literature should be an interesting case in point. Not only does literature seem to make us consider itself as a process rather than as a static monument – it moreover seems to suggest, quite literally, a "thinking in verbs." It also has the reflexive character of a *second-order technique*, always

---

[1] For two instructive examples, see Bernhard Siegert, *Cultural Techniques: Grids, Filters, Doors, and Other Articulations of the Real*, trans. Geoffrey Winthrop-Young (New York: Fordham University Press, 2015); Thomas Macho, "Zeit und Zahl: Kalender- und Zeitrechnung als Kulturtechniken," in *Bild – Schrift – Zahl*, ed. Sybille Krämer and Horst Bredekamp (Munich: Fink, 2003), 179–192.
[2] Actor-Network-Theory and Latour's approach serve as a major inspiration in this regard. For an instructive introduction, see Bruno Latour, *Reassembling the Social: An Introduction to Actor-Network-Theory* (Oxford: Oxford University Press, 2005).
[3] Cornelia Vismann, "Cultural Techniques and Sovereignty," *Theory, Culture & Society* 30, 6 2(2013): 83–93, here 83.
[4] Thomas Macho, "Second-Order Animals: Cultural Techniques of Identity and Identification," *Theory, Culture and Society* 30, 6 (2013): 30–47.

Open Access. © 2020 Katrin Trüstedt, published by De Gruyter. This work is licensed under a Creative Commons Attribution 4.0 International License.
https://doi.org/10.1515/9783110647044-017

also being in some way about literature – an operation of writing that turns in on itself, that is of or about writing. In this sense, literature can indeed be understood as a cultural technique. The question then would be, what it is that distinguishes literature from other cultural techniques. By way of re-reading Franz Kafka's *Die Verwandlung (The Metamorphosis)*, I want to suggest in the following that literature is not just self-reflective but also reflective of other types of cultural techniques and their own self-reflexivity. And this is so not just on the level of content, but on the level of the very form of its textual operations.

A crucial notion of operation with regard to the distinctive sense in which literature is a technique reflective of other techniques is the idea of literature as a "procedure" (*Verfahren*).[5] According to Victor Sklovskij, literature is a "procedure of estrangement." By producing what Sklovskij calls "impeded form," literature decelerates and interrupts our common routines in such a way that they first become fully apparent as cultural techniques.[6] If that is true, literature appears not just as a second-order technique (as can be said about many cultural techniques), but as a certain type of procedural reflection of cultural techniques. In order to develop this perspective and bring out the distinctive character of literary procedures, I will turn to a reading of a passage of one of Kafka's most-read literary texts.

In Kafka's *Metamorphosis*, there is a peculiar scene that seems to offer the prolonged elaboration of something like a procedural onset – an onset that, in addition, leads into a reversal. From outside his bedroom, Gregor Samsa is described as being imperatively addressed and pressured to open the door by his mother, his sister, his father, and the *Prokurist* of his employer. After prolonged hesitation and complex preparation, he is depicted as finally intending to make such an appearance: "He actually intended to open the door, actually present himself … ; he was eager to find out what the others, who were now so anxious to see him, would say at the sight of him."[7] When he is told to actually open the door

---

[5] See Rüdiger Campe, "Evidenz als Verfahren: Skizze eines kulturwissenschaftlichen Konzepts," in *Vorträge aus dem Warburg-Haus* 8 (2004): 105–134; Malte Kleinwort, *Kafkas Verfahren: Literatur, Individuum und Gesellschaft im Umkreis von Kafkas Briefen an Milena* (Würzburg: Königshausen und Neumann, 2004); Arne Höcker, "Literatur durch Verfahren: Beschreibung eines Kampfes," in *Kafkas Institutionen*, ed. A.H. and Oliver Simons (Bielefeld: transcript, 2007), 235–254.

[6] Viktor Sklovskij, "Die Kunst als Verfahren," in *Russischer Formalismus: Texte zur allgemeinen Literaturtheorie und zur Theorie der Prosa*, ed. Jurij Striedter (Munich: Fink, 1969), 3–35.

[7] Franz Kafka, *Oxforder Quartheft 17 (Die Verwandlung): Historisch-Kritische Ausgabe sämtlicher Handschriften, Drucke und Typoskripte*, Faksimile-Edition (Frankfurt am Main: Stroemfeld, 2003), 35: "Er wollte tatsächlich die Tür aufmachen, tatsächlich sich sehen lassen … ; er war begierig zu erfahren, was die anderen, die jetzt so nach ihm verlangten, bei seinem Anblick sagen würden." The translation is taken, with minor modifications, from Franz Kafka, *The Metamorphosis*:

to show and declare himself, however, the opposite of what the scene has built up to ensues. His appearance, expected and carefully prepared, fails to come to pass in the narration. It is not described but rather omitted by the text. Only the reaction of others is depicted – the mother as falling with her face on her breast, leaving her unable to see any appearance, the father as actively refusing to look at Gregor by covering his eyes with his hands, and the *Prokurist* as retreating from the scene. The father is then described as attempting to undo Gregor's entrance in driving him back into his room. The appearance, so urgently called for and prepared from different sides, is a misfire. As it is not described and not acknowledged as such by those who called for it, it seems to have resulted in a nonappearance that is then just as carefully undone. This scene as a whole recalls the way in which Benjamin has linked Kafka's writing to experimental settings: "Kafka's entire work constitutes a code of gestures which surely had no definite symbolic meaning for the author from the outset, rather the author tried to address such a meaning in ever-changing contexts and experimental arrangements [*Versuchsanordnungen*]."[8]

In this sense, Kafka's scene stages the onset, the beginning, and the reversal of what I would like to describe as a procedure of appearance, arranged around an omission of the actual appearance. It seems like a mirrored two-phase scene: an elaborate scene of preparation on the one hand, and a backward movement haltering, rewinding, and disintegrating, on the other. In what follows, I would like to propose a reading of this scene as staging a particular kind of proto-procedure going wrong. In doing so, I aim to show that literature is neither just a procedure like any other, nor just a description of a procedure. As I will indicate in the following, Kafka's *Metamorphosis* rather evokes other procedures – like those of the life sciences, the theatre, and the law – but evokes them by testing, confronting, and suspending them. This scene recalls other procedures and subverts them, showing how they are arranged around a meaning that is itself not given and yet to be attained, as Benjamin suggested. Instead of just being a second-order technique, it characterizes literature that the relation of materiality and meaning, of first and second order, is itself uncertain and constantly at play.

Translation, Backgrounds and Contexts, Criticism: The Norton Critical Edition, trans. and ed. Stanley Corngold (New York and London: Norton and Company, 1996), 10.
8 Walter Benjamin, "Franz Kafka: Zur zehnten Wiederkehr seines Todestages," in *Benjamin über Kafka. Texte, Briefzeugnisse, Aufzeichnungen*, ed. Hermann Schweppenhäuser (Frankfurt am Main: Suhrkamp, 1981), 18; my translation. The German original reads: "... daß Kafkas ganzes Werk einen Kodex von Gesten darstellt, die keineswegs von Hause aus für den Verfasser eine sichere symbolische Bedeutung haben, vielmehr in immer wieder anderen Zusammenhängen und Versuchsanordnungen um eine solche angegangen werden."

2 The Procedures of the Life Sciences: The Fruit Fly

The experimental arrangement of *The Metamorphosis* involves a configuration of rooms and devices that connect and separate them: doors, a familial order inhabiting and structuring these rooms, the various family members, the maid, and ultimately the *Prokurist*, organizing these rooms and their doors. In this way, the set up depicted in the narration presents itself as a form of what the German historian of science Hans-Jörg Rheinberger calls "assemblages."[9] Assemblages are, as he contends in view of the life sciences, configurations of various actors, materials, and processes, arranged in such a way, that they might produce a certain appearance: a deed that is at the same time an item of knowledge: a fact or result – in short: a *Tat-sache*.

Kafka's scene of appearance is, first of all, addressing a procedure of making something visible – but what, exactly? Gregor's appearance is, as the scene demonstrates, monstrous in more than one way. What is omitted and only readable in the reaction of the audience seems to preclude any form of straightforward presentation. This resonates with Kafka's claim regarding the possible depiction of Gregor's transformed state on the book cover: "The insect itself cannot be drawn."[10] This claim itself is somewhat paradoxical, since it, firstly, specifically identifies an insect, but secondly, claims that this insect *cannot* be made present ("*kann* nicht gezeichnet werden"). The "monstrous vermin" is characterized in its unpresentability by the narrator with two negative prefixes: "*un*geheures *Un*geziefer." Throughout the story, "it" is being designated by various actors as an "Untier," "ein[] solche[s] Tier," as well as a "Ding." These nondescriptions question not only the extent to which Gregor has an ascribable form, but also challenge his status: is "it" an object or a subject? Thing or living being? A mere brute or an intelligent being? Such "a generic species of vermin, a hybrid thing, a true *Mischling*,"[11] recalls model organisms that have played a crucial role in experimental procedures in the life sciences around Kafka's time. The respective model organisms are very often produced as *mutations* of given living things: organisms,

---

[9] Hans-Jörg Rheinberger, *An Epistemology of the Concrete: Twentieth-Century Histories of Life* (Durham, NC: Duke University Press, 2010), 5f.

[10] Kurt Wolff, *Briefwechsel eines Verlegers, 1911–1963*, ed. Bernhard Zeller and Ellen Otten (Frankfurt am Main: Fischer Verlag, 1966), 37; my translation. The German original reads: "Das Insekt selbst kann nicht gezeichnet werden."

[11] Simon Ryan, "Franz Kafka's *Die Verwandlung*: Transformation, Metaphor, and the Perils of Assimilation," in *Franz Kafka*, ed. Harold Bloom (New York: Infobase, 2010, new edition), 197–216, here 214.

such as the well-known fruit fly *Drosophila*, were used as *living models*[12]: living objects of manipulation that at the same time act as models and markers of the respective investigation. As Rheinberger describes them, these model organisms are hybrids in terms of their status as living beings or things, as subjects or objects of the procedure and ultimately in terms of what it is they make visible. Insofar as they function as living objects of knowledge, the evidence procedure is not located outside of them (an external observation of which these beings are the mere object) but actually runs through them: the mutations produced in the breeding of these model organisms is the main part of the procedure. Even if we want to locate them in terms of nature or culture, the status of these organisms seems blurred. The "wild" insect is not only tamed and domesticized as the house pet of laboratories. But through highly artificial procedures allowing for, giving rise to, and selecting mutations, their biological, "natural" make-up is technically modified. In this ambivalence between a (natural) object of research and an (artificial) instrument, the model organisms are supposed to bring knowledge of life to light:

> As Robert Kohler has forcefully argued by reference to the favorite of classical genetics, the fruit fly *Drosophila melanogaster*, model organisms function not just as exemplars, but also as research "instruments." ... As an instrument the model organism is one of the technical conditions of an experimental system in which an epistemic object acquires its characteristic contours. To stick to the example of classical genetics *Drosophila* mutants were not themselves *epistemic objects* when it came, say, to drawing up chromosomal gene maps, but rather *instruments* that helped pin down the relative position of genes – the object of scientific interest – on chromosomes. Indeed many of the *Drosophila* mutants identified in Thomas Hunt Morgan's laboratory were not interesting in and of themselves – because of their specific, "monstrous" defects – but, rather, served as mapping markers.[13]

The monstrous defects are the products of the mutations, but as such they are expected to make something else visible, over and above themselves. The experiment that arranges for these mutants to appear is not calling for their appearance

---

12 Rheinberger, *Epistemology of the Concrete*, 6: "That model organisms and the concept of model organism could emerge at all in this period presupposed the idea of a *general biology*, the notion that certain attributes of life were common to all living things and could consequently be experimentally investigated using *particular* organisms that were representative of all others [*konnten stellvertretend untersucht werden*]. In previous centuries, it was the *differences* between various living creatures that had commanded the interest of scientists, who in the natural history tradition had sought to account for life forms in all their diversity. ... If biology had asked, upon entering the ranks of the sciences around 1800, what distinguished living from nonliving things, it tended to ask, around 1900, what constituted life as such ... The concept and present-day meaning of the expression 'model organism' arouse under these epistemological conditions."
13 Rheinberger, *Epistemology of the Concrete*, 224; my emphasis.

as such but requires them as markers and instruments that are supposed to give access to underlying living processes at work in these creatures or their ancestors. The *Drosophila* mutants thus seem to hover between being an epistemic object to be observed, on the one hand, and a living actant and instrument that is supposed to bring an epistemic object to light, on the other. This also makes it unclear, what exactly is *to be seen*.

In his studies on experimental settings in the laboratory, Rheinberger has demonstrated that not only does general biology face the task of "creating" its singular models as *Tat-sachen*, but that those models actually have to remain indeterminate and in the process of becoming in order for them to be productive. He points to Gaston Bachelard and Georges Canguilhem to stress this point:

> Georges Canguilhem once quipped that models are distinguished precisely by a certain lack of knowledge [*Datenarmut*; so, more literally: "poorness in data"]: they are relevant to research only as long as they leave something to be desired. We can extend this idea: from the standpoint of the research process, models maintain their function only for as long as this representational relation remains somewhat hazy [*ein wenig unscharf*], only as long as we cannot say exactly what a particular model ultimately represents.[14]

So as part of a procedure (or "research process"), the *Drosophila* mutants are, in their existence as "natural objects," not only artificially manipulated, so that what is visible is produced and affected by the procedures of making it visible.[15] Moreover, they remain hazy with regard to their meaning. The procedure seems, similar to what Benjamin wrote with regard to Kafka, experimentally arranged around an open question of meaning.

---

[14] Rheinberger, *Epistemology of the Concrete*, 8. Analogous points can be made about the function of the example in philosophical judgment (Kant), the paradigm in scientific research (Kuhn), and life objects in laboratories in Bruno Latour's analysis: the singular "case" refers to something more general, and yet this relation of representation has to remain "somewhat hazy" if the example, paradigm or model is to have a productive, generative and genuine function beyond mere illustration of a general rule. On the underlying logic of exemplarity, see Giorgio Agamben, "What Is a Paradigm?" in G.A., *The Signature of All Things* (New York: Zone Books, 2009), 9–32.

[15] According to Rheinberger's studies on the experimentalization of life, general biology faces the task of "creating" its "cases" or singular entities that can be used as models. Following Bachelard's famous thesis that objects "do not *exist* in nature: they have to be technically produced," Rheinberger contends that "the sciences do not find their objects ready-made but have to constitute them using specific epistemic settings" (Rheinberger, *Epistemology of the Concrete*, 2). This is even more so the case with contemporary sciences: atoms or genomes need complex procedures in order to become apparent "objects of study." Rheinberger also describes paradoxical procedures of authentication in various forms of exhibitions that involve the hiding of the very procedures that made the exhibiting of bodies possible in the first place.

3 Theatrical Procedures: The Entrance

The scene in Kafka's *Metamorphosis* does not only evoke an experimental arrangement, it also invokes a theatrical setting with doors marking an *off-stage area* (Gregor's room), from where his entering of the stage (the living room) is expected by the various actors that are at the same time the designated audience of this appearance. "'Gregor,' *rief es*," as the German text says, highlighting the fact that the calling is attributed to an impersonal source and seems to transcend any personally defined being. This is strengthened by the multiplication of callings that arise from different sides and voices and resound in different registers and pitches: "'Gregor, Gregor,' [the father] called, 'what's going on?' And after a little while he called again in a deeper, warning voice, 'Gregor! Gregor!' At the other side door, however, his sister moaned gently, 'Gregor?...'"[16] The expectation is built up by the various actors who attempt to produce an entrance from different sides, calling upon Gregor to appear and to declare himself: to give "an immediate and precise explanation,"[17] as the *Prokurist* demands. It is this expectation by the others that moves Gregor to finally become eager to appear himself: "He actually [*tatsächlich*] intended to open the door, actually [*tatsächlich*] present himself and speak to the *Prokurist*; he was eager to find out what the others, who were now so anxious to see him, would say at the sight of him."[18]

The scene that he is said to initiate – "tatsächlich die Tür aufmachen, tatsächlich sich sehen lassen" – is at its core a theatrical one, like many of Kafka's scenes. When Benjamin claimed that Kafka addresses meaning in ever-changing contexts and experimental arrangements, he declared that the theatre is actually the matrix of such experimental set ups: "The theatre is the given place of such arrangements [*Das Theater ist der gegebene Ort solcher Versuchsanordnungen*]."[19]

---

16 Kafka, *Die Verwandlung*, 26–27; Kafka, *The Metamorphosis*, 5. The original reads: "'Gregor, Gregor,' rief er, 'was ist denn?' Und nach einer kleinen Weile mahnte er nochmals mit tieferer Stimme: 'Gregor! Gregor!' An der anderen Seitentür aber klagte leise die Schwester: 'Gregor? ... '"
17 Kafka, *Die Verwandlung*, 33; Kafka, *The Metamorphosis*, 9. The German original reads: "Ich spreche hier im Namen Ihrer Eltern und Ihres Chefs und bitte Sie ganz ernsthaft um eine augenblickliche, deutliche Erklärung."
18 Kafka, *Die Verwandlung*, 35; Kafka, *The Metamorphosis*, 10. The original reads: "Er wollte tatsächlich die Tür aufmachen, tatsächlich sich sehen lassen und mit dem Prokuristen sprechen; er war begierig zu erfahren, was die anderen, die jetzt so nach ihm verlangten, bei seinem Anblick sagen würden."
19 Walter Benjamin, "Franz Kafka: Zur zehnten Wiederkehr seines Todestages," in *Benjamin über Kafka. Texte, Briefzeugnisse, Aufzeichnungen*, ed. Schweppenhäuser, 18; my translation. Rainer Nägele describes the theatre machine as a "machine of phenomenalization." Against this background, theatre is revealed as the "model of presentation itself [*Darstellung überhaupt*]" – see

If the theatre is indeed the place of such experimental arrangements of gestures lacking symbolic meaning, the stepping on stage of one actor to expose himself is the theatrical gesture par excellence. It recalls the foundational scene constituting Greek theatre, when one actor steps out of the chorus to expose himself in front of an audience that is itself constituted as such and as separate from him by that very move.[20] What becomes clear in the instance of Gregor's expected stepping out is that making an entrance is not a simple and self-confined move but needs to be prepared.[21] The arrangement of doors, to be opened,[22] the knocking, calling, and waiting outside of the door is part of that preparation. The expectation of others, separated as viewers, is needed for an entry to be possible and noticeable as an entrance that has a certain significance invested in it and ascribed to it by these others. Since Gregor has been depicted as trying repeatedly and unsuccessfully to get out of bed, it is this constellation, not an act of a genuine will of his own, that gets him to actually make the first step towards appearing: "And more as a result of the excitement produced in Gregor by these thoughts than as a result of any real decision, he swung himself out of bed with all his might."[23] The assumed perception by the others, the insistent normative request of his appearance, and the curiosity as to how the others will react to his appearance, move him, as we are told, to finally "stand up" and open the door. The door, dividing and connecting inside and outside, is a central actor in this scene, preparing the stage with a dark offstage from which the entrance is possible. The gesture of opening of the door – "actually opening the door" – is necessary to appear – to "actually present himself" or, more literally, "indeed let himself be seen" (*tatsächlich sich sehen lassen*). The imagination of the others, who are all affixed to his door expecting his entrance, then causes Gregor to go to work on the door key. When he manages to unlock the door, it first opens without him being seen: this is the first step in the setting of the stage. Even once the door is finally open, Gregor is depicted as carefully maneuvering and continuing to prepare his entrance, so as to avoid falling into the room on his back, which would not count as properly entering the

---

Rainer Nägele, "Mèchanè: Einmaliges in der mechanischen Reproduzierbarkeit," in R.N., *Hölderlins Kritik der Poetischen Vernunft* (Basel: Engler, 2005), 133–149, here 137f.

20 See Hans-Thies Lehmann, *Theater und Mythos: Die Konstitution des Subjekts im Diskurs der antiken Tragödie* (Stuttgart: Metzler, 1991), 40ff.

21 See Bettine Menke, "Suspendierung des Auftritts," in *Auftreten. Wege auf die Bühne*, ed. Juliane Vogel and Christopher Wild (Berlin: Theater der Zeit, 2014), 247–274.

22 See Oliver Simons, "Schuld und Scham – Kafkas episches Theater," in *Kafkas Institutionen*, ed. Arne Höcker and O.S. (Bielefeld: transcript, 2007), 269–293.

23 Kafka, *Die Verwandlung*, 31; Kafka, *The Metamorphosis*, 8. The German original reads: "Und mehr infolge der Erregung, in welche Gregor durch diese Überlegungen versetzt wurde, als infolge eines richtigen Entschlusses, schwang er sich mit aller Macht aus dem Bett."

room: "Since he had to use this method of opening the door, it was really opened very wide while he himself was still invisible. He first had to edge slowly around the one wing of the door, and do so very carefully if he was not to fall flat on his back just before entering."[24] Although all kinds of careful preparations occur for him to actually enter the stage, his appearance is subsequently not narrated but rather omitted. It seems that Gregor's appearance itself remains unpresentable. Recall that Kafka held that "the insect itself cannot be drawn." Instead of Gregor, Kafka suggested to show the expectation of the others on the book's cover: "If I were to make suggestions for an illustration, I would choose scenes such as: the parents and the procurist in front of the closed door, or, even better, the parents and the sister in the lit-up room, while the door to the dark adjoining room stands open."[25]

The expectations of the others for Gregor to appear, their gazes and glances, are at the center of attention here, not his appearance itself. The scene upon which Kafka wants to base the book cover seems to be the exact moment when the door is already open, but Gregor's appearance does not (yet) take place. The contrast in illumination that informs this moment – the family in the well-lit living room facing the dark adjoining room – stresses the theatricality of this scene and alludes to the theatrical convention that the entering of the stage is supposed to take place from a dark back room to the stage and into the spotlight.

The carefully prepared appearance, however, is not only held in suspense for a while and not directly addressed and described by the narration; as we learn from the reaction of the others, however, the appearance is quite literally a non-appearance: it is not acknowledged by the others as Gregor entering the stage but as the appearance of a negativity. The response is described as one of not seeing, not wanting to see, retreating from and even repelling the (non)appearance. The reaction marks a theatrical situation in which any appearance precisely requires the acknowledgement of others. Put differently, theatrical appearances are marked by their to-be-seenness.[26] In this case, however, the opposite of a watched

---

24 Kafka, *Die Verwandlung*, 38; Kafka, *The Metamorphosis*, 12. The German original reads: "Da er die Türe auf diese Weise öffnen mußte, war sie eigentlich schon recht weit geöffnet, und er selbst noch nicht zu sehen. Er mußte sich erst langsam um den einen Türflügel herumdrehen, und zwar sehr vorsichtig, wenn er nicht gerade vor dem Eintritt ins Zimmer plump auf den Rücken fallen wollte."
25 Kurt Wolff, *Briefwechsel eines Verlegers*, 37; my translation. The German original reads: "Wenn ich für eine Illustration selbst Vorschläge machen dürfte, würde ich Szenen wählen, wie: die Eltern und der Prokurist vor der geschlossenen Tür oder noch besser die Eltern und die Schwester im beleuchteten Zimmer, während die Tür zum ganz finsteren Nebenzimmer offensteht."
26 See Stanley Cavell, "The Avoidance of Love: A Reading of *King Lear*," in S.C., *Disowning Knowledge in Seven Plays of Shakespeare* (Cambridge: Cambridge University Press, 2003), 39–123;

appearance and an appearance of acknowledgment ensues. That something or other seems indeed to be happening is only readable in the various ways in which Gregor cannot reach the gaze of his audience. The mother's face is "hidden on her breast," his father shields his eyes with his hands, and the *Prokurist* is repelled by an "invisible force":[27]

> But at Gregor's first words the procurist had already turned away and with curled lips looked back at Gregor only over his twitching shoulder. And during Gregor's speech he did not stand still for a minute but, without letting Gregor out of his sight, backed toward the door, yet very gradually, as if there were some secret prohibition against leaving the room. He was already in the foyer, and from the sudden movement with which he took his last step from the living room, one might have thought he had just burned the sole of his foot. In the foyer, however, he stretched his right hand far out toward the staircase, as if nothing less than an unearthly deliverance were awaiting him there.[28]

When the *Prokurist* retreats and watches Gregor only over his shoulder while leaving the scene, Gregor – in a comical reversal of the procedure in which the *Prokurist* had pressured him to step out of the door – actually now attempts in turn to go after him. The father then tries to undo the laboriously prepared entrance in driving Gregor back into his room, which turns out to be just as laborious. While Gregor is supposed to move into his room in reverse, as if actually some apparatus would allow us to rewind the movement, it becomes apparent that this movement is not only slow but that "in reverse he could not even keep going in one direction."[29] Gregor thus tries to turn around and, after further complications, it becomes apparent, with his body finally facing the door head-on, that he is too broad to get through anyway. Driven by his father, Gregor forces himself in the doorway and gets stuck. Finally, a hard shove by his father (*einen jetzt wahrhaftig erlösenden starken Stoß*) throws

---

and Michael Fried, *Why Photography Matters as Art as Never Before* (New Haven and London: Yale University Press, 2008).
27 Kafka, *Die Verwandlung,* 38; Kafka, *The Metamorphosis*, 12.
28 Kafka, *Die Verwandlung,* 40–41; Kafka, *The Metamorphosis*, 13. The German original reads: "Aber der Prokurist hatte sich schon bei den ersten Worten Gregors abgewendet, und nur über die zuckende Schulter hinweg sah er mit aufgeworfenen Lippen nach Gregor zurück. Und während Gregors Rede stand er keinen Augenblick still, sondern verzog sich, ohne Gregor aus den Augen zu lassen, gegen die Tür, aber ganz allmählich, als bestehe ein geheimes Verbot, das Zimmer zu verlassen. Schon war er im Vorzimmer, und nach der plötzlichen Bewegung, mit der er zum letztenmal den Fuß aus dem Wohnzimmer zog, hätte man glauben können, er habe sich soeben die Sohle verbrannt. Im Vorzimmer aber streckte er die rechte Hand weit von sich zur Treppe hin, als warte dort auf ihn eine geradezu überirdische Erlösung."
29 On the media conditions of time axis manipulation, see Friedrich Kittler, "Real Time Analysis, Time Axis Manipulation," *Cultural Politics* 13, 1 (2017): 1–18.

him back into the room and severely injures him. "Then at last everything was quiet."[30] The reaction of the "audience" that had so insistently approached his room and called for his appearance now negatively mirrors this preparation in the disintegration of the scene. The experiment has gone wrong and thereby shown retroactively that it has been an experimental arrangement to produce an entrance and to investigate what it takes for an entrance to count as such.[31] As a potential scene of subjectivation, it requires the call from an other: from the institutions of the family and the firm.[32] The way in which these institutions react to the way Gregor reacts to his interpellation makes it evident that something other than the interpellated subject must have "appeared" on the scene.

The peculiar address by the *Prokurist* demanding that Gregor steps out and declares himself, marks this as a procedure to make Gregor not just appear but appear in a certain expected way: "You ... cause your parents serious, unnecessary worry, and you neglect – I mention this only in passing – your duties to the firm in a really shocking manner. I am speaking here in the name of your parents and of your employer and ask you in all seriousness for an immediate, clear explanation."[33] Gregor is addressed by an institution to appear and explain himself as the person he is expected to be: as the worker and bread winner for the family. Appearing, as it is reflected in this scene *ex negativo*, always means appearing *as someone*. The failing of the entrance is due to the expectations of not just any but a particular entrance. Given that expectation, the nonappearance of a monstrous vermin seems not like an instance of just falling short of this expectation but as a radical negation of any determinate expectation. At any rate, one might ask whether a failure to meet the expectations of others is enough to make one appear as an "ungeheures Ungeziefer." What appears in this distinctive form of nonappearance cannot be just a certain trait or characteristic that is not expected or called for at this point; it is a more radical other of the expected

---

30 Kafka, *Die Verwandlung*, 44; Kafka, *The Metamorphosis*, 15.
31 See Paul North, *The Yield: Kafka's Atheological Reformation* (Stanford, CA: Stanford University Press, 2015), 64ff.
32 See Arne Höcker, "Literatur durch Verfahren," 244: "And just like in Louis Althusser's brief theoretical narrative, the subject finds its origin in the interpellation by the institution."
33 Kafka, *Die Verwandlung*, 33; Kafka, *The Metamorphosis*, 9. The German original reads: "Sie ... machen Ihren Eltern schwere, unnötige Sorgen und versäumen – dies nur nebenbei erwähnt – Ihre geschäftliche Pflichten in einer eigentlich unerhörten Weise. Ich spreche hier im Namen Ihrer Eltern und Ihres Chefs und bitte Sie ganz ernsthaft um eine augenblickliche, deutliche Erklärung."

person: the remainder of the un-person or im-person that is the correlate of any subjectivation.[34]

4 Legal Procedures: Procura

The nonappearance of a human subject as a "monstrous vermin," an "ungeheures Ungeziefer" has political resonances and raises questions not only of textual and literary form, but also of legal status. Appearing in the doubly negative mode of an "ungeheures Ungeziefer" is not only in tension with appearing in a certain role in the eyes of the others, but also in tension with being an actor with legal status.[35] If we attend to this register, it seems obvious that the theatrical scene of entrance is closely linked to the scene of an appearance *before the law*. The "monstrous" entrance challenges our ways of conceiving what it could mean to display personhood and to act as an agent, to appear as a being to which we could accord a certain legal status just as well as a certain narrative intelligibility. Gregor's failure to appear thus raises the question of what it could mean to resist or withdraw from such forms of appearance or performance and what it would mean for such forms of appearance to be withheld from someone. If it is indeed vital to our legal and political existence that we can make an appearance, and if certain political pathologies express themselves in extreme form of depravation of legal status, the nonappearance of the "ungeheures Ungeziefer" in facts raises political questions.[36]

---

34 See Niklas Luhmann, "Die Form 'Person'," in N.L., *Soziologische Aufklärung 6: Die Soziologie und der Mensch* (Opladen: Westdeutscher Verlag, 1995), 142–154, here 148: "Other things remain on the unmarked side, since one does not expect them to become the object of communication. What belongs to the unperson therefore is as indeterminate as the non-loop in knitting or the non-hole in billards."
35 The positive term "Geziefer" from medieval sacrifice did not survive in German, nor did "Geheuer" as a positive term; see "Geziefer," and "Geheuer," in *Deutsches Wörterbuch*, ed. Jacob Grimm and Wilhelm Grimm (Stuttgart: Hirzel, 1853ff.), vol. 7, col. 7045–7048, http://www.woerterbuchnetz.de/DWB?lemma=geziefer; and vol. 5, col. 2478–2480, http://www.woerterbuchnetz.de/DWB?lemma=geheuer (both webpages last visited on January 28, 2020).
36 In *The Origins of Totalitarianism*, Hannah Arendt made clear that the stateless are exposed to such an extreme form of rightlessness that they can only make themselves known legally and become visible again by violating the law, as that may secure them the rights and representation before the law: "As a criminal even a stateless person will not be treated worse than another criminal, that is, he will be treated like everybody else. Only as an offender against the law can he gain protection from it. ... The same man who was in jail yesterday because of his mere presence in this world, who had no rights whatever and lived under threat of deportation, or who was

The cultural history of the term "Ungeziefer" points to two important connotative meanings. As Stanley Corngold has shown, the noun "Ungeziefer" in Middle High German signifies an "unclean animal not suited for sacrifice."[37] Sander L. Gilman detailed the anti-Semitic connotations that the word "Ungeziefer" carried in Prague around the turn of the century. In German and Austrian anti-Semitic publications, Jews were referred to as "rats," "mice," "insects," and "vermin," and more specifically, "Ungeziefer der Menschheit."[38] In the context of the Dreyfus Affair, in which – as Simon Ryan has shown – Kafka was keenly interested, the Jew was described as "a stinking and dangerous animal, a plague, a centipede, a microbe, a mite, a cancer, an ugly spider and synagogue lice. ... 'Long live the sabre that will rid us of the vermin.'"[39] Reports of Dreyfus's trial and incarceration focused particularly on the alleged unfitness of male Jews for military service.[40]

When Ryan refers to Gregor as "a generic species of vermin, a hybrid thing, a true *Mischling*," he points to this anti-Semitic context of bio-politics.[41] Following this line of thought, one might ask whether the precarious character of this "monstrous vermin" is connected to a certain political and legal logic of exception and whether the appearance as an "ungeheures Ungeziefer" is the shadow thrown by a failure to appear as a legal person. Clearly, *The Metamorphosis* does not seem to address any distinct legal and political regimes especially prone to deprive people of their rights and to turn them into un-persons. But the fact that the failure to appear as an accountable agent is connected to the (non)appearance

---

dispatched without sentence and without trial to some kind of internment because he had tried to work and make a living, may become almost a full-fledged citizen because of a little theft. Even if he is penniless he can now get a lawyer, complain about his jailers, and he will be listened to respectfully. He is no longer the scum of the earth but important enough to be informed of all the details of the law under which he will be tried. He has become a respectable person." Hannah Arendt, *Origins of Totalitarianism* (Cleveland and New York: Meridian Books, 1962), 286–287.

37 Stanley Corngold, *The Commentator's Despair: The Interpretation of Kafka's "Metamorphosis"* (Washington: National University Publications, 1973), 10.

38 Stanley Corngold, *The Commentator's Despair*, 31, 80; the last phrase stems from Ernst Hiemer, *Der Jude im Sprichwort der Völker* (Nuremberg: Der Stürmer, 1942), 34–40.

39 Ryan, "Franz Kafka's *Die Verwandlung*," 11; see Jean Denis-Bredin, *The Affair: The Case of Alfred Dreyfus* (New York: George Braziller, 1986), 351.

40 See Ryan, "Franz Kafka's *Die Verwandlung*," 11. Kafka writes to Milena Jesenská: "Ist es nicht das Selbstverständliche, daß man von dort weggeht, wo man so gehaßt wird (Zionismus oder Volksgefühl ist dafür gar nicht nötig)? Das Heldentum, das darin besteht, doch zu bleiben, ist jenes der Schaben, die auch nicht aus dem Badezimmer auszurotten sind." Franz Kafka, *Briefe an Milena*, enlarged and newly arranged edition, ed. Jürgen Born and Michael Müller (Frankfurt am Main: Fischer, 2015), 288.

41 Ryan, "Franz Kafka's *Die Verwandlung*," 16.

of this remainder, suggests that *The Metamorphosis* has a latent connection to these issues.[42]

In order to bring out the text's legal and political implications, we can start from a certain "appearance, in the legal sense" that concerns Kafka and that Derrida has outlined by reference to *Before the Law* – an appearance that has a strong kinship with the theatrical sense of entering the stage and thus is both a matter of performative action and of legal procedure: "To appear before the law means in the German, French, or English idiom," as Derrida reminds us, "to come or to be brought before judges, the representatives or guardians of the law, for the purpose, in the course of a trial, of giving evidence or being judged. The trial, the judgment (*Urteil*), this is the place, the site, the setting – this is what is needed for such an event to take place: 'to appear before the law.'"[43]

Assuming, as a starting point, that the cited passage of *The Metamorphosis* also marks a place, a site, a setting of appearing in front of others, or before the law, a first moment that Kafka's text brings out is a *hesitation to appear*. This hesitation to appear is also present in *The Trial* and *Before the Law*, and it is connected to a certain procedural structure of Austrian law of the time. Austrian procedural reforms in the nineteenth century had led to an oral and public main trial but continued to rely on a secret and written pretrial. As Wolf Kittler has convincingly shown, Kafka's *Trial* seems to remain arrested, or lingers, in some important sense in the stage of the secret and written pretrial proceedings that precede the main oral trial.[44] Kafka's *Trial* never leaves those preliminary stages and thus never results in an actual and public "appearance before the law in order to be judged." Something similar can be said of Kafka's *Metamorphosis*, even though elements of legal procedure are of course much less prominent in this text. If we attend to these elements, however, it becomes clear that *The Metamorphosis* is

---

[42] The notion of a "minor literature" by Deleuze and Guattari also takes up these political implications; see Gilles Deleuze and Félix Guattari, *Kafka: Toward a Minor Literature* (Minneapolis and London: University of Minnesota Press, 1986). They do so not from the side of reflecting on a certain depravation of rights that manifests itself in Gregor's (non)appearance as a vermin but rather from the complementary perspective that recognizes a certain movement of resistance in this (non)appearance. By failing to appear in his institutional role and by appearing as this "Ungeziefer" instead, Gregor points to a certain dynamic of "becoming animal." According to this reading of a nonappearance of a monstrous vermin, the question of meaning and nonmeaning, of material and symbol, is itself constantly at stake.
[43] Jacques Derrida, "Before the Law," in J.D., *Acts of Literature*, ed. Derek Attridge (New York: Routledge, 1992), 188.
[44] Wolf Kittler, "Heimlichkeit und Schriftlichkeit: Das Österreichische Strafprozessrecht in Franz Kafkas Roman *Der Proceß*," *The Germanic Review. Literature* 78, 3 (2003): 194–222.

also troubled by attempts to appear before the law and lingers in a preliminary stage of a procedure.

If we think of the structure of legal proceedings in Kafka's time, we might say that the described scenes remain in the pretrial proceedings rather than entering the actual oral trial. The procedural reforms first initiated by the so-called *Code d'instruction criminelle* of the Code Napoléon transformed what used to be an inquisitorial procedure under the principle of secrecy and written procedure into a public and open procedure (*Öffentlichkeit*) that involved oral presentation of evidence (*Mündlichkeit*) and its immediate presentation before the deciding judges (*Unmittelbarkeit*).[45] The need for an immediate appearance before the law recalls the Prokurist's demand for an "*immediate* and precise explanation," and Gregor's attempts (*tatsächlich die Tür öffnen, tatsächlich sich sehen lassen*) suggest that we are here confronted with a scene of an investigation ultimately requiring the subject to expose itself to the gaze of the law, without mediation, as it were.

The goal of the reformed procedures was the appearance of a subject in its immediate, authentic, and actual presence before the law. The proceedings turned away from the priority of the professional representatives and instead focused on the accused subject and the project to make this subject, its intentions and deeds, appear in court. This far-reaching reform in the wake of the French Revolution emphasizes the presence of the subject and aligns such immediate subjective presence with orality.[46] This change of procedure was initiated in the name of the subject who was supposed to have the "right" to appear before the law and defend itself. And yet the reforms concerned only the last or main part of a trial. The pretrial procedures remained in line with the principle of secrecy and of written procedure. What "immediacy" meant in this reformed context, then, is that evidence established during the written and secret stage of the procedure had to be *reproduced* during the main trial in an oral and public manner. The supposed immediacy of the procedure therefore relied precisely on this mediated written procedure while also attempting to conceal the complex procedural

---

[45] Carl Joseph Anton Mittermaier, *Die Mündlichkeit, das Anklageprinzip, die Öffentlichkeit und das Geschworenengericht: in ihrer Durchführung in den verschiedenen Gesetzgebungen* (Stuttgart and Tübingen: Cotta, 1845).

[46] On the implications of that change, see also Cornelia Vismann: "As the criminal process in the nineteenth century shifted to the maxim of oral trials, the emphasis was no longer on the lawyer as orator but rather on the interrogated subject. Since then, the criminal process has aimed at taking the wrongdoer to task (*den Täter zur Rede zu stellen*) in order to hold him accountable. It is he who has to be heard, in order for him to be judged." Cornelia Vismann, *Das Recht und seine Mittel* (Frankfurt am Main: Fischer, 2012) 229; my translation.

entanglement of forms in order to enable an "actual, immediate appearance in person" before the law.

While one instance of calling upon Gregor to appear is one of urging him (the father "mahnte nochmals," urged again), another (the sister's) is one of lamenting, moaning, or suing ("klagte leise"), as the German term *klagen* also has this legal meaning ("klagen," "anklagen," "verklagen"). In this way, Gregor also seems to be – just as Josef K. was in the *Trial* – called upon to appear *in the legal sense* – as Derrida had it: "to come or to be brought before ... the representatives or guardians of the law, for the purpose, in the course of a trial, of giving evidence or being judged."[47] As the *Trial* never actually reaches the main part of an oral and public appearance that would be needed in order for K to be judged, but remains in the preliminary stages, *The Metamorphosis* offers a scene of Gregor being called to appear – to come out of his room – that nevertheless results in a scene of not-appearing to then remain, just like the secret pre-procedures, largely "behind closed doors."

This oscillating mode of being called to appear, attempting, hesitating, and failing to appear involves not only a dual constellation – Gregor facing an other, calling him – but a more complex scenario indebted to an older, Roman procedural tradition (we know Kafka studied Roman law as part of his law school curriculum): namely that which is encapsulated in the juridical genre of rhetoric where an orator, patron, or procurator speaks for another (the client) against an accusing or antagonistic party, and in front of a third party who is addressed so that they may judge.[48]

The nonappearance of the vermin that raises questions of form and legal status therefore also raises the question of (legal and literary) representation by others, to others. When the front doorbell rings, Gregor expects "someone from the office," a thought that paralyzes him – "'They're not going to answer,' Gregor said to himself, captivated by some senseless hope. But then, of course, the maid went to the door as usual with her firm stride and opened up. Gregor only had to hear the visitor's first word of greeting to know who it was – *the procurist himself*."[49] The *Prokurist* in Kafka's story is referred to not by his name, but by his

---

[47] On the comparison with Josef K. in the *Trial*, see Höcker, "Literatur durch Verfahren," 245.
[48] On Kafka's familiarity with Roman law, see Peter-André Alt, *Franz Kafka. Der ewige Sohn: Eine Biographie* (Munich: Beck, 2008), 128. On the rhetorical scene of judging, see Rüdiger Campe, "An Outline for a Critical History of Fürsprache: Synegoria and Advocacy," *DVjs* 82 (2008): 355–381; R.C., "Kafkas Fürsprache," in *Kafkas Institutionen*, ed. Arne Höcker and Oliver Simons (Bielefeld: transcript, 2007), 189–212.
[49] Kafka, *Die Verwandlung*, 30; my emphasis; Kafka, *The Metamorphosis*, 7. The German original reads: "'Sie öffnen nicht,' sagte sich Gregor, befangen in irgendeiner unsinnigen Hoffnung. Aber

function – which is to speak and act for an Other. The Latin *procuratio* designates an agent taking care (*cura*) of some task for (*pro*) another, as John Hamilton has developed in greater length.[50]

> [D]id the procurist himself have to come, and did the whole innocent family have to be shown in this way that the *investigation of this suspicious affair* could be entrusted only to the intellect of the procurist? And more as a result of the excitement produced in Gregor by these thoughts than as a result of any real decision, he swung himself out of bed with all his might."[51]

The situation of the *Prokurist* speaking for the company and acting as Gregor's adversary in front of the family in some kind of investigative procedure – evoked by the text speaking of this suspicious circumstance to be investigated – leads to the scene of (non)appearance.

What distinguishes Kafka's procedure in this scene from the legal procedures it recalls is that it never stabilizes. Rather, the various roles constantly change places and rearrange themselves, in what Benjamin called "ever-changing contexts" and "experimental arrangements." When the *Prokurist* addresses Gregor "himself," the constellation shifts and his mother – until then part of the "innocent" third party before whom the *Prokurist* speaks for the company vis-à-vis Gregor – now steps in and in turn speaks "for Gregor," in his place, defending him as "his *Prokurist*," so

---

dann ging natürlich wie immer das Dienstmädchen festen Schrittes zur Tür und öffnete. Gregor brauchte nur das erste Grußwort des Besuchers zu hören und wußte schon, wer es war – *der Prokurist selbst*."

50 A *"Prokurist"* is someone, "whose functions include both representation and execution, both speaking for the company and carrying out its policies," as Hamilton writes. "In accordance with Austrian business practice, the Prokurist is the agent who has been delegated 'full power of attorney' (*Prokura*), the *pleinpouvoir* or *Vollmacht*, to inspect the case on the firm's behalf." See John Hamilton, "Procuratores: On the Limits of Caring for Another," *Telos* 170 (2015): 7–22. It is in the presence of this *Prokurist*, reduced to his function of speaking not for himself but for another, that Gregor loses the ability to speak for himself. Moreover, the *Prokurist* speaks primarily not for another person with a genuine voice of his/her own, but for "the company." Gregor claims that the *Prokurist* himself (again: *Prokurist selbst*) has a better understanding of the organization than the boss. See also Doreen Densky, "Proxies in Kafka: *Konzipist* FK and *Prokurist* Joseph K.," in *Kafka for the Twenty-First Century*, ed. Stanley Corngold and Ruth Gross (Rochester, NY: Camden House, 2011), 120–135.

51 Kafka, *Die Verwandlung*, 31; my emphasis; Kafka, *The Metamorphosis*, 8. The German original reads: "[M]ußte da der Prokurist selbst kommen, und mußte dadurch der ganzen unschuldigen Familie gezeigt werden, daß die *Untersuchung dieser verdächtigen Angelegenheit* nur dem Verstand des Prokuristen anvertraut werden konnte? Und mehr infolge der Erregung, in welche Gregor durch diese Überlegungen versetzt wurde, als infolge eines richtigen Entschlusses, schwang er sich mit aller Macht aus dem Bett."

to speak, against an assumed accusation by the company, and thereby giving an account of him in a third person narration. "'He's not well,' said the mother to the procurist."[52] The mother makes a claim on Gregor's behalf that Gregor himself – despite the numerous speculations about travel-sickness – had denied in those exact same terms ("Gregor fühlte sich tatsächlich ganz wohl"), thereby shows that she regards herself to be in a better position to speak for him than he himself could.[53] The silenced conflict between these statements (*Gregor fühlte sich tatsächlich ganz wohl* and *ihm ist nicht wohl*) suggests that she at the same time blocks him by speaking in his place.

While the *Prokurist* speaks for the company, in his address to Gregor he turns the constellation yet again in claiming to speak for the parents first and foremost and for his company only in the second place: "I am speaking here in the name of your parents and of your employer and ask you in all seriousness for an immediate, clear explanation [*augenblickliche, deutliche Erklärung*]."[54] Claiming to speak not in his own but in the mother's name in demanding an explanation – even though Gregor's mother had just given him such an explanation, namely that Gregor is not well – the *Prokurist* claims the power to speak more convincingly for the mother than she can speak for herself, and ironically so, just as the mother with regard to Gregor. Demanding "quite seriously" an "*immediate* and precise explanation" stresses again that this is a scene of investigation that ultimately aims for an "immediate appearance" before the law, which is exactly what the procedural reforms of the nineteenth century envisioned with regard to the main trial, to wit, an oral presentation and the bodily presence of the subject itself.[55]

---

52 Kafka, *Die Verwandlung*, 32; my translation. The German original reads: "'Ihm ist nicht wohl,' sagte die Mutter zum Prokuristen."
53 On speculations about Gregor's travel sickness, see John Zilcosky, "'Samsa war Reisender': Trains, Trauma, and the Unreadable Body," in *Kafka for the Twenty-First Century*, ed. Stanley Corngold and Ruth Gross (Rochester, NY: Camden House, 2011), 179–206.
54 Kafka, *Die Verwandlung*, 33; Kafka, *The Metamorphosis*, 9. The German original reads: "Ich spreche hier im Namen Ihrer Eltern und Ihres Chefs und bitte Sie ganz ernsthaft um eine augenblickliche, deutliche Erklärung."
55 See Peter Horn, "Tier werden, um der Sprache, der Macht, zu entkommen: Zu Kafkas Verwandlung," in *Sprache und Macht*, ed. Walther Köppe (Frankfurt am Main: Peter Lang, 1993), 101–22, here 104: "Unsere Aussagen sind individualisiert und unsere Sprache subjektiviert nur in dem Maße, in dem ein unpersönliches Kollektiv das erfordert. Wir werden immer durch Sprache zum Sprechen gebracht und durch das Sprechen als Subjekte unterworfen; [...] Dieses unpersönliche Kollektiv interpelliert uns als Subjekte und befiehlt uns: Du wirst als Organismus organisiert sein und deinen Körper als Signifikant und Signifikat artikulieren, du wirst als Subjekt festgenagelt werden, und wenn du dich weigerst, bist du ein Ungeziefer."

In a final, yet crucial twist, the *Prokurist* suggests himself as a potential procurator for Gregor vis-à-vis the company's boss, but only by way of denying his willingness to actualize such a potential *procuratio*:

> The head of the firm did suggest to me this morning a possible explanation for your tardiness – it concerned the cash payments recently entrusted to you – but really, I practically gave my word of honor that this explanation could not be right. But now, seeing your incomprehensible obstinacy, I am about to lose even the slightest desire to *stick up for you in any way at all* [verliere ganz und gar jede Lust, *mich auch nur im geringsten für Sie einzusetzen*].[56]

"*Mich für Sie* einsetzen" means "taking your side," "sticking up for you," but also "taking your place" (setting *me* in for *you*, positing me instead of you). This declaration by the *Prokurist* seems to be a threat of a double bind: the threat made by the *Prokurist* to take Gregor's place or part, just as much as the threat to not take his place or part. This double bind seems to be inherent to the practice of *procuratio* as speaking or acting on behalf of another and stepping into another's place (in German: *Stellvertretung*).

Instead of his appearance, we see the image of Gregor in the picture hanging on the wall. It depicts Gregor in his former shape, which is characterized as "carefree" (*sorglos*) and which displays Gregor as someone of rank. As his attempt to appear has just failed, the picture shows him as a proxy, namely a lieutenant: "Right opposite Gregor on the wall hung a photograph of himself in military service, as a lieutenant, hand on sword, a carefree smile on his face, inviting respect for his uniform and military bearing."[57] *Lieu-tenant* – a participle of *lieu tenir* – in French means "place holding." Gregor's appearance is not being described, only the reactions to it; in place of his appearance, we only find the description of this picture of Gregor as a place holder, itself occupying the place and blocking the central omission of his appearance.[58]

---

56 Kafka, *Die Verwandlung*, 34; my emphasis; Kafka, *The Metamorphosis*, 9. The German original reads: "Der Chef deutete mir zwar heute früh eine mögliche Erklärung für Ihre Versäumnisse an …, aber ich legte wahrhaftig fast mein Ehrenwort dafür ein, daß diese Erklärung nicht zutreffen könne. Nun aber sehe ich hier Ihren unbegreiflichen Starrsinn und verliere ganz und gar jede Lust, *mich auch nur im geringsten für Sie einzusetzen*."
57 Kafka, *Die Verwandlung*, 39.
58 This scene could be read as a "Vexierbild" in Rainer Nägele's sense; see Rainer Nägele, "Vexierbilder des Andern. Kafkas Identitäten," in R.N., *Literarische Vexierbilder: Drei Versuche zu einer Figur* (Eggingen: Isele, 2002), 9–29; on the importance of this picture, see also Ryan, "Franz Kafka's *Die Verwandlung*," 5: "As Gilman's evidence for the cultural significance of Jewishness in relation to fitness for military service will suggest, a direct link may be made between the suggestion of loss when Gregor catches sight of his portrait and his experience, only a few moments before struggling to open the door to the living room, of the loss of his normal human

5 Literature as Procedure: *Hope and Fear Alike*

That Gregor's metamorphosized attempt to appear results in a retreat and dissolution of the scene seems to be in line with the inherent ambivalence of the procedure. On the one hand, it urgently aims to produce an actual and immediate appearance; and yet such an "im-mediate appearance" without the imprint of the procedure is impossible in the complex artificial process that necessarily involves many actors, positions, movements and operations informing the *Tatsachen* they are treating. The complexity and abstraction of the procedure even seem to further the need for an authentic, actual, and immediate appearance (that is, an appearance without media) untouched by the procedure that aims at producing it. So when such an appearance is actually attempted, the procedure seems to implode. The text not only exposes the inherent tension in this project of appearance and traces the transformation that one undergoes in the process of such a failed appearance. It somehow manages to make this appear and to produce the evidence of something that is just the opposite of something presentable: ein "ungeheures Ungeziefer." The immense political implications are latent but readable in the scene.

We can see both the initiation of such procedures of representation in *The Metamorphosis* that seem to aim at the subject finally appearing, as well as moments of resistance against this process: procedures of hovering, halting, and ultimately breaking off. The procedures never come to the stage of a main trial where a person appears, a case is made and debated, and a verdict reached. This hesitation finally recalls one other tradition of legal procedure: the tradition of rabbinic interpretation. Benjamin, among many others, connected Kafka's texts to a particular aspect of that tradition when he emphasized the preliminary character of these "procedures" that are experimentally arranged around an open meaning:[59]

> We may remind ourselves here of the form of the Haggadah. ... Like the haggadic parts of the Talmud, these books, too, are stories; they are a Haggadah that constantly pauses, luxuriating in the most detailed descriptions, in the simultaneous hope and fear that it might encounter the halachic order, the doctrine itself, en route [*immer in der Hoffnung und Angst zugleich, die halachische Order und Formel, die Lehre könnte ihr unterwegs*

---

voice." Hartmut Binder calls Gregor in the picture a "Stellvertreter der Staatsmacht" and a borrowed authority; see Hartmut Binder, *Kafkas "Verwandlung": Entstehung, Deutung, Wirkung* (Frankfurt am Main, Basel: Stroemfeld, 2004), 449f.

59 As Judith Butler has pointed out, Kafka read "– 'greedily' as he puts it – *L'Histoire de la littérature Judéo-Allemande* by Meyer Pines, which was full of Hasidic tales, followed by Fromer's *Organismus des Judentums*, which details rabbinic Talmudic traditions." Judith Butler, "Who owns Kafka?" *London Review of Books* 33, 5 (2011): 3–8, here 8.

zustoßen]. ... Novels are sufficient unto themselves. Kafka's books are never that; they are stories pregnant with a moral to which they never give birth [*die mit einer Moral schwanger gehen, ohne sie je zur Welt zu bringen*].[60]

In *The Metamorphosis*, it is the hesitation before stepping out of the dark room in order to appear before the representatives (*Prokurists*) of the institution (the family, the firm, the law). Just as the Haggadah "constantly pauses ... in the simultaneous hope and fear that it might encounter the halachic order, the doctrine itself, en route," so too does Gregor seem to be in hope and fear alike of actually appearing. The scene of appearance in Kafka's *Metamorphosis* is, as I have tried to show, not just a procedure like any other, nor is it a mere description of one. Rather, it evokes procedures of evidence production constitutive of the life sciences, the theatre, and the law, but it does so in an experimental arrangement in and of itself. The laboratory, theatrical, and legal operations are estranged, diffracted, and decelerated by the arrangements of the literary text. In drawing upon and transforming, subverting or interrupting procedures in this manner, the ways of literature ultimately allow us to reflect upon them in their essence: we gain a sense both of their operativity and also their aporias. This sense is dependent upon a specific way in which the evoked operations are arranged – neither in the form of a detached observation, nor of a teleological and practical engagement, but instead in a distinct form of playful experiment that I would call a "literary procedure."

---

[60] Walter Benjamin, "Franz Kafka: Beim Bau der Chinesischen Mauer," in *Benjamin über Kafka*, ed. Hermann Schweppenhäuser, 39–46, here 42.

Christiane Lewe
Collective Likeness: Mimetic Aspects of Liking

1 Like/Unlike

When Facebook introduced six versions of the like button in 2016 (Facebook Reactions), it placated long-standing calls for a dislike button without directly complying with these requests. Ever since the like button had been introduced in 2009, users had come together in Facebook groups to demand a way of expressing "thumbs down" to complement the regular "thumbs up." Many saw the absence of a dislike button as an exclusion of negative points of view. And even just a few years ago, there were still complaints that a lack of possibilities for expressing criticism, disapproval, and dissent on Facebook exerted pressure to conform.[1] The vivid desire for a dislike button indicated that negative affects, too, could be a motivation for liking. Not only adorable kittens can become viral successes but also content that triggers passionate hate, outrage, and dismay. Because the like button carries connotations of a positive judgment, however, it did not seem to be the right operation for reacting, for example, to images of a natural disaster. Instead of introducing a dislike button, Facebook added a range of various emojis symbolizing love, laughter, amazement, sadness, and anger, in addition to thumbs up. Facebook thus multiplied the opportunities available to react to all different kinds of content.

Just why the dislike button never had a chance becomes apparent if we understand the like button not as a value judgment or an expression of emotion but rather as an act of connecting, as a basic operation of social networking. Regardless of whether liking is given symbolically positive or negative connotations, it is always an act of affirmation, networking, or engagement. Accordingly, the opposite of the like button would not be a pejorative version of itself but an act of dis-connection, or dis-engagement. And indeed, a previously activated like can

---

[1] Neil Strauss, "The Insidious Evils of 'Like' Culture," *The Wall Street Journal*, July 2 (2011), https://www.wsj.com/articles/SB10001424052702304584004576415940086842766 (visited on October 26, 2018); Sam Fiorella, "The Social Media Borg: A Culture of Likes," *The Huffington Post*, April 29 (2014), https://www.huffingtonpost.com/sam-fiorella/facebook-likes_b_3175615.html (visited on October 26, 2018); Lea Z. Singh, "The Facebook Culture of 'Like,'" *Culture Witness*, May 14 (2014), http://www.culturewitness.com/2014/05/the-facebook-culture-of-like.html (visited on October 26, 2018).

Translated by Michael Thomas Taylor

Open Access. © 2020 Christiane Lewe, published by De Gruyter. This work is licensed under a Creative Commons Attribution 4.0 International License.
https://doi.org/10.1515/9783110647044-018

be undone even without a dislike button. The difference articulated by the like is thus not: like/dislike but rather connected/disconnected. To put it pointedly: a user's opinion about something is completely irrelevant for the practice of liking as long as it affects them – be it positively or negatively.

2 Habit

But what does it mean to connect in social networks with content or other users? What is connection? The media studies scholar Wendy Chun has posed this question while rejecting models for explaining networks that she finds much too simplistic.[2] How has the network become such a sweeping, universal model that it equally reduces phenomena as diverse as global financial flows, public transportation infrastructure, food chains in the animal kingdom, and social media platforms to an utterly ridiculously simplified diagram of nodes and edges (connections)? According to Chun, what makes network models so irresistible is that they make visible what would otherwise remain invisible: complicated, confusing social and physical movements. Network models translate dynamic interactions and events in time into a spatial structure of nodes and lines of connection and thus create an overview, an order, and a predictability of future events. But this always makes them, as Chun writes, "too early and too late."[3] Networks are historiographic because they record previously occurring incidents of interaction as a line. And they are also predictive because they anticipate future interactions on the basis of previous, repeated interactions, which they project as lines of connection. This is why Chun also describes them as "imagined networks": "They describe future projections as though they really existed; they relay past events as if they were unfolding in the present."[4] In this way, networks spatially represent events in time. And the basis for this spatial connection of nodes is the (potential) repetition of past events, which is what makes it possible to anticipate future events:

> Networks ... spatialize the temporal by rendering constant repetition – or the possibility of repetition – into lines. To be able to repeat, then, is the basis of connection, or the basis for

---

[2] Wendy Hui Kyong Chun, "Networks NOW: Belated Too Early," in *Postdigital Aesthetics: Art, Computation and Design*, ed. David M. Berry and Michael Dieter (Basingstoke: Palgrave Macmillan, 2015), 289–315.
[3] Chun, "Networks NOW," 299.
[4] Wendy Hui Kyong Chun, *Updating to Remain the Same: Habitual New Media* (Cambridge, MA: MIT Press, 2017), 50.

the elucidation/imagining of connection. To be able to repeat is what links the machinic and the human.[5]

Understanding liking as an act of networking, as a link between a user profile and a piece of content, means counting on its repetition. Only if a like does not remain a singular event but is a repetitive practice carried out by the user-subject, can it be captured as a connection and hence as information. Facebook's newsfeed algorithms, which select and arrange content of a user's newsfeed based on their past likes, count on them liking things similar to what they have already liked. Wendy Chun thus suggests understanding connections that are projected based on repetition as habits: "Imagined connections are habits."[6] With recourse to a number of well-known theorists of habit – ranging from Felix Ravaisson, Henri Bergson, and Gilles Deleuze to William James, Gabriel Tarde, and Pierre Bourdieu – Chun understands habits as conditions for predicting events, that is to say, for projecting networks. Habits are repetitive practices that can be acquired and learned through imitating others. For this reason, habits connect those who practice them, making these practitioners similar to one another.

Through embodiment, culturally acquired habits become "second nature." They thus transcend dichotomies such as nature/culture, inside/outside, conscious/automatic, human/nonhuman, and individual/collective. They furthermore oscillate between stability and change. Through repetition, they give permanence to what is ephemeral, and at the same time, habits are also open to change. For one thing, this is because they are only acquired through external impulses, which means that different habits can also be learned. And for another, this is because every repetition produces difference. Habits form the subject who practices them in relation to an environment; they generate a "way of being"[7] or a "habitus."[8]

Habitualized liking is the social media subject's way of being. One cannot use Facebook or other social media apps in a passive manner. Without likes, friends, or connections, a profile remains amorphous, empty, and isolated. Only through repeated liking (or other acts of networking) do user-subjects take shape. And only through repeated liking can subjects' newsfeeds become their habitat,

---

[5] Chun, "Networks NOW," 300.
[6] Chun, *Updating to Remain the Same*, 53.
[7] Félix Ravaisson, *Of Habit*, trans. Clare Carlisle and Mark Sinclaire (London and New York: Continuum, 2008), 25.
[8] Pierre Bourdieu, *Outline of a Theory of Practice*, trans. Richard Nice (Cambridge: Cambridge University Press, 1977 [1972]).

which supports their existence by encouraging them to express even more likes. "Like breeds like."[9]

3 Homophily

Others use the terms "echo chamber" or "filter bubble" to denote what I am here calling "habitat." Discussions about echo chambers or filter bubbles are predominantly shaped by concerns about the loss of democratic values, i.e., access to heterogeneous content and diverse perspectives and spaces for difference and controversy. These discourses accuse the algorithms of the dominating platforms of only showing us what we already know and like, that is, of merely reflecting back what we ourselves shout out, just like an echo chamber. They argue that the culture of likes has herded us into gated communities with people like ourselves by filtering out everything that disrupts this unity. Eli Pariser, who coined the term "filter bubble," writes: "the filter bubble confines us to our own information neighborhood, unable to see or explore the rest of the enormous world of possibilities that exist online."[10]

Wendy Chun follows up on this criticism by scrutinizing the networking mechanisms that allegedly produce echo chambers.[11] According to Chun, a basic assumption for explaining the formation of clusters in networks has been incredibly influential in network research for years: homophily, the love of the same. In a 2001 sociological study that has proven authoritative for network science,[12] homophily is defined as the idea that people prefer connections with people like themselves. *Birds of a feather flock together*. This can be seen, the study argues, in all kinds of personal networks: "Similarity breeds connection."[13] The supposed

---

[9] Wendy Hui Kyong Chun, "Difference and Discomfort: Intervening in Habits and Homophily: Make a Difference! An Interview with Wendy Hui Kyong Chun by Martina Leeker," in *Interventions in Digital Cultures: Technology, the Political, Methods*, ed. Howard Caygill, Martina Leeker, and Tobias Schulze (Lüneburg: meson press, 2017), 75–85, here 75.
[10] Eli Pariser, *The Filter Bubble: What the Internet Is Hiding from You* (New York: Penguin Press, 2011), 222.
[11] Wendy Hui Kyong Chun, "Queering Homophily: Muster der Netzwerkanalyse," *Zeitschrift für Medienwissenschaft* 10, 1 (2018): 131–148. An expanded English version of this article appeared in *Pattern Discrimination*, ed. Clemens Apprich, et al. (Lüneburg and Minneapolis: meson press/ University of Minnesota Press, 2018), 59–97.
[12] Miller McPherson, Lynn Smith-Lovin, and James Cook, "Birds of a Feather: Homophily in Social Networks," *Annual Review of Sociology* 27 (2001): 415–444.
[13] McPherson et al., "Birds of a Feather: Homophily in Social Networks," 415.

similarity structuring connections is found here in the usual sociodemographic categories, although ethnic homophily caused the strongest differences in the formation of personal networks, closely followed by age, religion, education, profession, and sex. Chun argues that the axiom of homophily represents a grossly simplified notion of the influence between individuals, as well as a problematic reversal of cause and effect that attributes and thereby also justifies real injustices – such as racist or classist segregation of residential areas in modern cities – to a naturalized love of the same. According to this point of view, Muslim migrants, low-wage workers, or white middle class families simply like being together in their respective groups, and therefore in different parts of the city. Yet actual political causes for housing segregation remain hidden.

Homophily has subsequently become a foundational concept in network science – a simple answer to the question of how stable social networks are constituted and maintained. One reason is that the principle enables predictions: when network analysis identifies similarities, it is assumed that they will reproduce themselves, and thereby create connections. As a result, in all areas in which network analyses are applied (for example, as machine learning algorithms), historically developed forms of social discrimination are reproduced without the need for racist, classist, or sexist categorizations to explicitly come into play. Networks project past connections into the future and thereby reproduce what they claim to only describe. This makes them self-fulfilling prophecies. Chun names a self-learning prediction algorithm used by the Chicago police for crime prevention that predicts possible wrongdoing on the basis of social proximity and sociodemographic similarity, thus exposing minorities, and in particular African-Americans, to a higher risk of being arrested or murdered, without effectively lowering the crime rate.[14] This bias problem in network science is enormous, and Chun is right to point it out. Yet despite her criticism of the one-sided axiom of homophily, she herself appears to remain wedded to thinking in dichotomies of homogeneity/heterogeneity, similarity/difference, or control/freedom in a way that does not adequately capture the transgressive, mimetic dimensions of network operations.

If we follow the arguments of Chun and other critics, the supposed love of the same leads precisely to the unity, homogeneity, and harmony of segregated filter bubbles. Pariser calls this "friendly world syndrome."[15] The experience of everyday social media use, however, is of anything but harmony. Facebook, Twitter, and their ilk make opposing viewpoints and opinions, as well as modes of perception that are completely different from one's own, more available than

---

14 Chun, "Queering Homophily," 133f.
15 Pariser, *The Filter Bubble*, 147.

ever. In comments to posts, one can take a seat at the kitchen table of political opponents.[16]

If filter bubbles are a new phenomenon in times of social media, then this is only because they become perceptible in the first place in colliding and overlapping with one another. These kinds of encounters between different filter bubbles often erupt in brutal verbal hostilities and hate speech. Does this hate, then, truly lead to mutual rejection and a withdrawal into the homogeneous bubble, as Wendy Chun claims for the phenomenon of white flight?[17] In social media, at least, other mechanisms are apparently at work. Hate and outrage are also emotions that create connections. *Even hate breeds likes.* Just think about the torrential power of shitstorms. According to the logic of the network, these waves of negativity are also breeding grounds of positive connections, just like any cute cat meme. The most bitter enemies find themselves in close proximity. Is Chun's claim, then, really true that networks do not model conflicts?[18] Does the supposed love of the same actually eradicate all difference?

We get a different idea if we look beyond the duality of sameness and difference by instead focusing on the mimetic dimension of repetitive liking. In doing so, it is important to distinguish between sameness and similarity. Theoreticians of homophily, as well as Wendy Chun, imprecisely use the term "similarity" when they actually mean "homogeneity." Yet "the like is not the same" ("le semblable n'est pas le pareil"), as Jean-Luc Nancy writes in his reconceptualization of the concept of community.[19] Things that are similar are also different. In a reversal of the naturalized causality of the axiom of homophily, the hypothesis would thus be: *connection breeds similarities, and therefore differences.*

4 Collectivity

The mimetic dimension of habitualized liking lies in the repetitive structure of habit, which produces similarities and differences with each repetition. But it also exists in the term's polysemy: "to like something" and "to be like something."

---

[16] Seth Flaxman, Sharad Goel, and Justin M. Rao, "Filter Bubbles, Echo Chambers, and Online News Consumption," *Public Opinion Quarterly* 80, 1 (2016): 298–320, DOI: 10.1093/poq/nfw006.
[17] Chun, "Queering Homophily," 144. "White flight" designates the tendency of middle-class whites to leave residential areas when these areas become too diverse and whites are no longer clearly in the majority.
[18] Chun, "Queering Homophily," 146.
[19] Jean-Luc Nancy, *The Inoperative Community* (Minneapolis and Oxford: University of Minnesota Press, 1991), 33.

The word "like" demonstrates an etymological relation between attraction and similarity. The cultural studies scholar Jonathan Flatley takes up this idea in his book *Like Andy Warhol* (2017). Flatley offers an in-depth investigation of practices of liking from the perspectives of art history, queer theory, and affect theory. Although he completely excludes the field of social media from his considerations, he provides helpful approaches for understanding networking practices in social media. Flatley shows that the practice of liking plays a central role in Andy Warhol's pop art. He takes Warhol at his word when Warhol identifies liking, in his famous 1963 interview with Gene Swenson, "What is PopArt?," as a basic operation of the genre:

Warhol	Well, I think everybody should like everybody.
Swenson	You mean you should like both men and women?
Warhol	Yeah.
Swenson	Yeah? Sexually and in every other way?
Warhol	Yeah.
Swenson	And that's what Pop art's about?
Warhol	Yeah, it's liking things.
Swenson	And liking things is being like a machine?
Warhol	Yeah. Well, because you do the same thing every time. You do the same thing over and over again. And you do the same … [ellipses in original]
Swenson	You mean sex?
Warhol	Yeah, and everything you do.
…	
Warhol	Well, I want everybody to think alike.
Malanga	Well that's a communistic attitude.
Warhol	Is it?
…	
Warhol	I mean, Russia is sort of doing it under government, and we're doing it … [ellipses in original] it's happening without even being under government here. Everybody looks alike and acts alike and we'll be getting more and more that way, you know. And it will just sort of happen.
Swenson	And you like it?
Warhol	Yeah. *[laughing]* Everybody should wear the same uniform. *[laughing]*[20]

For Warhol, liking is a repetitive, machinic practice that makes people more alike. His interlocutor worries that this might hide a communist idea threatening individuality. But Warhol argues that the dissolution of individuality in similarity has already happened right within North America's liberal, capitalistic consumer

---

[20] Jennifer Sichel, "'What is Pop Art?' A Revised Transcript of Gene Swenson's 1963 Interview with Andy Warhol," *Oxford Art Journal* 41, 1 (2018): 85–100, DOI: 10.1093/oxartj/kcy001, here 88–99.

culture – and it's something he likes. For Warhol, liking is a queer relationality that undermines the opposition of sameness/difference, homo/hetero, or friend/foe, thus making possible collectives that harbor both similarities and differences. "Warhol is encouraging us to forget the sense that we must relate to others by way of *either* identification ('being') or desire ('having'), which itself relies on the opposition of same and different."[21]

Warhol makes a habit of liking things. This can be seen in his serial works of visual art and film (such as the silkscreen prints of Marilyn Monroe or his *Screen Tests*), his imitation gags (instead of accepting invitations to give talks, he sent friends dressed as himself), or his intense practice of collecting all kinds of things: from everyday objects and photos to expensive pieces of furniture and works of art. In these practices, Warhol shows a particular interest for flawed, imperfect objects. These flaws allow the objects within the collection to remain singular; they are not identical, but similar.

For Warhol, liking is a practice of perceiving similarities, producing similarities, and becoming similar. Liking constitutes a fundamental mode of being and perceiving that Flatley describes, in recourse to Walter Benjamin, as "a collector's relationship to the world."[22] This mode of being is characterized by an affective openness and a search for correspondences. "[E]verything concerns"[23]: everything gains the potential to affect and become part of a collection, to relate to other things and people, to become like other things and people. Something becomes part of a collection because it is in some way similar to the things that have already been collected. But it also becomes similar to them by becoming part of the collection. The collection is slightly changed by every new item. Collecting is thus not only conservative but transformative. Together with the collection, it also forms and changes the collector.

Flatley summarizes this affective openness of the collector with a series of questions that could also be posed by social media users: "How am I like this? How is this like other things? How can I relate to this thing as somehow imitable? In what way are we alike? How do we (mis)fit together?"[24] Social media users can also be characterized as possessing an affective openness and a collector's relationship to the world. Everything concerns, everything has the potential to affect and become part of a collection. Liking is collecting and the collection of likes constitutes the likeness of the user. The user composes themselves as a collection that transforms with every like. At the same time, the user is also an object of

---

21 Jonathan Flatley, *Like Andy Warhol* (Chicago and London: University of Chicago Press, 2017), 11.
22 Jonathan Flatley, "Like: Collecting and Collectivity," *OCTOBER* 132 (2010): 71–98, here 80.
23 Flatley, *Like Andy Warhol*, 57.
24 Flatley, *Like Andy Warhol*, 43.

other collections. By being liked or befriended they become part of the likeness of others.

The practice of liking forms a complex network of relations of similarity that are constantly being renewed and changed. Hence, we could speak of a "collective likeness": first, in the sense of an individually attributable appearance as a collection; and second, as a collective similarity, i.e., a collective whose elements are like one another.

Liking is a practice that erodes and transcends the distinctions between same/different, self/other, individual/collective, thereby shifting the focus from nodes (individuals) to connections. As the legal scholar Antoinette Rouvroy has shown, the form of government of social media platforms – which she calls "algorithmic governmentality" in recourse to Michel Foucault – operates precisely through the avoidance of individuality. The "object ... of algorithmic governance" is no longer individuals but "precisely relations: the data shared are relations and only subsist as relations; the knowledge generated consists of relations of relations; and the normative actions that derive from it are actions on relations (or environments) referred to relations of relations."[25]

Facebook is not interested in individuals. The identity and personality of user-subjects, what they like, and what they believe, play no role in the implementation of Facebook's business model. Although the social media giant processes enormous quantities of personal data, which users knowingly share and leave behind as digital traces, the company's power is not based on an in-depth knowledge of its users – of their characteristics and personalities. Facebook sells targeted advertising that is much more differentiated thanks to network analysis than target groups determined by means of conventional demographic information. But this is more of a microsegmentation of the market than an individualization. Users are not addressed individually but rather assigned to certain, nuanced target groups (profiles) based on their habits. For example, Facebook offers advertisers the service of identifying "lookalike audiences." Lookalike audiences are Facebook users who are like, or act like, the advertiser's customer base and who – it is assumed – will probably also become customers. Individual users are targeted only inasmuch as they have connections to other users. By means of network analyses, correlations are determined that make it possible to

---

[25] Antoinette Rouvroy and Thomas Berns, "Gouvernementalité algorithmique et perspectives d'émancipation," *Réseaux* 177, 1 (2013): 163–196, translated by Elizabeth Libbrecht as "Algorithmic Governmentality and Prospects of Emancipation: Disparateness as a Precondition for Individuation through Relationships," DOI: 10.3917/res.177.0163, https://www.cairn-int.info/load_pdf.php?ID_ARTICLE=E_RES_177_0163 (visited on September 23, 2019), II–XXXI, here XX.

construct the profile of a collective even based on interests that seem to be completely singular and particular. *People like me.*

5 Regulation

In the logic of the network, liking and hating are connected by the same mimetic operation. Both affirmation and rejection are expressed through an act of repetition, i.e., imitation, thus becoming part of one's own profile. Any user can take a look at the diverse patchwork that is their online likeness, since Facebook allows users a limited degree of insight into their own profile, which is automatically created based on their online behavior and serves as the basis for the selection of targeted ads.

In the settings of every account, it is possible to browse through and alter the categories and interests assigned to one's profile – categories such as "frequent traveler," "iPhone user," or "engaged shopper" and interests such as "Bauhaus," "Adidas," or "music videos." The latter form a disparate collection of certain fields of interest that are quite specific and accurate and others that appear to be completely arbitrary.[26] In some instances, Facebook's assessments seem to directly contradict one's own inclinations and convictions. For example, the profiles of liberal atheists might contain categories such as "Mormonism" or "Republican Party."[27] This is evidence for the fact that, as was explained above, networks are indifferent to the symbolic meanings of connections. Both love and hate are infectious, provoking likes and fostering engagement, resulting in similarities not only with friends and neighbors but also with enemies.

In the preface to the second edition of his *Laws of Imitation*, Gabriel Tarde recognizes the necessity of supplementing his social theory – which he bases on

---

[26] These considerations are based on an examination of my own Facebook profile. My impressions are nevertheless supported by a series of journalistic articles that similarly share analyses of their authors' own Facebook account: Julia Glum, "I Found Out Everything Facebook Knows About Me – And You Can Too," *Money – Time*, March 23 (2018), http://time.com/money/5212501/how-facebook-tracks-me (visited on October 26, 2018); Todd Haselton, "How to Find Out What Facebook Knows About You," *CNBC*, November 19 (2017), https://www.cnbc.com/2017/11/17/how-to-find-out-what-facebook-knows-about-me.html (visited on October 26, 2018); Elle Hunt, "Facebook Says All I Want Is Babies and Caviar: What Else Does It Think It Knows About Me?" *The Guardian*, August 4 (2017), https://www.theguardian.com/culture/2017/aug/04/facebook-says-all-i-want-is-babies-and-caviar-what-else-does-it-think-it-knows-about-me (visited on October 26, 2019); Andrew Hutchinson, "Wanna' Know What Facebook Thinks Your Interests Are? Here's How to Find Out," *Social Media Today*, June 5 (2016), https://www.socialmediatoday.com/social-networks/wanna-know-what-facebook-thinks-your-interests-are-heres-how-find-out (visited on October 26, 2018).
[27] Haselton, "How to Find Out What Facebook Knows About You."

mimetic practices – with a negative form of imitation. He notes that societies are constituted by two forms of imitation. People imitate models by doing the same thing or the opposite.

> In counter-imitating one another, that is to say, in doing or saying the exact opposite of what [people] observe being done or said, they are becoming more and more assimilated, just as much assimilated as if they did or said precisely what was being done or said around them.[28]

Following Tarde, we can use the term "counter-imitating" to designate repeated acts of connection that are motivated by hate, as well as the targeted provocations of trolls. Counter-imitation makes conservatives and liberals resemble each other. It connects misogynists and feminists. It joins together climate change deniers and climate-protection advocates in a collective. "But both kinds ... have the same content of ideas and purposes. They are assimilated, although they are adversaries, or, rather, because they are adversaries."[29]

Yet this portrayal is not meant to suggest a harmonious picture. Hate directed toward people in social networks regularly results in existential threats for groups and individuals making it impossible for them to participate and sometimes even leads to physical violence, mass shootings, mental health issues and suicides. Platforms find themselves compelled to intervene by instituting regulations – deleting content and blocking users – in order to remain hospitable for as many users as possible.

Mimesis possesses two poles, as Friedrich Balke demonstrated, as do mimetic operations in social networks: one that regulates and one that is excessive. As a principle of regulation, mimetic practices support an already existing order. Pierre Bourdieu's concept of habitus can be located at this regulatory end of the mimetic spectrum.[30] According to Chun, habitualized liking as an embodied mode of being is both the basis for predicting user behavior and the condition for modeling networks. Driven to an extreme, however, regulative mimesis turns into mimetic excess that destroys systems of order. This polarity is also found in the mimetic structure of habit. When they are excessive, good habits become addictions and obsessions.[31] In 2014, Mat Honan (an editor for Wired magazine) conducted an experiment in which he liked everything he encountered on Facebook. Very quickly, not only his newsfeed changed but also that of his Facebook friends. Postings from his friends disappeared. Instead, his screen was flooded

---

28 Gabriel Tarde, *The Laws of Imitation*, trans. from the second French edition by Elsie Clews Parsons (New York: Henry Holt and Company, 1903), xvii.
29 Tarde, *The Laws of Imitation*, xviii.
30 Friedrich Balke, *Mimesis zur Einführung* (Hamburg: Junius, 2018), 17.
31 Catherine Malabou, "Addiction and Grace: Preface to Félix Ravaisson's *Of Habit*," in F.R., *Of Habit*, trans. Clare Carlisle and Mark Sinclaire (London and New York: Continuum, 2008), vii–xx, here viii.

with politically provocative content, advertisements, clickbait, and banal gossip. Obsessive liking made him become unlike himself, leading friends to contact him and ask if he had been hacked. After two days of compulsive liking, Honan had destroyed his Facebook habitat: "By liking everything, I turned Facebook into a place where there was nothing I liked."[32]

Platforms such as Facebook must protect the order of the network, which is based on consistent, predictable user behavior, from excessive hate and excessive liking, because excessive mimesis destroys so-called good order by transforming it into a hostile environment and undoing predictability. To accomplish this, platforms must carry out policing measures that institute regulations and enforce their own community standards. Users are admonished to accept good social networking habits and let go of bad habits in order to achieve a "healthy" and satisfactory user experience and remain predictable for the platform.[33]

For example, since 2015 Facebook has been carrying out a large-scale multimedia campaign encouraging users to claim sovereignty over their own Facebook profile: "Mach Facebook zu deinem Facebook!" (Make Facebook Your Facebook!), is the call.[34] This includes a series of didactic videos informing users about how they can control and shape their own account. The first versions of the campaign in 2015 and 2016 aimed, in particular, to encourage reserved German users to become more active in their behavior, and to communicate the idea that repeated liking, following, and sharing – as well as unliking (hiding a post, unsubscribing to a page, blocking users) – could make Facebook into a personal habitat: "Mit mehr von dem, was dir gefällt. Und weniger von allem anderen" (With more of what you like, and less of the rest).[35] Since 2017, the campaign has aimed at communicating options for purging the results of bad habits. In the advertising

---

[32] Mat Honan, "I Liked Everything I Saw on Facebook for Two Days: Here's What It Did to Me," *Wired*, November 8 (2014), https://www.wired.com/2014/08/i-liked-everything-i-saw-on-facebook-for-two-days-heres-what-it-did-to-me/ (visited on November 6, 2018).
[33] In his critique of the one-sided content of filter bubbles, Eli Pariser refers to a talk that danah boyd gave in 2009 in which she compares the consumption of highly stimulating content with obesity to suggest instead a form of balanced, healthy consumption from which, she argues, individuals as well as society as a whole will benefit: danah boyd, "Streams of Content, Limited Attention: The Flow of Information through Social Media," talk *at Web2.0 Expo*, November 17 (2009), http://www.danah.org/papers/talks/Web2Expo.html (visited on June 11, 2018).
[34] Facebook, *Wiederaufnahme der Kampagne "Mache Facebook zu deinem Facebook,"* Facebook Newsroom (2017), https://de.newsroom.fb.com/news/2017/07/wiederaufnahme-mache-facebook-zu-deinem-facebook/ (visited on October 17, 2017).
[35] Facebook, *Mach Facebook zu deinem Facebook*, video from October 24 (2015), https://www.facebook.com/FacebookDeutschland/videos/mach-facebook-zu-deinem-facebook/10153568987925932/ (visited on October 26, 2018).

videos, Facebook users complain, for instance, that they are being shown too many cat videos and not enough art and culture, and Facebook shows them how liking and unliking can help. The goal is to animate users to create an environment for themselves through connective activities that foster the production of more connections, making users more valuable for Facebook.

The mimetic structure of liking forms the basis for the permanence and predictability of network order. At the same time, it possesses the power to threaten the existing order. Facebook reacts to this danger with regulatory measures. But critics of homogeneous filter bubbles, too, act out of concern for an order of sameness (control, censorship) and difference (freedom, democracy) that the mimesis of liking questions. In doing so, they ignore the collectivizing function of liking. Contrary to what Pariser claims,[36] one is not alone in a filter bubble but part of a collective likeness with others. Particular niche interests establish a relationship of similarity between users who otherwise share nothing with one another. Acts of connection motivated by affinity or rejection bring us closer together – to our friends as well as our enemies.

---

36 Pariser, *The Filter Bubble*, 9.

List of Figures

Tom Ullrich

Fig. 1	Bibliothèque nationale de France, Paris, https://gallica.bnf.fr/ark:/12148/btv1b53014004p (visited on December 17, 2019) —— **25**
Fig. 2	Ville de Paris, Bibliothèque historique de la Ville de Paris, https://bibliotheques-specialisees.paris.fr/ark:/73873/pf0001196016 (visited on December 17, 2019) —— **34**
Fig. 3	Bibliothèque nationale de France, Paris https://gallica.bnf.fr/ark:/12148/btv1b1200033h/f2.item (visited on December 17, 2019) —— **36**
Fig. 4	Anne S.K. Brown Military Collection, Brown Digital Repository, Brown University Library, Providence, RI, https://repository.library.brown.edu/studio/item/bdr:235843/ (visited on December 17, 2019) —— **39**
Fig. 5	Bibliothèque nationale de France, Paris, https://gallica.bnf.fr/ark:/12148/bpt6k6223591q/f16 (visited on December 17, 2019) —— **40**
Fig. 6	Bibliothèque historique de la Ville de Paris, Fonds Claretie, Ms 1122 —— **41**
Fig. 7	Bibliothèque nationale de France, Paris, https://gallica.bnf.fr/ark:/12148/bpt6k57466170/f324.image (visited on December 17, 2019) —— **43**

Wolfgang Struck

Fig. 1	Scheme by the author —— **63**
Fig. 2	*The Nautical Magazine and Naval Chronicle for 1852* (London, 1852). Forschungsbibliothek Gotha, Perthesforum, Pol 8° 00584/06 —— **64**
Fig. 3	Forschungsbibliothek Gotha, Perthesforum, SPB 4° 8010.00135 —— **65**
Fig. 4	Forschungsbibliothek Gotha, Perthesforum, SPB 4° 8010.00135 —— **67**
Fig. 5	Bundesamt für Seeschifffahrt und Hydrographie, Hamburg, Sammlung Neumayer, No. 394 —— **67**
Fig. 6a	Bundesamt für Seeschifffahrt und Hydrographie, Hamburg, Sammlung Neumayer, No. 418, recto —— **69**
Fig. 6b	Bundesamt für Seeschifffahrt und Hydrographie, Hamburg, Sammlung Neumayer, No. 418, verso —— **69**

Gabriele Schabacher

Fig. 1	Hermann Haeder, *Die kranke Dampfmaschine und erste Hülfe bei der Betriebsstörung: Praktisches Handbuch für Betrieb und Wartung der Dampfmaschine* (Duisburg: Selbstverlag von Hermann Haeder, 1899, second edition), 252 —— **83**
Fig. 2	Marshall M. Kirkman, *Maintenance of Railways* (Chicago: Trivess, 1886), 73 —— **84**

∂ Open Access. © 2020 Jörg Dünne, Kathrin Fehringer, Kristina Kuhn, and Wolfgang Struck, published by De Gruyter. This work is licensed under a Creative Commons Attribution 4.0 International License.
https://doi.org/10.1515/9783110647044-019

Christoph Eggersglüß

Fig. 1 Photo by the author (2012) —— 87
Fig. 2 Kent Stainless Ltd. http://www.kentstainless.com/our-products/stainless-steel-street-furniture/studs/ (snapshot/screenshot of July 29, 2016 available via Wayback Machine). Selection of bolts and deterrents for street furniture offered by Kent Stainless on this website until 2016 —— 89
Figs. 3–5 Stephane Argillet and Gilles Paté, *Le Repos du Fakir* (Paris: Production Canal Marches, 2003), video, 06:20 min —— 93

Hannah Zindel

Fig. 1 https://loon.com/ (x.company/loon; visited on April 13, 2017) —— 109
Fig. 2 https://loon.com/ (x.company/loon; visited on April 13, 2017) —— 110
Fig. 3 https://loon.com/ (x.company/loon; visited on April 13, 2017) —— 110
Fig. 4 "Ballon-sonde en papier pétrolé important le premier enregistreur," in Wilfrid de Fonvielle, *Les ballons-sondes de MM. Hermite et Besançon et les ascensions internationales* (Paris: Gauthier-Villars et fils, 1898), 12 —— 112
Fig. 5 "Return of the Aeronauts," woodcut, in Wilfrid de Fonvielle and Gaston Tissandier, *Windy Ascents and Dragging*, trans. T.L. Phipson, in *Travels in the Air: By James Glaisher, Camille Flammarion, W. de Fonvielle and Gaston Tissandier*, ed. James Glaisher (London: Bentley, 1871 [1868]), 351 —— 113
Fig. 6 "A route-tracking telegram by the balloonist and photographer Nadar," in Estate Nadar, *Bibliothèque Nationale de France*, Cabinet des estampes et photographies, Documents Imprimes, Notes XIX (photographed by the author; Paris, no year) —— 114
Fig. 7 "William Blair's first radio-tracking transmitter," in William R. Blair and H. W. Lewis, "Radio Tracking of Meteorological Balloons," *Proceedings of the Institute of Radio Engineers* 19, 9 (September 1931): 1532 —— 117
Fig. 8 Julius Maurer, "Wetterkarte vom 3. Oktober 1898, vormittags," map, in Albert Heim, Julius Maurer, and Eduard Spelterini, *Die Fahrt der Wega über Alpen und Jura am 3. Oktober 1898* (Basel: Schwabe, 1899), 101 —— 121
Fig. 9 "An arrangement of meteorological stations designed to fit with the chief mechanical properties of the atmosphere," in Lewis Fry Richardson, *Weather Prediction by Numerical Process* (Cambridge: Cambridge University Press, 1922), IV —— 123
Fig. 10 Source: François Schuiten, reproduced by permission of the artist —— 124

Bernhard Siegert

Fig. 1 Heneage Jewel (or Armada Jewel), ca. 1595. © Victoria & Albert Museum, London —— 140

Michael Cuntz

Fig. 1 Valerie Hansen, *The Silk Road: A New History* (Oxford: Oxford University Press, 2015), 20 —— **149**
Fig. 2 *Meisterwerke im J.Paul Getty Museum. Illuminierte Handschriften*, ed. Christopher Hudson (Los Angeles, CA: J. Paul Getty Museum, 1997), 102 —— **152**
Fig. 3 Barry Cunliffe, *By Steppe, Desert, and Ocean: The Birth of Eurasia* (Oxford: Oxford University Press, 2015), 234 —— **153**

Jörg Paulus

Fig. 1 Three labeled slipcases – in German, Schuber: (1) Papillones, (2) Bombices et Noctu[ae], (3) Spinges, Geometridae, Tortices, Pyralides, Tinides, Allucides (presumably around 1850, private collection) —— **166**
Fig. 2 Sheet 112r from slipcase 2 —— **168**
Fig. 3 Sheet 75r from slipcase 1 —— **168**
Fig. 4 Slipcase 1, sheet 31r (detail) —— **169**

Jürgen Martschukat

Fig. 1 Wellcome Library, London https://wellcomecollection.org/works/s9sbg97g CC-BY-NC 4.0, https://creativecommons.org/licenses/by/4.0/ —— **177**
Fig. 2 *Animal Locomotion: an Electro-Photographic Investigation of Connective Phases of Animal Movements* by Eadweard Muybridge, J.B. Lippincott Co., Philadelphia (1887), UPT 50 M993, Box 62 FF, https://archives.upenn.edu/digitized-resources/docs-pubs/muybridge/catalogue —— **182**

Nicolas Pethes

Fig. 1 Jean Paul, *D. Katzenberger's Badereise*, first edition (Heidelberg: Mohr & Zimmer, 1809), vol. 1 and vol. 2 (last page). Reproduction by Universitäts- und Landesbibliothek Bonn —— **246**
Fig. 2 Jean Paul, *Herbst-Blumine*, first edition (Tübingen: Cotta, 1810). Reproduction by Universitäts- und Landesbibliothek Bonn —— **252**
Fig. 3a *Zeitung für die elegante Welt*, No. 45 (March 4), 1809, table of contents. Reproduction by Universitäts- und Landesbibliothek Bonn —— **255**
Fig. 3b *Zeitung für die elegante Welt*, No. 45 (March 4), 1809, title page. Reproduction by Universitäts- und Landesbibliothek Bonn —— **256**

Kristina Kuhn

Fig. 1   *Bilder-Atlas: Ikonographische Encyklopädie der Wissenschaften und Künste: Ein Ergänzungswerk zu jedem Conversations-Lexikon*, vol. 1–8, *completely revised second edition*, ed. Karl Gustav von Berneck et al., *Fünfhundert Tafeln in Stahlstich, Holzschnitt und Lithographie* (Leipzig: F. A. Brockhaus, 1868–1875), vol. 7, *Ethnographie* (1875), plate 13, origin of digital copy: Klassik Stiftung Weimar, Herzogin Anna Amalia Bibliothek, shelfmark: Th Q 2: 11 (g) —— **267**

Fig. 2   *Bilder-Atlas zum Conversations-Lexikon: Ikonographische Encyklopädie der Wissenschaften und Künste*, vol. 1–10, ed. Johann Georg Heck, *500 Tafeln nebst Text und Universal-Register, fifth edition* (Leipzig: F. A. Brockhaus, 1860), vol. 4, *Völkerkunde der Gegenwart, 42 Tafeln nebst Text*, plate 110 (1), origin of digital copy: Klassik Stiftung Weimar, Herzogin Anna Amalia Bibliothek, shelfmark: N 15667 (f) —— **271**

Fig. 3   *Bilder-Atlas: Ikonographische Encyklopädie der Wissenschaften und Künste: Ein Ergänzungswerk zu jedem Conversations-Lexikon*, vol. 1–8, *completely revised second edition*, ed. Karl Gustav von Berneck et al., *Fünfhundert Tafeln in Stahlstich, Holzschnitt und Lithographie* (Leipzig: F. A. Brockhaus 1868–1875), vol. 6, *Culturgeschichte* (1875), plate 8, "Altägyptisches Leben," origin of digital copy: Klassik Stiftung Weimar, Herzogin Anna Amalia Bibliothek, shelfmark: Th Q: 11 (f) —— **275**

List of Contributors

Michael Cuntz has been Professor of Media Philosophy at Bauhaus-Universität Weimar until 2020.

Jörg Dünne is Professor of Romance Literatures at Humboldt-Universität zu Berlin.

Christoph Eggersglüß has been Researcher at the IKKM (Internationales Kolleg für Kulturtechnikforschung und Medienphilosophie), Bauhaus-Universität Weimar, until 2020.

Kathrin Fehringer has been Postdoctoral Researcher at the Cultural Techniques Research Lab and Assistant Professor of Romance Literatures at Universität Erfurt until 2020.

Stephan Gregory is currently pursuing a research project on "Treason and Subjectivity" at Bauhaus-Universität Weimar.

Kristina Kuhn has been Postdoctoral Researcher and Coordinator of the Cultural Techniques Research Lab at Universität Erfurt until 2019. She is Referee for Collection Management and Quality Standards in Museums (MUSEALOG) and works as a Project Manager for the Novalis Museum Schloss Oberwiederstedt since 2020.

Christiane Lewe is Doctoral Researcher and Coordinator at the DFG-Graduiertenkolleg Medienanthropologie, Bauhaus-Universität Weimar.

Jürgen Martschukat is Professor of North American History at Universität Erfurt.

Bettine Menke is Professor of Comparative Literature at Universität Erfurt.

Jörg Paulus is Professor of Archival and Literary Studies at Bauhaus-Universität Weimar.

Nicolas Pethes is Professor of German Literature at Universität zu Köln.

Gabriele Schabacher is Professor of Media and Culture Studies at Johannes-Gutenberg-Universität Mainz.

Bernhard Siegert is Professor of History and Theory of Cultural Techniques at Bauhaus-Universität Weimar.

Wolfgang Struck is Professor of German Literary Studies at Universität Erfurt.

Katrin Trüstedt is Assistant Professor of Germanic Languages and Literatures at Yale University.

Tom Ullrich is Doctoral Researcher at the Institute of Film, Theater, Media and Culture Studies at Johannes Gutenberg-Universität Mainz.

Hannah Zindel is Postdoctoral Researcher at the Institute for Advanced Study on Media Cultures of Computer Simulation at Leuphana Universität Lüneburg.

Open Access. © 2020 Jörg Dünne, Kathrin Fehringer, Kristina Kuhn, and Wolfgang Struck, published by De Gruyter. This work is licensed under a Creative Commons Attribution 4.0 International License.
https://doi.org/10.1515/9783110647044-020

Name Index

Adorno, Theodor W. 62
Aira, César 193f., 203–215
Anna Amalia (Duchess of Weimar) 260
Antrechaus, Jean d' 284–286
Aragon, Louis 23
Argillet, Stéphane 91

Bachelard, Gaston 300
Baggesen, Jens 236, 239
Bahr, Hans-Dieter 131
Bakhtin, Mikhail 184, 257
Balke, Friedrich 327
Barthes, Roland 229 f., 235, 243
Baudrillard, Jean 131
Bayle, Pierre 228
Becher, Alexander 63–66
Benedetti, Alessandro 283
Benedict, Barbara 244, 250
Benjamin, Walter 30n, 297, 300f., 311, 314, 324
Berend, Eduard 239
Bergson, Henri 319
Besançon, Georges 111f., 114–116
Bjerknes, Vilhelm 122
Blanqui, Auguste 25, 32, 37f.
Boehm, Gottfried 265
Borges, Jorge Luis 226
Bourdieu, Pierre 319, 327
Bouterwek, Friedrich Ludewig 225
Bowker, Geoffrey 77n
Brecht, Bertolt 61, 63, 70–72
Bühler, Karl 196
Buonarroti, Philippe 37
Bureau, Robert 117f.

Calfucurá 53
Callou, G. 32
Campe, Rüdiger 227, 237n
Canetti, Elias 30
Canguilhem, Georges 300
Canler, Louis 29
Cassirer, Ernst 196
Caussidière, Marc 23f., 45
Celan, Paul 62

Cerebotini, Luigi 117
Certeau, Michel de 211
Cervantes, Miguel de 206
Charles X 29
Chun, Wendy 318–322, 327
Cooper, Charlotte 185
Corday, Charlotte 247f., 260
Corngold, Stanley 307
Cunliffe, Barry 150, 156n, 161

d'Alembert, Jean-Baptiste le Rond 229n, 269
D'Hondt, Jacques 283f.
Danius, Sara 243
Darwin, Charles 207n
Daumier, Honoré 36
Davies, John 139
Davis, Mike 100–102
De Landa, Manuel 11 f.
Debord, Guy 42
Defoe, Daniel 1 f., 5, 8, 245
Déjacque, Joseph 31f., 40
Deleuze, Gilles 12, 53, 290, 308n, 319
Dercum, Francis X. 186
Derrida, Jacques 131, 220n, 308, 310
Diderot, Denis 225n, 228f., 269, 281f., 289f.
Digard, Jean-Pierre 145, 155f.
Dreyfus, Alfred 307
Du Camp, Maxime 41
Dumas, George 194
Durkheim, Émile 194, 197n

Edward III 153, 158
Elizabeth I 139
Engels, Friedrich 27, 32, 37
Ersch, Johann Samuel 229

Fitzpatrick, Matthew P. 146
Flatley, Jonathan 223f.
Flaubert, Gustave 41, 198n
Fonvielle, Wilfrid 115
Foucault, Michel 13n, 33, 35, 92, 136, 225, 238, 257, 290, 325
Fourier, Charles 31, 40
Fracastoro, Girolamo 238

Frankopan, Peter 150–153
Freud, Sigmund 163

Gaillard, Napoléon ("Gaillard père") 39–42
Gamman, Lorraine 104
Garve, Christian 287
Gazagnadou, Didier 158, 162
Gellert, Christian Fürchtegott 223
Gibbon, Edward 289
Gilman, Sander L. 177, 307, 313n
Girard, René 138, 279
Glaisher, James 111f.
Goethe, Johann Wolfgang von 198, 260, 281, 290
Goody, Jack 150
Gracq, Julien 154
Grillon, Edme Jean Louis 32
Grimm, Jakob und Wilhelm 78
Gruber, Johann Gottfried 229
Guattari, Félix 12, 53, 308n

Hamilton, John 311
Hartmann, Karl 84f.
Haudricourt, André-Georges 141, 156n
Haussmann, Georges-Eugène 25, 33, 35, 37, 42, 44
Hebel, Johann Peter 247
Hederich, Benjamin 228
Hegel, Georg Wilhelm Friedrich 224, 235f., 277–294
Heidegger, Martin 135
Heinse, Gottlob Heinrich 229
Hennion, Antoine 75n, 139
Hergesell, Hugo 118n, 122
Hermite, Gustave 111–116, 120n
Herodotus 54f.
Hilliard, Nicholas 140
Holbach, Paul Thiry d' 280
Honan, Mat 327f.
Humboldt, Alexander von 194, 204
Hutchin, Edwin 76

Ingold, Timothy 135

Jacobi, Friedrich Heinrich 247
Jacobs, Jane 100
Jacoubet, Théodore 32

James, William 319
Janet, Pierre 194
Jean Paul 219–241, 243–261
Jeanne, Charles 29
Joyce, James 243

Kafka, Franz 191, 295–315
Kant, Immanuel 225, 231n, 300n
Kapp, Ernst 163
Kircher, Athanasius 283
Kirkman, Marshall M. 83
Kittler, Friedrich 3f., 131–133, 135, 219
Kittler, Wolf 308
Kleist, Heinrich von 118–120
Kluge, Alexander 131
Knigge, Adolph Freiherr von 284
König, René 197
Krünitz, Johann Georg 228

Lacan, Jacques 137, 292
Latour, Bruno 12, 14, 49, 94-96, 136, 156n, 173n, 249f., 300n
Lecouturier, Henri 32
Leroi-Gourhan, André 6n, 10n, 75n, 145, 156n, 159
Lévi-Strauss, Claude 196f.
Lichtenberg, Georg Christoph 221n
Livy 136f.
Luxemburg, Rosa 292

Macho, Thomas 295
Marat, Jean Paul 261
Marco Polo 150
Marville, Charles 36
Marx, Karl 37
Maurer, Julius 121f.
Mauss, Marcel 2f., 6n, 9f., 13, 75n, 131, 156n, 159f., 187–215
Maye, Harun 189, 194, 212f.
McLuhan, Marshall 163
Meckel, Johann Friedrich 257
Merck, Klemens 174
Merruau, Charles 35
Mitre, Bartolomé 53
Monroe, Marilyn 324
Montaigne, Michel de 226f., 231
More, Thomas 53

Morgan, Thomas Hunt 299
Muybridge, Eadweard 177–186

Nadar, Félix 116
Nadaud, Martin 30f., 36
Nancy, Jean-Luc 322
Napoleon III 32, 37
Negt, Oskar 131
Neumann, Peter Horst 254
Neumayer, Georg von 68
Newman, Oscar 100
Newton, Isaac 259
Nietzsche, Friedrich 289
Norman, Nils 103f.

Paré, Ambroise 279
Pariser, Eli 320f., 328n, 329
Paté, Gilles 91
Paviot, Jacques 153
Pethes, Nicolas 187n, 199, 203, 212
Pliny (the Elder) 146f., 223n
Plutarch 136f.
Pope, Alexander 244, 258n
Price, Leah 249

Rabelais, François 184–186
Ravaisson, Felix 319
Reimarus, Johann Albert Heinrich 285f.
Rheinberger, Hans-Jörg 298–300
Richardson, Lewis Fry 122–125
Richter, Johann Paul Friedrich (see Jean Paul)
Robida, Albert 42
Rosas, Juan Manuel de 51
Ross, Sir John 64
Rousseau, Jean-Jacques 5
Rouvroy, Antoinette 325
Rugendas, Georg Philip 207
Rugendas, Moritz 194, 204–215
Rustichello da Pisa 150
Ryan, Simon 307

Sarmiento, Domingo Faustino 47–59
Schäfer, Armin 238, 254, 257
Schivelbusch, Wolfgang 131
Schlegel, Friedrich 225n

Schlegel, Gustaaf 152–154
Schmidt-Hannisa, Hans-Walter 234
Schmitt, Carl 50, 52f.
Scholl, Ewald Friedrich 82
Schott, Gerhard 68
Schuiten, François 124
Schupbach, William 185
Schüttpelz, Erhard 75n, 187f., 190, 192, 199
Seckendorff, Leopold von 206
Seneca 146f., 154
Serres, Michel 136–138
Shaftesbury (Anthony Ashley Cooper, Earl of Shaftesbury) 244
Siegert, Bernhard 79n, 172
Silbermann, Albert 117
Sklovskij, Victor 296
Star, Susan Leigh 75f.
Starr, S. Frederick 150
Sterne, Laurence 245
Stockhammer, Robert 232n, 265
Swenson, Gene 323
Swift, Jonathan 245

Tarde, Gabriel 290, 319, 326f.
Teisserenc de Bort, Léon-Philippe 116

Vanhöffen, Ernst 170, 172, 174
Vespucci, Amerigo 54
Viollet-le Duc, Eugène 153
Vismann, Cornelia 6n, 11, 23, 309n
Voigt, Johann Gustav 250f.

Warhol, Andy 323f.
Wheatstone, Charles 116
White, Lynn 157f.
Whyte, William H. 99–102, 105
Widlok, Thomas 76
Winthrop-Young, Geoffrey 134, 187

Yersin, Alexandre 283

Zedelmaier, Helmut 264n
Zischler, Hans 243
Žižek, Slavoj 282, 287, 292
Zola, Émile 36

Subject Index

actor-network theory / ANT 7, 9, 74f., 88, 135, 139, 295n
actors (human/nonhuman) 4, 11, 13, 17f., 28, 31, 44, 75, 91f., 94–96, 111, 135, 142–145, 149, 292, 295, 298, 301f., 306, 314
adapt 91, 98, 118, 161f., 167, 172
agency 28, 44, 75, 136, 213, 295
agriculture 3, 7f., 24, 47–49, 58, 86, 121n
almanac 243–261
anecdote 187–215
animals 3f., 11, 13, 74, 79, 86, 133, 134, 141–164, 168, 178–183, 190f., 200, 208, 211, 246, 253f., 286, 307f., 318
anthology 196, 242–261
anthropology / anthropological difference 3f., 9, 72, 78, 90n, 135n, 138, 156n, 178f., 185, 190, 193, 212, 232n, 270n, 273
appearing 54, 66, 71, 137, 139, 165, 172, 208, 211f., 270, 284f., 295–315
arranging 14–16, 27, 62, 71, 88, 91f., 95, 100, 118, 158, 168, 172, 178, 180, 186, 190, 195, 203, 205, 215, 241, 243, 247, 264f., 268–276, 297–302, 305, 311, 314f., 319
articulation 9f., 18, 49, 133f.
assembling 9–18, 27f., 38, 68, 71, 138, 143, 156, 166, 172, 195, 207f., 213, 241, 264–266, 273
assemblage 9, 17, 95n, 136, 140, 143f., 147, 157n, 162, 164, 165, 167, 172, 208f., 229, 236, 241, 274f., 298
assembly (human) 17, 100, 102, 135f., 274
athlete 198–202, 211
atlas 72, 263–276
attaching/attachment 94n, 105, 112, 131–140, 147, 157, 158n, 161, 164f., 167

ballooning/balloon 18, 107–127
barricades 5, 18, 23–45
battle 40, 53, 143, 201, 206n, 207, 290–294
biography 70–72, 201, 205, 212, 247, 260
biopolitics/biopolitical 179, 183, 307
blocking/block 24, 45, 92, 94, 98, 100, 222, 289, 312f., 327f.

body techniques 2f., 9–11, 13f., 18, 44, 94, 100, 102, 105, 131, 141, 144f., 156n., 160f., 178, 180, 187–215
boulevard 23–45
butterfly 165–175

calculating 38, 80, 83, 96n, 100, 107f., 111, 115, 120, 122–125, 133
caption 177f., 184f., 269, 272, 274
caring 38, 40, 78–82, 85, 180
carrying 137f., 143, 148, 157, 159–164
case study 18, 193, 263
categorize 44, 168, 177–186, 194, 222f., 252, 267, 321
chance 219–241
charting/chart 2, 8, 61–72, 108, 115, 122, 266
chronophotography 177–186
cite 224–226, 232–237, 241, 269
civilize/civilization 1, 5f., 51–59, 145, 150, 158n, 164, 190, 273
classify 10, 76, 85, 91, 105, 173, 178, 185, 191, 228, 230n, 233, 252, 254, 257f.
clothing/cloth 13, 79, 119, 141–164, 209, 273, 283
collect/collection 12n, 13, 15f., 68f., 89, 108, 111, 119f., 122, 126, 167n, 168, 170, 172–174, 178, 181, 184f., 221, 223f., 240, 243–261, 263f., 270, 274, 276, 324–326
collective 9–18, 27f., 30f., 40, 62, 72, 126, 131–140, 145, 174, 195, 229, 263f., 270, 273f., 276, 319, 324–327
community 17, 31, 62, 76 (of practice), 104, 132, 141, 279f., 320, 322, 328
compiling 220–223, 238, 243f., 250f., 254, 261
compression 24, 136–140, 205, 215, 270
connect 17, 33, 45, 52, 55f., 58, 63, 66, 73, 75f., 78, 81, 86, 104, 107, 110, 133f., 138–140, 141–145, 153, 155, 174, 179, 209, 213, 215, 223, 230, 236, 241, 244, 269, 273, 295, 298, 302, 307, 317–329
consciousness 16, 104, 111, 278–282, 286f.

Subject Index

constructing/construction site 5, 23–45, 97, 104, 109, 125, 137, 172, 263
control 33, 35, 45, 49, 66, 80, 87, 97, 100, 103–107, 114–120, 126, 136, 154, 179, 246, 291, 321, 328f.
cultivate 3, 7, 24, 48, 55, 58
cultural techniques
– and body techniques 10, 13f., 18, 102, 156n, 178, 187, 190, 194, 197, 210, 212–215
– and collectives 15, 17, 132f., 260
– and infrastructure 73–77, 126
– and interspecies studies 141, 156n
– and literary studies/philology 15f., 47, 59, 72, 194, 219f., 243, 261, 263, 295f.
– and media studies 44, 131–132
– and spatiality 8f., 18, 48, 58, 125, 167, 171
– and temporality 49, 72, 138, 145
– elementary 6–8, 134,
– primitive 47–49, 58, 133f.
– first/second order 47, 134–136, 295f.

deterrent 87–106
digression 223, 232, 238f., 245, 254, 258
dimensioning/dimensionality 14f., 17f., 49, 61, 63, 66, 71, 115, 167f., 171, 173, 204, 210f., 266n
disappearing 91, 116, 137, 168, 211, 327
discourse 32f., 35, 42, 59, 85, 88, 91, 133, 163, 186, 191, 257f., 283, 288, 291, 320
discourse network 131f., 136, 220, 236, 274
disperse/dispersal 15, 103, 136f., 237
displace 96–106, 161, 238, 240f., 279
disseminate/dissemination 16, 244, 279, 288f., 294
dissolve 5, 12, 17, 70, 135, 261, 271
divide 27, 54f., 63, 122, 142, 276
domesticate/domestication 3, 133, 141–164
door 77, 103, 210f., 296–298, 301–304, 310, 313n
drawing 9, 13, 27, 48, 62–65, 98, 116, 133f., 162, 173, 193, 204–215, 226, 239, 241, 281, 298, 303
drift 63, 66–68

eating 2, 132, 185
encyclopedia/the encyclopedic 16, 223n, 226–230, 232n, 234f., 257, 263–276
engineering 7, 8, 55f., 75, 88, 104f.
entanglement 3, 7, 11, 73, 92, 140, 246, 265, 310
environment 3f., 24, 52n, 55, 77, 84, 99, 100–102, 105, 108, 118, 122, 125, 162, 164, 319, 325, 328f.
epidemics/epidemiology 36, 277–294
ethnography/ethnology 190–198, 201, 204f., 227, 266–276
excerpting/excerpts 15f., 219–241, 248n, 263
exclude 17f., 27, 87–106, 237n, 257, 317, 323
experiment 1, 16, 42, 61f., 71f., 111f., 118, 127, 167n, 193, 201–205, 215, 229n, 234n, 238n, 245, 257, 297–302, 305, 311, 314f., 327

fashion 80, 151–154, 160, 188, 195, 203f., 244
fat 177–186
filter 100f., 105f., 134, 263n
filter bubble 320–322, 328f.
fitness 177–186, 313n
folding 14, 140, 165f., 168, 171f.
forecasting 118f., 121n, 122–125
founding act/fiction/narrative 1, 3, 6, 7, 47–59, 137f.
fragment 5, 137f., 141, 197f., 204, 209, 211, 213–215, 229n, 244
furniture 14, 29, 79, 91f., 100–105, 267, 324

garment 141–164
geopolitics/geopolitical 18, 48–52, 57, 59, 126
governmentality 32f., 325
grotesque 16, 183f., 186, 205–211, 224f., 244, 254, 257–260
ground-laying/ground 7, 9, 48–50, 52, 57–59, 66, 88, 90f., 98, 103, 110, 117f., 133, 136, 172, 180f., 278, 281
grouping 11, 244, 263–276

habit/habitualization 9, 13, 27, 76f., 91, 145, 151, 153, 156, 159–164, 178, 195, 200f., 205, 209, 213, 232n, 287, 318–320, 322, 324–328
homogenize/homogeneity 12, 66, 101, 258f., 261, 322, 329
horse-riding 13f., 55, 113, 141–164, 206, 208, 210f.

identify 102, 177–186, 267, 273, 325
illustrate/illustration 171, 266, 268f., 274, 276, 303
imitate/imitation 16, 92, 152, 195, 229n, 287, 319, 324, 326f.
immutable mobiles 14, 49
index 18, 222n, 230n, 240, 263, 272–274
infection 16, 153, 277–294, 326
infrastructuring/infrastructure 11, 18, 52, 56, 58, 73–86, 91, 97, 105, 107–111, 125–127, 150, 295, 318
inscription 49–51, 54, 165, 167, 249
insect 165, 168, 171, 236, 298f., 303, 307
internet access 107, 109f–111, 119, 126f.
invent/invention 16, 54–56, 59, 66, 78, 117, 155f., 158, 164, 184, 189–193, 199, 204, 209, 212, 223f., 231–233, 235, 239f., 289f.

judgment 202, 225n., 227n., 236n., 300n., 308, 317

knight 144, 156–159

labeling 92, 166, 172, 189, 222, 243f., 253, 279
law 16, 49f., 96n, 121–123, 199, 297, 306–310, 312, 315
leaf through 15, 230–232, 240, 269
liking 317–329
list/list 2, 45, 82, 89, 181, 189, 220n, 222f., 226, 230, 235, 238, 246f., 251, 263, 269
literary procedure 72, 236n, 296, 315

maintaining/maintenance 24, 73–86, 150, 300, 321
making visible 65, 92, 264, 299, 318
making flat 49, 51, 61, 71f., 168, 274
map 2, 15, 27, 29, 34, 42, 49, 66, 68, 123, 211, 266, 272f., 299
margins 44, 91, 98, 102, 148, 183, 219, 239, 276
materiality 44, 54, 71, 73, 75, 135, 227, 237, 297
measuring 97n, 98–100, 108, 111, 115–119, 122, 124f., 180, 204, 210, 214

media techniques/technical media 3f., 18, 26, 96, 132, 211, 213, 220
media theory/studies 3, 4, 6n, 8, 15, 73f., 92, 95, 126, 132f., 135, 187–189, 212, 263
meteorology 107–127
mimetic/mimesis 279, 317–329
miscellaneous/miscellaneity 198, 221, 243–261
monster/monstrosity 16, 153, 204f., 208f., 212, 225, 254, 257f., 298f., 305–308
monture/Montur 142–145

nature 1–4, 6, 12, 14, 53, 55–58, 125, 135n, 142
nature self-imprinting 165–175
navigate 76, 107, 118–120, 125, 272
networking/network 12f., 24, 28, 38, 52, 76f., 91, 95, 103, 107–110, 119, 121, 126f., 135, 147, 164, 229n, 239, 245, 274, 282, 317–329
niche 87f., 91f., 96, 98, 100, 329
nomadism 141–164
normalization 87–106

obesity 177–186
operational chains 1, 5f., 8, 13, 24, 28, 38, 48, 75n, 133f., 167, 172n, 175
orthopedics 87–106
overlap 120, 142, 192, 264, 269, 272, 274, 322

painting 27, 77, 168, 170, 172, 203–215, 268n
paper 1, 5, 14, 16, 23, 41, 61, 70f., 108, 112, 117, 119, 139, 165–172, 204, 210–213, 220n, 236, 240n, 243, 263, 275, 281, 291
paperwork 49, 249
pass through 98, 103, 108, 272
patience 277–294
pedantry 226f., 237n, 283
philology 15f., 191, 219f.
photography 10, 27, 36, 38, 42, 94n, 116, 121n, 173, 177–186, 313
plague 279, 282–288, 293, 307
planning/plan 25–29, 32–35, 38f., 42, 44, 80, 84, 86, 88, 97, 107, 230f.

planting/plants 11, 24, 54, 56n, 86, 168, 170, 282
poisoning 247, 280, 284, 293n
politics 91f., 95, 105, 185, 269, 273, 287, 290, 307, 235
postal techniques 108, 112–114, 133n, 150
practices
– material/symbolic 3–5, 11, 13n, 15, 25–28, 33, 44, 62f., 108, 133, 171, 221, 226f., 231, 243, 251, 254, 263, 281, 283, 313, 318
– operationality of 9, 12, 14, 71f., 78f., 125, 164, 167f., 210, 234f., 288, 323f.
– priority of 1, 5f., 48, 55, 86, 295
– practice theory 74–77, 131–133, 163, 195n, 319, 327
precipitation 277–294
predicting 114–127, 212, 290, 318–321, 327–329
printing 14f., 27, 41, 153, 165–175, 237n, 239, 250, 253f., 259–261, 264, 266, 269, 314, 324
procedure/proceeding 7, 14–16, 23, 72, 74, 76, 78, 80, 94f., 92, 100, 103, 171, 191n, 200, 204, 208, 214, 232n, 234, 236n, 263, 266, 277, 279, 295–315
programming/program 92, 96n, 102–105, 120, 131–133, 287
proliferate 16, 227n, 233, 291

quasi-object /quasi-subject 133, 136–140

radio (-sonde/-tower) 107–127
reading 6f., 15f., 62f., 77, 107, 131, 133f., 220, 223, 226, 230–234, 237–239, 241, 245, 251, 258, 261, 263–276
reflexivity/self-reflexivity 7, 16, 47, 149, 295f.
register/register 118, 185, 205, 221–223, 226, 240, 273
regulate/regulation 9, 16, 75, 81f., 88, 91f., 95f., 98–105, 126, 233, 326–329
repairing 25, 29, 31n, 32f., 45, 77–80, 85f., 166
repeat 30, 32, 80, 134, 143, 206, 210, 219, 288f., 318f., 327f.
reprint 250–254, 259–261
revolution 5, 23–45, 260f., 287–293, 309
riding (see *horse-riding*)
river delta 53–58

sacrificing/sacrifice 132–134, 138, 306n, 307
scene
– of appearance/entrance 295–315
– of begining/primal scene 1–6, 48, 50, 54, 57, 59, 153, 302
– writing 211f., 215, 235, 263
security 33, 35, 100–102
self-imprinting 165–175
seriality 13, 191, 198f., 249, 253, 258f., 324
shipwrecking 1, 5
similarity 17, 317–329
simulate/simulation 108, 111, 119, 125
sleeping 88, 102, 198, 202, 253
slipcase 165–175
social media 317–329
space/spatiality
– airspace 107f., 114, 119, 125f.
– aquatic/maritime 52, 63, 66, 70
– dimensions of (see dimensionality)
– geographical 49, 52, 204, 272
– mathematical 18, 68, 125, 171f.
– of connectivity (see *connect, networking*)
– of control 45, 87–92, 100
– of inscription 49, 51f., 168, 173, 230, 238n, 241, 271, 274, 276
– of travel/transit 17, 63, 65f., 108, 114, 125, 193, 204, 268n
– public 88, 92, 94f., 100–103, 189
– symbolic 3, 8, 15, 288
– techniques of spatialization 6, 8–11, 17f., 45, 48, 318
– urban 28f., 35, 88, 92, 95, 99f., 103
speaking for/on behalf of someone else 295–315
sports 180, 202f.
stabilize/destabilize 8f., 13, 15, 17f., 25, 28, 33, 44f., 74, 95n, 134, 138n, 157f., 167, 173, 200, 205–213, 311, 319
staging 18, 38, 62, 193, 204f., 209–212, 214, 292, 297, 301–303, 308–310
steam engine 66, 73, 81–86, 200f., 213
steering 107f., 112, 114, 119, 131, 157
subjectivation 8f., 15, 71f., 75, 78, 133–136, 139, 305f., 236, 295, 298f., 305f., 309, 312, 314, 319
suffrage 137f.
superstition 277, 280, 291, 294

supplement 5n, 15f., 68, 82, 121, 145, 147, 245, 254, 263f., 269, 273, 276
symbolic order 1, 3, 8, 11, 15, 24, 27f., 58, 71f., 133, 144, 156n., 167, 173, 265, 282, 297, 302, 317, 326

table of contents 220–222, 246–251, 268f., 272, 274
techniques of the body (see body techniques)
telling stories 57, 61, 187–215, 245, 266n
temporality 8f., 11, 18, 57n, 71, 78f., 86, 141, 277f., 282, 285–287, 290, 318
territorialization/deterritorialization 12, 33, 47–59, 139, 148
textile 14, 48, 135n, 141f., 145f., 152
throwing dice 239f., 247
tracking/tracing 62, 68, 71, 94n, 96, 113–118, 125, 138, 141, 145, 148, 152f., 160, 170, 172, 181, 184, 210, 240f., 269, 272, 314, 325
transmitting 110, 116–118, 283, 285
transporting/transport 28, 52, 55–58, 74, 110, 148, 156, 162, 164, 173, 288, 318

uprighting 2n, 3, 144, 160, 202
urban design/planning 26, 32, 35, 44, 80, 100–104
user 91f., 94, 185, 317–329

verbs (thinking in) 6, 11, 13, 295
vermin 295–315

waiting 73–86, 282, 292f., 302
walking 9f., 15, 144, 160, 177–186, 188, 200, 203f., 206, 285
warfaring/waging war 149, 157f., 162, 200–203, 206, 289, 291
wearing 79, 138, 141–164, 274
wrapping 139, 147f., 159
writing 1f., 5–8, 13, 15f., 45, 47, 51, 61f., 71f., 77, 82, 108, 112, 131–134, 193, 207, 212f., 219–241, 250, 254, 263, 266, 268, 296f., 309
writing scene (see scene of writing)